ARCTIC ECOLOGICA RESEARCH FROM MICROWAV SATELLIT OBSERVATIO

Arctic Ecological Research from Microwave Satellite Observations

Gennady I. Belchansky

CRC PRESS

Boca Raton London New York Washington, D.C.

Library of Congress Cataloging-in-Publication Data

Belchansky, Gennady I.
Arctic ecological research from microwave satellite observations / Gennady I. Belchansky.
p. cm.
Includes bibliographical references (p.).
ISBN 0-415-26965-2 (alk. paper)
1. Ecology--Research--Arctic regions. 2. Microwave remote sensing--Arctic regions. I. Title.
QH84.1B45 2004
577‘.0911’3—dc22 2003065517

This book contains information obtained from authentic and highly regarded sources. Reprinted material is quoted with permission, and sources are indicated. A wide variety of references are listed. Reasonable efforts have been made to publish reliable data and information, but the author and the publisher cannot assume responsibility for the validity of all materials or for the consequences of their use.

Neither this book nor any part may be reproduced or transmitted in any form or by any means, electronic or mechanical, including photocopying, microfilming, and recording, or by any information storage or retrieval system, without prior permission in writing from the publisher.

The consent of CRC Press LLC does not extend to copying for general distribution, for promotion, for creating new works, or for resale. Specific permission must be obtained in writing from CRC Press LLC for such copying.

Direct all inquiries to CRC Press LLC, 2000 N.W. Corporate Blvd., Boca Raton, Florida 33431.

Trademark Notice: Product or corporate names may be trademarks or registered trademarks, and are used only for identification and explanation, without intent to infringe.

Visit the CRC Press Web site at www.crcpress.com

© 2004 by CRC Press LLC

No claim to original U.S. Government works
International Standard Book Number 0-415-26965-2
Library of Congress Card Number 2003065517
Printed in the United States of America 1 2 3 4 5 6 7 8 9 0
Printed on acid-free paper

PREFACE

Active (imaging radar) and passive (radiometer) microwave satellite systems are increasingly widely used in diverse fields of Arctic ecological research. Nevertheless, ecologists interested in remote sensing often have limited access to the full suite of physical and analytical techniques of microwave systems, data processing and ecological applications because a suitable reference book has not been produced.

This book provides a summary of main microwave satellite missions and applications for Arctic ecological research. It will be useful to undergraduate and postgraduate students, specialists with a background in microwave techniques and ecologists interested in applications of microwave active and passive remote sensing for tundra, boreal forest, and Arctic marine mammal studies.

Chapter 1 presents a brief introduction to Arctic ecological problems, the role of satellite remote sensing for systematic monitoring of Arctic ecosystems, elements of microwave remote sensing, data processing and applications.

Chapter 2 provides a summary of main characteristics and applications of Russian KOSMOS–OKEAN (real aperture radar, multispectral optical and passive microwave radiometer instruments) polar-orbiting satellite series and ALMAZ, RESURS–ARKTIKA (synthetic aperture radar) satellite series. To a lesser extent, we also treat the main national and international microwave and multispectral optical satellite systems that are of particular interest to the Arctic research community.

Chapter 3 represents some results of Arctic sea-ice habitat studies using remotely sensed and tracking satellite data. The sea-ice types, concentration and surface temperature are analyzed based on passive and active microwave OKEAN-01 satellite data.

Chapter 4 includes an example of Arctic sea-ice variability studies in the Barents–Kara Seas and adjacent parts of the Arctic Ocean using radar and passive microwave satellite measurements. Trends and sea-ice concentration are studied based on KOSMOS, OKEAN-01 and ALMAZ SAR satellite and historical data.

Chapter 5 gives the comparative analysis of multisensor satellite monitoring of Arctic sea-ice habitat using OKEAN-01, SSM/I and AVHRR satellite instruments. OKEAN-01 sea-ice type and concentration algorithms utilize radar and passive microwave information and *a priori* knowledge about the scattering and emission parameters of the basic sea-ice types.

Chapter 6 presents an example of boreal forest habitats studies using OKEAN-01 satellite data. Data processing and classification algorithms are based on the minimum loss criterion. These algorithms are used to classify satellite multispectral and microwave data into groups corresponding to different terrain cover of boreal forest habitat, and to evaluate information content of satellite data for discriminating boreal forest habitats.

Chapter 7 describes the evaluation of relative information content of ALMAZ-1, ERS-1 and JERS-1 SAR and Landsat-TM multispectral satellite data for identifying wet tundra habitats. Classification algorithms are based on minimum loss criteria to classify satellite data into groups corresponding to the different terrain covers of tundra habitats.

Chapter 8 summarises the results of investigating the influence of SAR data-focusing parameters on the efficiency of tundra habitat classification. The special software is used for synthesizing images from the raw SAR data. The focusing algorithm allows control of a spatial resolution by means of the multi-look technique, and provides a set of window functions for flexible adjustment of the synthesized gain pattern.

Chapter 9 describes some aspects of polar bear regional ecology studies using simultaneous satellite telemetry and microwave remote sensing data collected by OKEAN-01 and SSM/I instruments. These aspects include regional and seasonal changes in habitat parameters; daily, seasonal and annual variability of movement rates; individual and group-specific direction and migration patterns, and a characteristic of using a particular type of habitat.

Chapter 10 presents some results of sensitive boreal forest type detection to monitor and assess potential impacts of climate change on the boreal forest structure. The primary objectives and analytical approaches employed by this study included a syntaxonomic classification of nature reserve forest community types, an evaluation of the biodiversity among the forest types and a comparative micro-climatic analysis to investigate bio-physical sensitivity.

ACKNOWLEDGMENTS

This book was prepared with the participation of I.N. Mordvintsev (Chapters 4, 5) and V.G. Petrosyan (Chapters 3, 6, 9, 10), Institute of Ecology and Evolution, Russian Academy of Sciences. Chapters 3–10 are based on the results of research collaboration between the Alaska Science Center, U.S. Geological Survey; and the Institute of Ecology and Evolution, Russian Academy of Sciences in the framework of activity 02.05-7105 of the U.S. – Russia Environmental Agreement (Area V). These results were received with the principal participation of D.C. Douglas (Chapters 3–10) and G.W. Garner† (Chapter 9), Alaska Science Center, U.S. Geological Survey.

ERS-1 and JERS-1 imagery from the Alaska SAR Facility (ASF), Fairbanks received funds from the National Aeronautics and Space Administration. OKEAN-01 imagery from the Scientific Research Centers for Natural Resources Studies (NITS IPR), Moscow Region, Dolgoprudny and ALMAZ-1 SAR imagery from the NPO Mashinostroenia, Moscow was provided with funds from the Russian Ministry of Science and Technology. The National Snow and Ice Data Center, University of Colorado, provided the DMSP SSM/I Daily Polar Gridded Sea Ice Concentrations, the Global Hydrology Resource Center provided SSM/I brightness temperature data sets, and U.S. Geological Survey provided AVHRR data. Some studies included in this book were carried out with the support of The International Arctic Research Center, University of Alaska, Fairbanks; and USGS Alaska Science Center.

Contents

1

Introduction to Arctic Ecological Research and Microwave Remote Sensing

1.1 ARCTIC ECOLOGICAL PROBLEMS AND REMOTE SENSING

The global ecological situation is characterized by a steady population growth and by an unprecedented threat to the biosphere by man's activities. Negative consequences of these processes are ozone depletion, acid precipitation, pollution of the ocean, contamination of the ground and subsoil waters, decreasing the fertility of the soil and rapid degradation of biodiversity and climatic change.

Arctic regions play an important role in the global ecology, having an influence on physical-chemical processes occurring in the atmosphere. Differences between conditions at the equator and at the poles are the main driver of the large-scale atmospheric and oceanic circulation systems that redistribute heat, water, gases and nutrients around the world and determine the global ecology and climate (Aagaard et al., 1985; Aagaard and Carmack, 1989). The effects of climate change are amplified in the Arctic. A particularly significant consequence of climate warming is an expected modification of the floristic composition of tundra and boreal forest and the transformation of Arctic ecosystem functions. Over longer periods, the effects of warming will feed back into the climate system through sustained changes in the extent of snow and sea ice, in the structure of Arctic ecosystems and in global ocean circulation (Gloersen and Campbell, 1988; Hall, 1988; Kondratyev, 1995). This offers an opportunity for early detection of changes by systematic monitoring of the amount of sea ice and snow, the duration of summer melting and Arctic ecosystems. The main goals of Arctic research are understanding the role of the Arctic Ocean and ice cover in the heat and moisture transfer between the ocean and the atmosphere, and heat redistribution from low to high latitudes. These goals include identifying and quantifying the responses of the Arctic Ocean, ice cover, permafrost and Arctic ecosystems in the context of climate change (Thomas, 1992).

In the framework of these problems, detection and investigation of sensitive tundra vegetation and boreal forest communities, as well as ecological studies of marine mammals in Arctic environments are important in an assessment of how they will be impacted by global change. Unfortunately, Arctic marine mammal status, natural habitat parameters, seasonal distributions, extensive movement patterns, sea ice relationship, and adaptation to environmental changes are not well understood because ecological information remains lacking.

Consequently, understanding the mechanism of the Arctic climate changes is critical to an understanding of associated natural patterns, including a variability of sea ice habitats, a correlation between areas of open water in ice-covered seas and biological productivity, and others. It is known that the areas of open water surrounded by ice (*polynyas*) are zones of intense biological activity. The large Arctic polynyas play a very important role in the conservation of the high numbers of marine mammals and birds because the migration of overwintering birds and mammals depends on their existence at critical times during the period when the sea is largely ice covered (Stirling, 1996).

Among different Arctic ecological problems, the radioactive contamination of the Arctic Ocean from nuclear tests and industrial centers, and subsequent monitoring of these contaminants in Arctic environments, are both difficult and topical. This contamination may impact critical biological resources of Arctic and sub-Arctic regions through transport by currents and through bioaccumulation in apical predators of the marine ecosystem. Arctic marine mammals (polar bears and walruses) can function as a unique biological indicator of radioactive contamination in the Arctic because of their extensive movement pattern, their potential for accumulating radioactive pollutants and their potential to act as biological transport mechanisms (Garner and Chadwick, 1997).

The dramatic expansion of the spatial and temporal scales at which Arctic ecological problems must be considered has presented another difficult quantitative challenge. Ecologists now face a series of problems that require regional, continental and global levels of data and an understanding of both short- and long-term consequences. In this context it is so vital to intensify multidisciplinary investigations to improve our understanding of processes that occur at the Arctic by developing new methods of obtaining data based on advance satellite and information technologies. Satellite technologies are one of the most effective data sources for Arctic ecological studies because wide-ranging movements of Arctic biota, environmental extremes, geographic vastness and logistical expenses greatly complicate more conventional data collection and analysis methods. Contemporary ecological monitoring of the Arctic includes the satellite telemetry and remote sensing of the tundra, boreal forest and sea ice habitats (Fancy et al., 1988; Garner et al., 1989, 1994; Belchansky et al., 1993, 1994).

The satellite telemetry for tracking and obtaining physiological and other data from Arctic animals began in the mid-1970s with studies of the efficiency of tracking polar bears using the Nimbus satellite system (Fancy et al., 1988). Since that time, numerous technological advances have made it possible to develop accurate and reliable systems for satellite tracking of Arctic marine mammals.

Long-term Arctic satellite measurements began in the early 1970s with the polar-orbiting weather Landsat and Nimbus satellites series, which used passive sensors measuring the intensity of natural radiation from the earth's surface and the atmosphere. Great volumes of data have been acquired to provide continuous time series measurements. Multispectral optical systems such as Advanced Very High Resolution Radiometer (AVHRR) are widely used for Arctic ecological research. However, periods of prolonged darkness or persistent cloud cover in the Arctic often preclude data acquisition on desired dates. The appearance of microwave sensors essentially increased the scope of Arctic ecological studies. Unlike optical systems, microwave sensors are not dependent on the light from the sun or restricted by clouds, fog and precipitation. Understanding and interpretation of microwave data by ecologists requires some knowledge of microwave sensor design, mission operations and image processing. Basic principles of imaging microwave sensors and applications are presented in a series of special publications and books (Lillesand and Kiefer, 1979; Curlander and McDonough, 1991; Fung, 1994). In this chapter we describe only the main principles of microwave remote sensing that will be useful for ecologists to understand image data processing in the context of some Arctic ecological studies that used the active and passive microwave data presented in this book.

1.2 ELEMENTS OF MICROWAVE REMOTE SENSING

Coincident multispectral measurements can only provide an accurate description of Arctic ecosystem parameters only because each part of the electromagnetic spectrum contains certain information about the surface properties. For example, remote measurements in infrared (3–30 μm), visible/near infrared (0.4–3 μm) and thermal infrared (10–15 μm) regions of the electromagnetic spectrum characterize the earth's surface properties in terms of chemical, physical and biological parameters. Microwave remote sensing utilizes the microwave portion of the electromagnetic spectrum within the approximate range of 1.0 GHz to 300.0 GHz (wavelengths within the range of 1 m to 1 mm) and is characterized by two distinctive features: first, within this frequency range there are several frequency regions where microwave signals are capable of penetrating the atmosphere under virtually all conditions (clouds, smoke, light rain, snow etc.), and second, microwave reflection and emission from surface objects contain information other than in the visible or thermal portions of the electromagnetic spectrum. Therefore, microwave imaging instruments are capable of continuously monitoring biophysical and geophysical parameters related to the structural and electrical properties of the earth's surface and subsurface, independent of day/night and weather conditions.

For microwave measurements, both active (radars) and passive (microwave radiometers) sensors are widely used. The radars supply their own source of energy, transmitting the microwave electromagnetic waves and measuring the reflected signals with real or synthetic aperture. The microwave radiometers use only real apertures and measure the microwave energy emitted and/or reflected from terrain features. Radar systems are divided into radar-altimeters, scatterometers (measuring the radar backscatter from wind-induced capillary waves), and imaging radars providing high (synthetic aperture radar — SAR) and middle (real aperture radar — RAR) resolution day/night and all-weather images of ocean, land and ice surfaces.

There are three main principles of frequency selection for all microwave instruments (Curlander and McDonough, 1991). The first is governed by minimizing of atmospheric absorption (attenuation) properties, and the microwave frequency is selected such that its attenuation by the atmosphere (cloud cover or precipitation) is small. In the 1–10 GHz region, the transmissivity of atmosphere approaches 100%. This region is used in imaging radars because it permits all-weather observations. The second principle is governed by image space resolution; and the third, by the character of surface reflection and emission.

The passive multispectral sensors (radiometers) operate with an aperture that defines the angular resolution as the ratio of the radiation wavelength to the aperture size. The spatial resolution is the angular resolution times the sensor distance from the earth's target. Therefore, at visible and infrared wavelengths, it is not a problem to obtain a

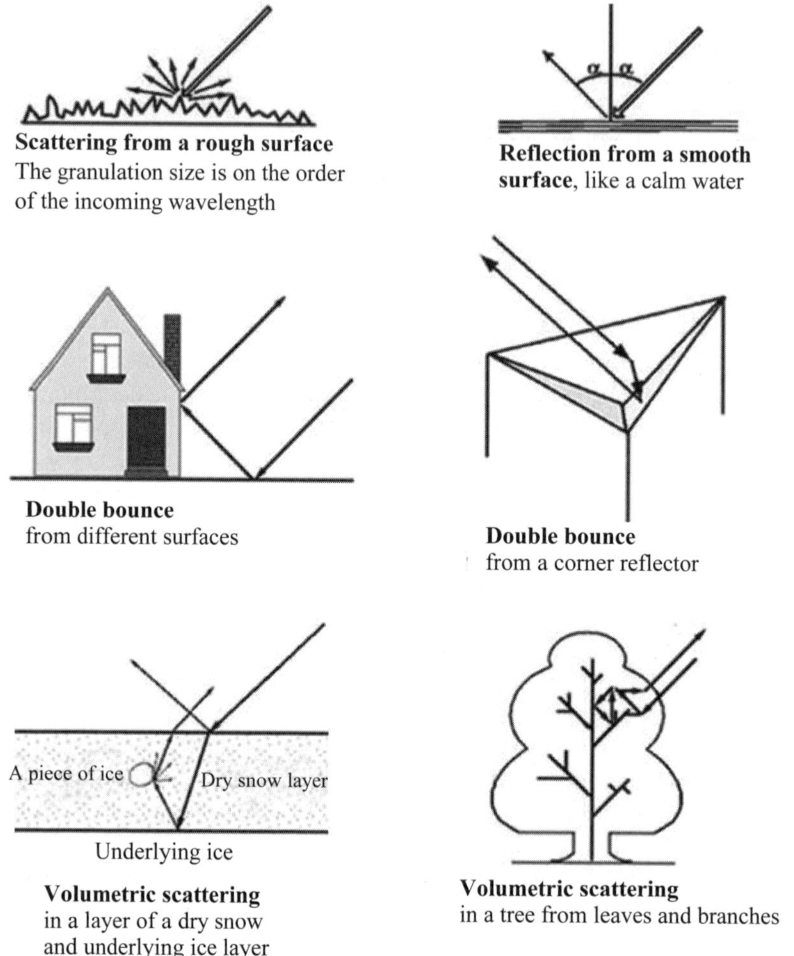

Figure 1.1 Scattering mechanisms.

high-resolution image for real instrument aperture size even at the altitude of a satellite. For microwave wavelengths (10×10^5 times longer than the wavelength of light), to obtain a high-resolution image from the physical antenna size (real aperture) is not possible. To improve this resolution without increasing the frequency and the physical size of the antenna, a signal processing (synthesizing aperture) is used by processing raw radar data on the satellite board or on the earth. When the radar frequency is increased within the microwave spectrum, an attenuation of the transmission is increased. Therefore, the SAR systems operate in the microwave region of 1–10 GHz (the wavelengths within the range of 3 cm to 30 cm) where the transmissivity of the atmosphere is 100%. The RAR systems (scatterometers, altimeters, imaging RAR) operate in the 10–20 GHz region where the resolution can be bigger, but a transmissivity of the atmosphere is not 100% in all weather conditions. For example, at a frequency of 22 GHz, there is a water vapor absorption band that reduces the transmission to about 85%. There are several windows in the atmospheric absorption bands in the spectral region of 30–300 GHz. The frequencies of 35 GHz, 90

GHz and 135 GHz are important for studies of surface properties, provide a narrower radiation beam to obtain a high resolution in the framework of a real aperture and are used in passive microwave radiometers.

The third principle of frequency selection is governed by the mechanism of electromagnetic wave reflection and emission. This mechanism is highly wavelength- and surface-property dependent. Some examples of scattering mechanisms are presented in Figure 1.1.

The signals transmitted from the radar are characterized by the power, frequency and polarization of the electromagnetic waves. These signals can be transmitted and/or received with horizontal (H) or vertical (V) polarization. Thus, it is possible to have four different combinations of signal transmission and reception: HH, HV, VH, VV. The parameters of the scattered wave (power, phase, polarization) are complex functions of the electrical properties of the surface (dielectric constant) and the surface roughness. These functions are usually derived from empirical observations. When the radar pulse interacts with the ground surface, it can be reflected, scattered, absorbed or transmitted (and reflected). The reflection is due to a high dielectric constant of an object (for example, a high water content). A very smooth surface encourages reflection (for example, calm water reflects the radar pulse at the same angle as the angle of incidence) (Figure 1.1). Such reflection returns a very small signal back to the radar. When reflections can again bounce off other objects and be redirected back, this is called a volumetric scattering (Figure 1.1). A volumetric scattering often occurs in vegetation, where leaves have a high water content. When the radar signal is transmitted through the surface, it will be refracted depending upon the density of the substance according to the index of reflection. The index of reflection is the velocity of an electromagnetic wave in a vacuum divided by the velocity of an electromagnetic wave in a particular substance. This index is used to determine how a signal is altered when it passes into a different material. The transmitted signal can be absorbed by the material or have another surface interaction (reflection, transmission, etc.) when it hits an object with different properties. For example, a radar signal is often transmitted through cold, dry snow. If the signal then hits a patch of smooth ice under the snow, it will reflect back up through the snow. The reflected signal can hit small rocks or denser ice pockets in the overlying snow and bounce yet again. This reflection is complex (combination of transmission, reflection and scattering).

Microwave radiometers operate in the low-energy region of the radiant emittance from terrestrial features in much the same manner as thermal radiometers. Therefore, blackbody radiation theory is central to the conceptual understanding of passive microwave sensing. The intensity of measured microwave radiation over any given object is dependent on the object's temperature and the incidence radiation, and the emittance, reflectance and transmittance properties of the object. These properties are a complex function of the electrical, chemical and textural characteristics of the object and the angles from which it is viewed. As mentioned, the frequencies of 35 GHz, 90 GHz and 135 GHz are important for studies of several surface properties. They provide a narrower radiation beam to obtain high resolution in the framework of real aperture and are used in passive microwave radiometers.

Thus, the selection of real aperture radar and microwave radiometer frequency is influenced by resolution, atmospheric effects and an object's scattering and emittance characteristics. Selection of the synthetic aperture radar frequency is influenced primarily by atmospheric effects and target scattering parameters (roughness and dielectric constant etc.). In addition, a number of other system parameters (the imaging geometry, wave polarization etc.) are used to further characterize the surface properties.

1.3 ELEMENTS OF IMAGING RADAR SATELLITE SYSTEMS

Imaging radar satellite systems provide high- (synthetic aperture radar) and middle- (real aperture radar) resolution all-weather images of ocean, land and ice surfaces. The SAR and RAR are side-looking radar (SLR) systems (Curlander and McDonough, 1991). This arises through variations in the relative sensor/terrain geometry for various terrain orientations. Local variations in terrain slope result in varying angles of signal incidence. The SAR and RAR systems use the following main principles: emission by an antenna of a brief electromagnetic pulse in a precise direction, detection of the echo scattered from a target with directional precision, measuring the time delay between emission and detection and scanning with the directional beam. A simplified geometry and the general scanning parameters of a right-looking RAR with a rectangular antenna are presented in Figure 1.2. To image terrain, SLR is carried on a satellite platform moving at uniform speed, V_s, and altitude, H. The forward motion provides scanning in the along track (azimuth) direction because a microwave beam is directed perpendicular to the velocity vector of a satellite and down toward the surface. The look angle, α, relative to the vertical, is the same as the incidence angle, η, which is the angle between the radar beam and the normal to the earth's surface at a point of interest (Figure 1.3). The beam is wide in the vertical direction and narrow in the horizontal direction. The continuous swath of the earth's surface is parallel to the flight direction. Microwave energy is emitted by the antenna in very short pulses with duration T_p. The emitted pulse moves outward from the antenna, and the reflected wave (echo) reaches the antenna (Figure 1.2).

By measuring the time between pulse transmission and echo reception, T_r, the slant range, SR, (direct distance between transmitter and target) is determined. Since the energy propagates in the air at approximately the speed of the light wave $C = 3.0\times10^8$ m/sec, the slant range to any given object is

$$SR = CT_r/2, \tag{1.1}$$

where SR — slant range, C — speed of light, T_r — time between pulse transmission and echo reception.

The beamwidths, θ_h and θ_v, are defined by the ratio of the wavelength, λ, of the electromagnetic radiation to antenna sizes, L_a, D_a. The azimuth beamwidth, θ_h, is

$$\theta_h = \lambda/L_a, \tag{1.2}$$

and the elevation beamwidth, θ_v, is

$$\theta_v = \lambda/D_a. \tag{1.3}$$

The continuous strip of the earth's surface is parallel to the flight direction. The swath width, W_g (the ground swath width), is defined by the formula

$$W_g = \lambda R_m/D_a\cos\eta. \tag{1.4}$$

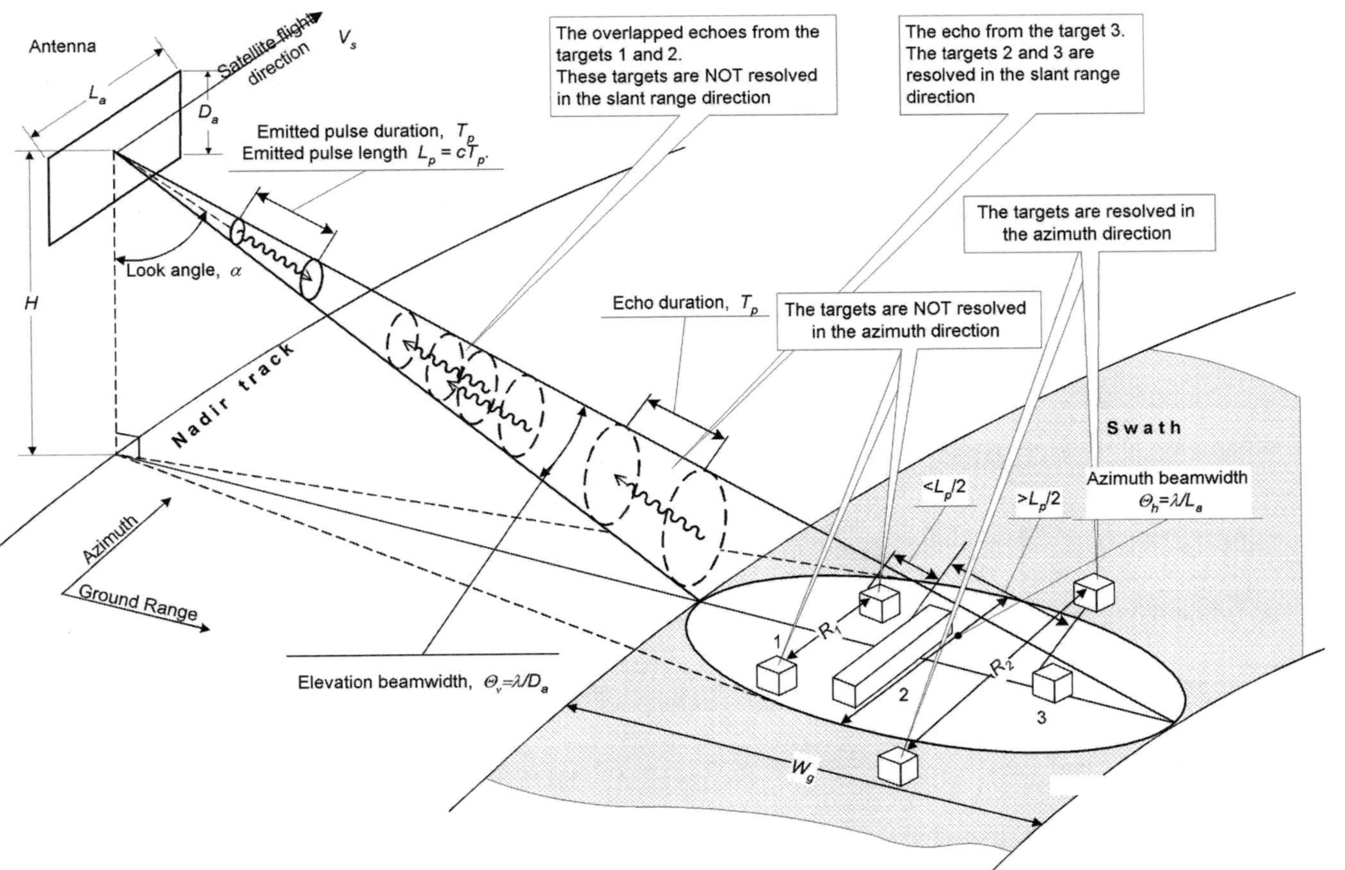

Figure 1.2 Simplified geometry of a right-looking real aperture radar (RAR) and the relationship between the slant range resolution and the azimuth resolution.

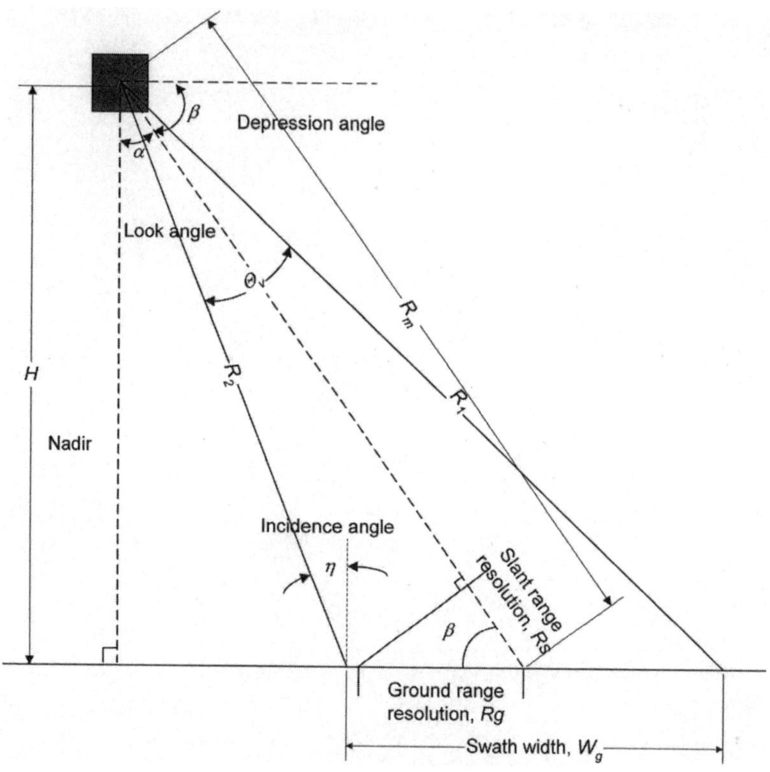

Figure 1.3 Ground range (cross-track) resolution.

In RAR systems, the spatial resolution is controlled by the pulse length, T_p, antenna size, D_a, and look angle, α. The pulse length, T_p, determines the spatial resolution in the direction of energy propagation (slant range resolution). This resolution does not depend on angular resolution and on the track scanning mechanism, since the microwave pulses scan the "echoes" returning at the same rate. Imaging radars achieve resolution dimensions by measuring the relative time delays of each echo component radar pulse. The governing factors for the ground range resolution (cross-track resolution) are the emitted pulse duration, T_p, and the incidence angle, η. The theoretical resolution in the slant range direction is half the pulse length to be detected. The ground range resolution, R_g (Figures 1.2 and 1.3), is

$$R_g = CT_p/(2\sin\eta). \tag{1.5}$$

To achieve the high ground range resolution without changing the pulse duration, T_p, and the pulse energy, the pulse compression technique (matched filtering) is commonly used, which includes appropriate processing of the received pulse. In this case the ground range resolution is

$$R_g = C/(2B\sin\eta), \tag{1.6}$$

where B is the frequency bandwidth of the transmitted pulse (Curlander and McDonough, 1991).

The azimuth resolution (along-track resolution), δX, (Figure 1.4) of the RAR is determined by the width of the radiated radar beam, θ_h, which is characterized by antenna aperture L_a and range, R.

$$\delta X = R\lambda / L_a. \tag{1.7}$$

Two targets on the ground separated by an amount R_1 (Figure 1.2) in the azimuth direction and at the same slant range, can be resolved only if they are not both in the radar beam at the same time.

To improve azimuth resolution without increasing the physical antenna size, and to achieve a resolution independent of the sensor altitude, SAR technology is employed. SAR technology is based on spectral analyzing of precisely phase controlled signals. By means of detection of small Doppler shifts in the signal from a target in motion relative to the radar, it is possible to obtain imaging resolutions on the order of 3 arc seconds for satellite SAR (Olmsted, 1993). This technology depends on precise determination of the relative position and velocity of the radar with respect to the target and on integrating the return signal information over a time period (or look) that is long compared with the time between pulses. For example, the ERS SAR look is characterized by 1000 pulses and, therefore, by a specific target's radar backscatter response about 1000 times. The response to each of the 1000 pulses depends on changes in target-sensor geometry and Doppler effects. This data allows a target's backscatter to be analyzed as though it had been seen by 1000 different antennas, so high resolution is achieved by synthesizing in the signal processor an antenna with the length equal to the distance the satellite passed through (Olmsted, 1993).

To understand how imaging radar measurements are used in remote sensing a study of the basic radar equation is required. The radar equation expresses a conversion of transmitted power into received power in terms of the ratio of received power (P_r) due to a target reflection to receiver power due to noise (P_n), together with some system and target parameters (Curlander and McDonough, 1991). The radar equation for an isolated target is

$$SNR = P_r / P_n,$$

where SNR is a receiver signal-to-noise ratio, received power is $P_r = P_t G^t \sigma S_e / (4\pi R^2)^2$, and noise is $P_n = F_{op}\, kT_s B_n$, or, in terms of main radar parameters, the radar equation is

$$SNR = P_t G^t \sigma S_e / [(4\pi R^2)^2 F_{op} k T_s B_n], \tag{1.8}$$

where P_t is the power intensity flowing across a spherical surface at range R, G^t is an antenna gain (it characterizes the extent to which the radar antenna concentrates the power delivered to it by the transmitter into a beam aimed in the target direction), G^t is called an effective isotropic radiated power, σ is the radar cross section for an isolated target, S_e is a surface area, R is a range from the receiver to the point target, F_{op} is a noise factor (it expresses the extent to which internal receiver noise increases apparent receiver input noise), $k = 1.38 \times 10^{-23}$ J K^{-1} is Boltzmann's constant, T is a temperature and B_n is noise power.

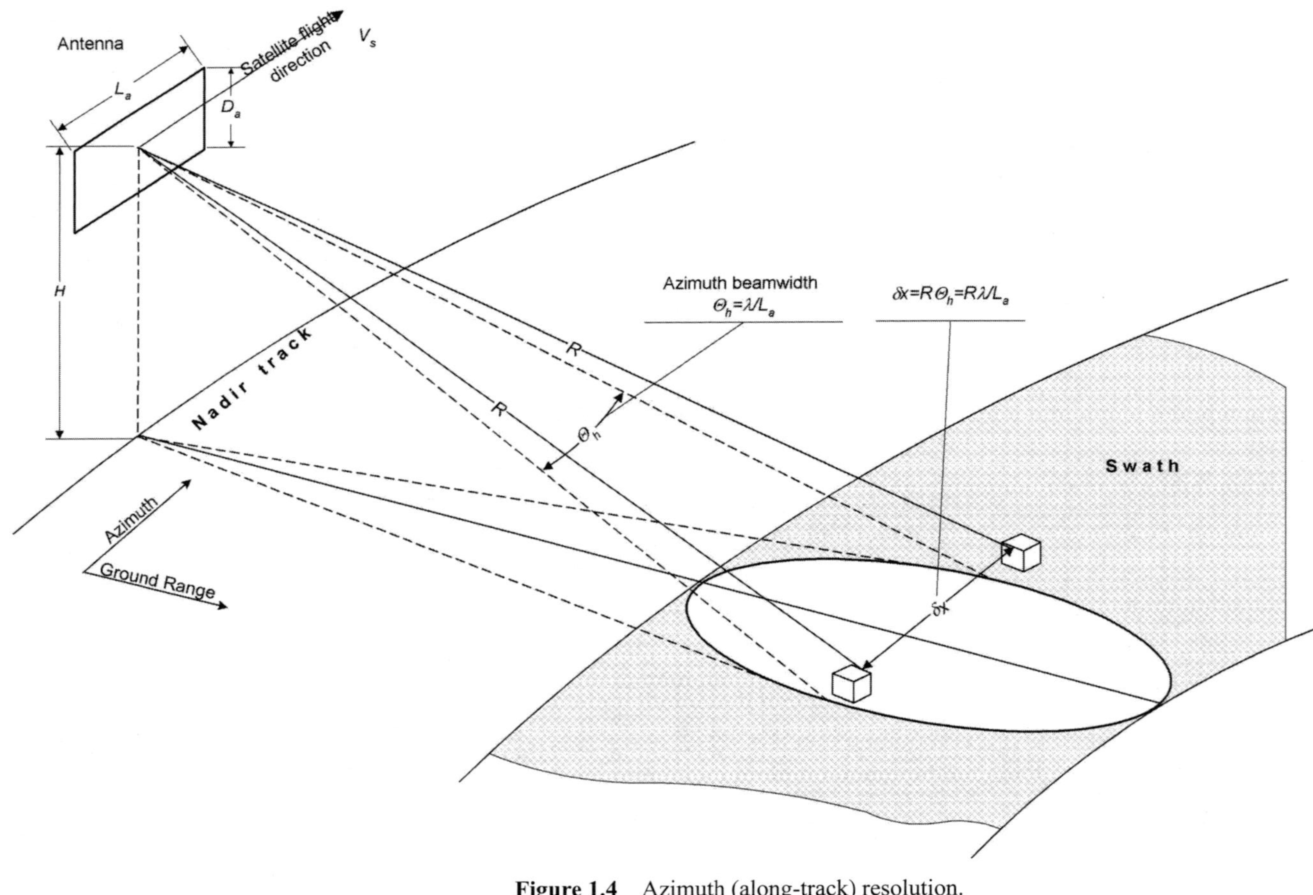

Figure 1.4 Azimuth (along-track) resolution.

The radar equation for an area-extensive target is

$$SNR = P_t \int [G^t(\theta, \phi)\sigma^0(\theta, \phi)A_e(\theta, \phi)(4\pi R^2)^2]\mathrm{d}S/F_{op}kT_sB_n, \tag{1.9}$$

where θ and ϕ are polar coordinates relative to the antenna, σ^0 is the backscatter coefficient for an area-extensive target.

In remote sensing radar measurements the quantity measured is the radar cross section for an isolated target (σ) (Eq. 1.8) or the backscatter coefficient for an area-extensive target (σ^0) (Eq. 1.9). The radar cross section (σ) of an isolated target observed in a given direction is the cross section of an equivalent isotropic scatterer that generates the same scattered power density as the target in the observed direction. To measure the radar cross section of a target, the size of the target must be smaller than the coverage of the radar beam, and the converse is true in measuring the backscatter coefficient (σ^0). The backscatter coefficient (σ^0) can be interpreted as follows: Each element dS of the extended target (sea ice surface, tundra, boreal forest etc.) can be assigned a local value of σ. This implied target area σ relative to the geometrical area dS is the specific backscatter coefficient at the particular point in question on the extended target $\sigma^0 = \sigma/\mathrm{d}S$. It is dimensionless. σ^0 is defined with respect to a nominally horizontal plane, and, in general, has a significant variation with incidence angle, wavelength and polarization, as well as with the properties of the scattering surface itself. The classification of imaging radar data is based on comparisons between the image σ^0 and direct measurements of σ^0 made at test sites. Before any radar image-to-radar image comparisons can be performed, the data values must be converted to σ^0. This process, known as radiometric calibration, converts image data values into scene-independent σ^0 values. In defining the backscatter coefficient (Eq. 1.9) the effects of antenna pattern and range are removed so that these quantities are influenced only by the target and the exploring electromagnetic wave parameters independent of the particular sensor system used to do the measurements (Fung, 1994).

To retrieve surface parameters, a very detailed description of the interaction of the electromagnetic waves with the earth's terrain being imaged is required (the scattering models). Both ground and radar data are used to validate direct models, i.e. predict the backscatter coefficient as a function of the surface parameters, and retrieval algorithms to derive biophysical parameters from the radar data.

Compared with radiometers radar systems have unique sensitivity to terrain surface, soil composition and moisture regime (Harris, 1978; Curlander and McDonough, 1991). However, despite the unique capabilities of the RAR and SAR systems to measure the properties of a surface, their operating range is limited to a small portion of the electromagnetic spectrum. In this context RAR and SAR data in conjunction with passive microwave data and visible and infrared (IR) radiometer data are required to provide a complete description of the surface biophysical parameters.

1.4 ELEMENTS OF PASSIVE MICROWAVE SATELLITE SYSTEMS

The passive microwave satellite systems (microwave radiometers) measure emitted and reflected microwave radiation affected by surface, subsurface and atmospheric conditions, and thus provide a range of biophysical and geophysical information for global ecology and climate studies. They operate in the same spectral region as the shorter-wavelength radar but do not emit their own radiation and measure the naturally radiated power within their antenna field of view. Therefore, in many aspects microwave radiometers operate as thermal radiometers. In accordance with quantum theory, all objects at an absolute temperature above zero radiate energy in the form of electromagnetic waves. Intensity B (brightness) emitted by area dS at a distance R along a given direction (θ, ϕ) is

$$B = \mathrm{d}p/[(dS\cos\theta)/R^2], \tag{1.10}$$

where p is the emitted power density from area dS. The brightness, B, of an object is related to its physical temperature. Under thermal equilibrium, the physical temperature of an object is a constant (Fung, 1994). This means that an object must absorb and emit at the same rate to keep its temperature unchanged. A perfect absorber is an idealized target called a blackbody, which radiates uniformly in all directions. At microwave frequencies the brightness, B_{bb}, of a blackbody follows the Rayleigh–Jeans law (Ulaby et al., 1981) and is

$$B_{bb} = 2k(T/\lambda^2)/\Delta f, \tag{1.11}$$

where $k = 1.38\times10^{-23}$ J K^{-1} is Boltzmann's constant, T is the blackbody physical temperature in Kelvin, λ is the radiation wavelength in meters in the medium in which the brightness of the blackbody is measured, Δf is a narrow bandwidth in Hz. A real target radiates less than the blackbody at the same physical temperature. Its radiation is generally not uniform and brightness $B(\theta, \phi)$ as the function of the angular variables, θ, ϕ, in a spherical coordinate system is

$$\varepsilon(\theta, \phi) = B(\theta, \phi)/B_{bb}, \tag{1.12}$$

where $\varepsilon(\theta, \phi) < 1.0$ is a coefficient called emissivity.

The brightness temperature T_b of a target is the product of its emissivity, $\varepsilon(\theta, \phi)$, and physical temperature, T

$$T_b = \varepsilon(\theta, \phi)T. \tag{1.13}$$

Therefore, it is always less than its physical temperature T. The definition of T_b allows one to write down (Fung, 1994) the brightness $B(\theta, \phi)$ of the target in terms of its brightness temperature in the form similar to that for a blackbody

$$B(\theta, \phi) = B_{bb}\varepsilon(\theta, \phi) = (T_b/T)[2k(T/\lambda^2)/\Delta f] = 2k(T_b/\lambda^2)/\Delta f. \tag{1.14}$$

As follows from the last equation the brightness $B(\theta, \phi)$ of the target is expressed by the same formula as brightness B_{bb} of the blackbody. Thus, it can be interpreted as the blackbody-equivalent radiometric temperature. Since the Rayleigh–Jeans approximation holds in the microwave frequency region, the radiated power is proportional to the temperature of the radiator to a first order. Therefore, the brightness (the intensity) is synonymous with temperature at these frequencies and in passive microwave remote sensing a radiometer measures the brightness temperature, T_b, which is derived from radiance and has the unit of degrees Kelvin.

A simplified geometry of a right-looking passive microwave scanning radiometer with a rectangular antenna and the scheme of the energy reflection, surface and subsurface emission and transmission to the radiometer are presented in Figure 1.5. The microwave radiation detected by the sensor is characterized by the power, frequency and polarization of the electromagnetic waves. This radiation can be received with horizontal (H) or vertical (V) polarization. Thus, it is possible to have two combinations of signal reception: H, V. Main radiation parameters (power, frequency, polarization) are the complex functions of the surface electrical properties and structure. In passive microwave remote sensing in estimating the brightness temperature the effects of antenna pattern and range are removed. Therefore, the brightness temperature measurements are influenced only by the physical and electrical parameters of an object and by the parameters of an electromagnetic wave independent of the particular passive microwave system. The surface brightness temperature in conjunction with radiation emission models allows one to compute the important surface parameters. Let us consider an example of a parameter inversion problem based on the brightness temperature measurements. The radiative transfer equation and emission models are usually used to retrieve the sea ice parameters (a parameter inversion problem) from satellite microwave measurements. In an integral form, this equation can be presented as (Swift et al., 1985)

$$T_b = T_c(1-\varepsilon)e^{-2\tau_1} + T_a(1-\varepsilon)e^{-\tau_1} + T_s\varepsilon e^{-\tau_1} + T_a, \tag{1.15}$$

where T_b is the upwelling brightness temperature in K as measured from space, T_a is the apparent temperature of the atmospheric column, T_c = 2.7 K is the cosmic background radiation reflected from the earth and attenuated by the atmosphere, ε is the surface emissivity, T_s is the true surface temperature in K, and τ_1 is the total atmospheric opacity. The surface emissivity, ε, is usually presented by an empirical model as a function of the parameter retrieved based on the ground measurements. To quantify the fractional areas occupied by water, multiyear sea ice, and first-year sea ice in the Arctic regions the composite emissivity, ε, is assumed to be given by

$$\varepsilon = S_W\varepsilon_W + S_{FY}\varepsilon_{FY} + S_{MY}\varepsilon_{MY}, \tag{1.16}$$

S_W, S_{FY}, and S_{MY} are the respective area fractions of open water, first-year ice, and multiyear ice; and ε_W, ε_{FY}, and ε_{MY} are the emissivities of each of the three components that are estimated based on the ground measurements. If there is no land within the microwave antenna footprint, the following constraint is used

$$1 = S_W + S_{FY} + S_{MY}. \tag{1.17}$$

Equations 1.15, 1.16, and 1.17 are used for computing the parameters S_W, S_{FY} and S_{MY}

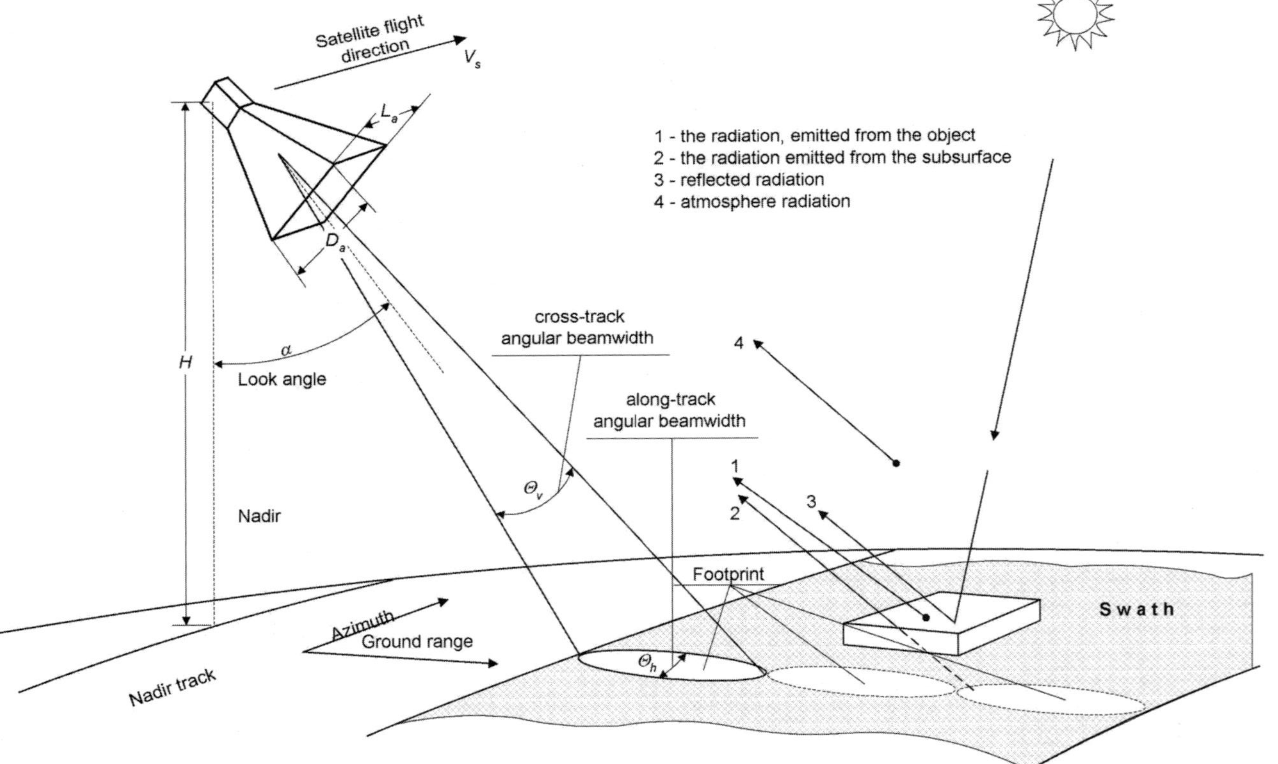

Figure 1.5 Simplified geometry of a right-looking passive microwave scanning radiometer and the scheme of the energy reflection, surface and subsurface emission and transmission to the radiometer.

when two microwave channel data are used (for example, one frequency with two polarizations or two frequencies).

In many ecological applications developing the inversion algorithm is quite a difficult problem, due to the many nonlinear and poorly understood factors involved. Therefore, in the last few years the neural network has received considerable attention as an alternative method to perform the inversion of microwave satellite data. The combination of emission models and neural network makes it possible to perform inversion with higher accuracy. The neural network does not require that the relationship between the inputs to the network and the outputs from the network be known, which determines this relationship directly from the training data (Fung, 1994; Belchansky and Korobkov, 1998).

The passive microwave radiometers operate with an aperture that defines the angular resolution as the ratio of the wavelength of the measured radiation to the antenna sizes. The spatial resolution is the angular resolution times the sensor distance from the earth's target.

To image terrain, the side-looking scanning microwave radiometer is carried on a satellite platform moving at uniform speed, V_s, and altitude, H (Figure 1.5). The forward motion provides scanning in the along track (azimuth) direction because a microwave beam is directed perpendicular to the velocity vector of the satellite and down toward the surface. The continuous swath of the earth's surface is parallel to the flight direction. The microwave radiometer has real aperture, which is defined by antenna sizes, L_a, D_a. The beamwidths, θ_h and θ_v, are defined by the ratio of the wavelength, λ, of received electromagnetic radiation to antenna sizes, L_a, D_a. The along-track angular beamwidth, θ_h, is

$$\theta_h = \lambda / L_a, \qquad (1.18)$$

and the cross-track angular beamwidth, θ_v, is

$$\theta_v = \lambda / D_a. \qquad (1.19)$$

The azimuth (along-track) resolution, δX, of the passive microwave radiometer is determined by the angular beamwidth, θ_h, (Figure 1.5) and the range, R,

$$\delta X = R\theta_h = R\lambda / L_a. \qquad (1.20)$$

The ground range (cross-track) resolution, R_g, is determined by the angular beamwidth, θ_v, and the range, R

$$R_g = R\theta_v = R\lambda / D_a \sin\eta. \qquad (1.21)$$

Modern passive microwave systems are conical scanning devices. They are multi-frequency and dual-polarized (vertical and horizontal polarization) sensors. For example, the Special Sensor Microwave Imager (SSM/I) aboard the Defense Meteorological Satellite Program (DMSP), is a conical scanning device sweeping an approximately 1400-km-wide swath as it looks forward (or backward) at a constant angle (of about 45 degrees from vertical). This instrument samples through 102 degrees for each scanning revolution and detects a microwave radiation at four frequencies: 19.35 GHz, 22.235 GHz, 37.0 GHz, 85.5

GHz, with dual polarization. These frequencies, which provide a narrower radiation beam to obtain high resolution in a framework of real aperture, are important for studies of sea ice surface properties (Cavalieri et al., 1997).

1.5 ELEMENTS OF MICROWAVE DATA INVERSION, CLASSIFICATION AND APPLICATIONS

The inversion and classification of the microwave satellite data are commonly employed to retrieve the biophysical and geophysical information. To date, the measured brightness temperatures (T_b) and the backscatter coefficient (σ^0) are used in conjunction with electromagnetic emission models and scattering models to determine the surface parameters (a parameter inversion problem) or which of several types of surface could be responsible for the measurements (a classification problem) (Fung, 1994). The parameter inversion is based largely on the empirical models, and the classification is based on the statistical classifiers. Developing the inversion algorithm is quite a difficult task due to the many nonlinear and poorly understood factors involved. The drawback of the statistical methods is that the underlying probability density functions must be assumed and that the classifier is theoretically optimal if this assumption is correct. In this book we will discuss these problems for some Arctic ecological applications. But, it is useful to mention that, in the last few years, the neural network has received considerable attention as an alternative method to perform inversion and classification of microwave satellite data. The combination of scattering models and neural network makes it possible to perform inversion and classification with higher accuracy. The neural network does not require that the relationship between the inputs to the network and the outputs from the network be known and determines this relationship directly from the training data (Fung, 1994; Belchansky, 1998). In a comparison of the neural network and the statistical classification of the remote sensing data, the neural network is superior to the statistical method in terms of classification accuracy, however, the time necessary for training is a problem. Recent work shows that optimized training methods can reduce the training time. Fung (1994) presents examples of how electromagnetic scattering and emission models can be used in conjunction with neural networks and unsupervised classification methods to perform retrieval of parameters and classification from remote sensing data.

Let us consider some examples of microwave satellite data applications (Figure 1.6) for Arctic ecology studies based largely on some empirical models and on the statistical classifiers (Belchansky and Pichugin, 1991).

Open Water Study — The scattering of radio waves in the 1-cm range because of the waves of ripples varying in length (Bass et al., 1968) is rather selective. In its turn, the spectral density of the ripples depends on the wind velocity. By using the dependencies of the specific backscatter coefficient between the velocity of the driving wind and the direction of the surface radiation in relation to the direction of movement of large waves it is possible to determine the vector of the driving wind velocity (Chan and Fung, 1977). To define the driving wind velocity by means of real aperture radar, the data are filtered with regard to the spatial structure of the image. Then the specific backscatter coefficient values are calculated with accounts of the through calibration of receiving–transmitting tract of the radar. Since the water surface gets irradiated only in one direction, additional information

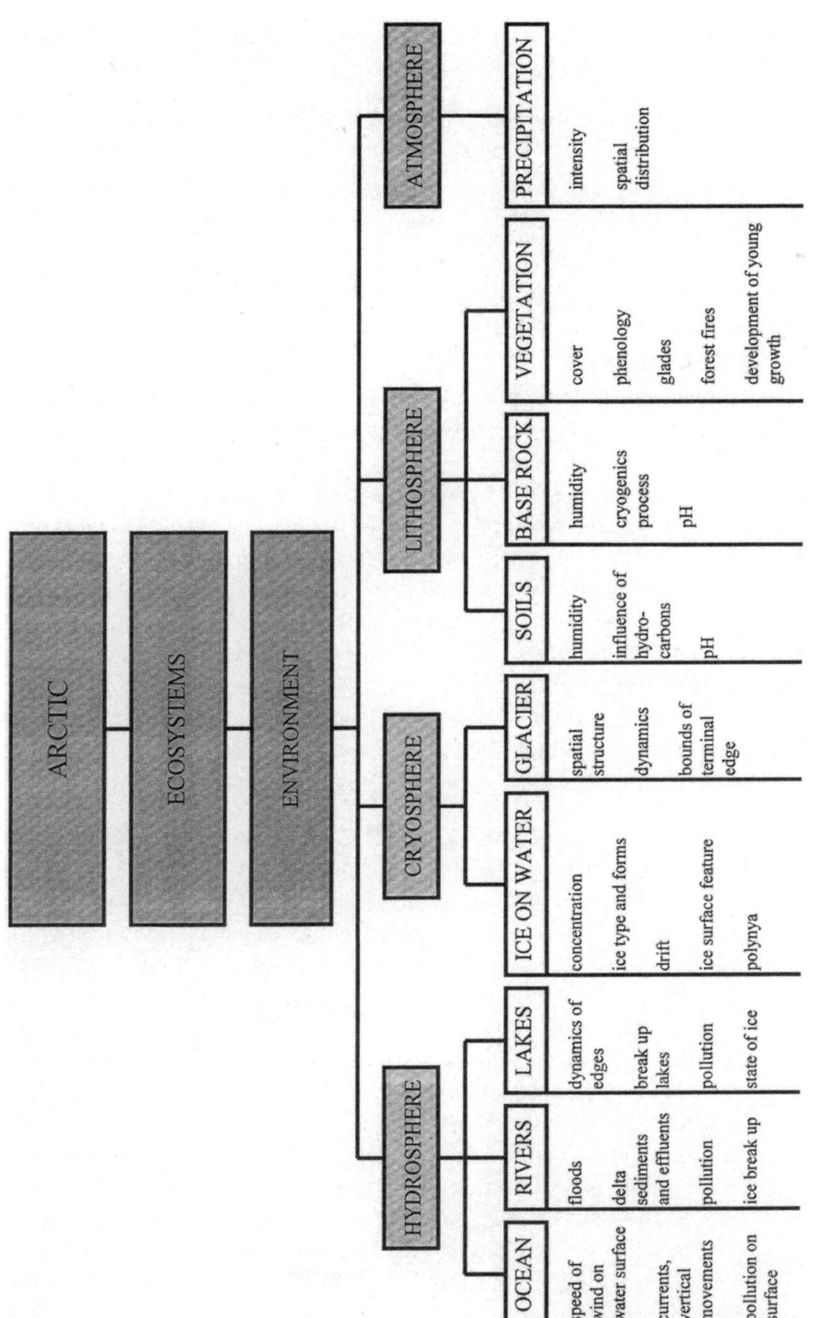

Figure 1.6 Objects of studies and parameters under measurement.

is used to specify the orientation of the driving wind. To this end, one considers the peculiar dislocation of clouds in the optical images, the shading behind islands and the coastline, as well as data from ships. Then, taking into account the direction of the driving wind, the measured specific backscatter coefficient values are corrected. After that, the radiation angle regarding the earth's curvature is found for each section of surface.

Pollution Observation — Pollution observation by means of radar remote sensing is based on the following principles:

- Radar images of water surface reveal different heterogeneities caused by numerous processes in the atmosphere, on the surface and in the aqueous medium.
- There is a special kind of heterogeneity due to the oil pollution of the water surface.
- An oil film drastically diminishes the spectral density of the scattering ripples, which is manifested in the reflected signal (Krishen, 1973).
- The reflected signal can decrease in an oil spot down to 20 dB, permitting reliable detection of oil pollution irrespective of the weather conditions.

Rivers Observation — Among the tasks of hydrology is the study of river inundation regime during river floods that elevate the water table in the watershed swamps. River inundation renders specific effects upon various ecological processes. Therefore it is very important to apply radar sensing to investigate the river inundation dynamics. The main task of river inundation control is to define the area of a territory involved in flooding. To automatically compute these areas after filtering, one specifies sections with different extents of flooding. It should be mentioned that the brightness threshold is to be chosen as a function of actual results of observations, since different territories of the earth's surface, which can be involved in river inundation, reveal their own scattering properties. A correct choice of the value of brightness threshold governs the accuracy of the flooded area estimation.

Sea Ice Cover Study — The results of satellite radar remote sensing have shown that the main ice cover parameters can be effectively measured using radar data. Such parameters include the ice cover structure, the ice age and thickness associated with it, the sea ice boundary and its variability in time. However, it is not simple to elaborate a model of radar reflection from the ice cover. There are certain difficulties, because the ice cover is very complicated. In addition, the problem becomes even more difficult to tackle because the conditions of ice generation are extremely diversified and can affect the process of scattering. Therefore, the methods for ice cover sensing and processing the data in such investigations are based on the summarized experimental data on the function of ice age (thickness), wavelength, radiation angle, polarization etc. Onstott et al. (1982) showed that the specific backscatter coefficient value within the centimeter range increases with the age of ice and depends very little on polarization. It is worth mentioning that, while in the centimeter range the contrast between young and old ice is quite marked, in the decimeter range this difference is practically negligible. The first experiments have shown that satellite imaging radar data can be used to determine multiyear and 1-year sea ice, channels and gaps in packs of perennial ice, offshore and coastal polynyas and openings, the position and configuration of gigantic ice fields and the speed and direction of ice drift. Satellite radar images are effectively used to study the dynamics of ice cover, since they offer a stable representation of its structural heterogeneity.

Soil Study — Among the serious problems in the Arctic study is the control of melting of soils and ground permafrost terrain. The results of investigations have demonstrated that the radar contrasts are a function of humidity at different angles of radiation. Therefore, it is possible to obtain integral data on the spatial distribution of soil moisture in the surface layer of soils over vast areas (Belchansky et al., 1990).

Northern Sea Route Application — The Russian Northern Sea Route (RNSR) extends over a distance of 2000–3100 nautical miles along the northern coast of Russia, offering a route between ports in Europe and Asia that is as much as 60% shorter than the traditional southern canal routes. The passage extends from the Barents Sea to the Bering Strait across four seas in the Arctic Basin (the Kara, Laptev, East Siberian and Chukchi Seas). Seasonally varying ice conditions, shallow waters and fog restrict sea transportation to the western part during the winter months, with the entire route open during the summer months. Access is available to ocean-going cargo vessels from July to the end of October, and requires ice class vessels as well as icebreaker escort during the remaining months. During the winter, these regions are characterized by the development of shore-fast ice that varies in thickness between 120–200 centimeters along the route. Eight polynyas have been observed to form along the fast ice–pack ice interface. During the freeze-up period from August to May, navigation along the RNSR exploits the regions of thinner ice and avoids the regions where convergence has taken place and are under pressure. The current monitoring of the RNSR involves the use of the OKEAN-01 satellite series with passive and active microwave sensors. Satellite active and passive microwave remote sensing in the Arctic has advantages over optical systems because data collection is not constrained by darkness or cloud cover.

1.6 CONCLUSION

The dramatic expansion of the spatial and temporal scales at which ecological problems must be considered has presented another difficult quantitative challenge. Ecologists now face a series of problems that require regional, continental and global levels of data and an understanding of both short- and long-term consequences. The Arctic plays an important role in the global ecological situation. But Arctic ecological studies are difficult, given the harsh conditions, frequently inaccessible habitats and often wide-ranging movements of Arctic biota. In this context, it is vital to carry out multidisciplinary investigations to improve our understanding of ecological processes that occur in the Arctic by developing and using new methods of obtaining the ecological data based on advance satellite and information technologies. Satellite remote sensing provides important data for scientific investigations in many areas of earth science. There is a clear need to use microwave remote sensing as part of ongoing research in Arctic ecological studies.

Microwave satellite systems essentially increase the possibilities of Arctic ecosystems monitoring. Unlike optical systems working in the visible and near-infrared bands, surface studies using radar and passive microwave systems are not limited by illumination or restricted by clouds and fog, thus permitting continuous day/night operation. Microwave systems are capable of continuously monitoring biophysical and geophysical parameters related to the structural and electrical properties of the earth's surface and subsurface. Compared with multispectral radiometers, microwave satellite data have unique sensitivity to terrain surface, soil composition and moisture regime. But, to retrieve biophysical parameters from microwave remote sensing data, a very detailed description of the interaction of electromagnetic waves with the terrain being imaged is required. Both ground data and microwave data are needed to validate direct models, i.e., predict the backscatter coefficient and emissivity as a function of the surface parameters, and retrieval algorithms to derive biophysical parameters from the microwave data.

Microwave radiometers operate in the low-energy region of the radiant emittance from terrestrial features in much the same manner as thermal radiometers. The intensity (brightness) of measured microwave radiation over any given object is dependent on the object's temperature, the incidence radiation and the emittance, reflectance and transmittance properties of the object. These properties are complex functions of the electrical, chemical and textural characteristics of an object and the angles from which it is viewed.

The SAR and RAR satellite systems operate in the microwave region of 1–10 GHz where the characteristics of the scattered wave (power, phase, polarization) depend mainly on the electrical properties of the surface and the surface roughness. Compared with multispectral sensors, imaging radars have unique sensitivity to terrain surface, soil composition and moisture regime.

Each part of the electromagnetic spectrum contains certain information about the properties of the earth's surface. Despite the unique capabilities of microwave systems to measure surface parameters, their operating range is limited to a small portion of the electromagnetic spectrum. In this context, only coincident multispectral multisensor measurements can provide an accurate description of Arctic ecosystem parameters. The imaging radar, in conjunction with microwave radiometer and visible and infrared (IR) radiometer, are needed to provide a complete description of the surface characteristics.

Owing to the increasing number of operational and proposed microwave satellite systems, the potential for microwave data to address Arctic ecological questions warrants thorough investigation. Some aspects associated with sea ice, tundra and boreal forest habitats have been identified in a series of publications. However, more studies are required to evaluate the diversity of methods for processing and analyzing imaging radar and passive microwave data in the context of specific ecological applications and system characteristics. The next chapters include some results of microwave data applications for Arctic ecology studies.

REFERENCES

Aagaard, K., Swift, J. H. and Carmack, E. C., 1985. Thermohaline circulation in the Arctic mediterranean seas. *J. Geophys. Res.*, Vol. 90, No. C3, pp. 4833–4846.

Aagaard, K. and Carmack, E. C., 1989. The role of sea ice and other fresh water in the Arctic circulation. *J. Geophys. Res.*, Vol. 94, No. C10, pp. 14485–14498.

Bass, F. G., 1968. Very high frequency radiowave scattering by a disturbed sea surface. *IEEE Trans.*, AP-16, pp. 554–559.

Belchansky, G. I., Vasiliev, A. I., Zhuravel, N. E., and Pichugin, A. P., 1990. Indicators of ecological soil transformation and its remote sensing observations. *Reports of USSR AS*, Vol. 313, No. 5, p. 1082–1085.

Belchansky, G. I., and Pichugin, A. P., 1991. Radar Sensing of Polar Regions. Proceedings of the International Conference on the Role of the Polar Regions in Global Change, Edited by Weller, G., Wilson, C. L., and Severin, B. A., Geophysical Institute, University of Alaska, Fairbanks, and Center for Global Change and Arctic System Research, University of Alaska Fairbanks, pp. 47–57.

Belchansky, G. I., 1993. The Ecological Situation and Problems Associated with Ecological Monitoring. *Soviet Journal of Remote Sensing*, Harwood Academic Publishers, Vol. 10(4), pp. 659–675.

Belchansky, G. I., Douglas, D. C. and Ovchinnikov, G. K., 1994. Processing of Space-monitoring Data to Document Parameters of the Habitat of Arctic Mammals. *Soviet Journal of Remote Sensing*, Harwood Academic Publishers, Vol. 11, No. 4, pp. 623–636.

Belchansky, G. I. and Korobkov, N. V., 1998. Neural network application for analysis of satellite remote sensing data. *Earth Research from Space, Russian Academy of Sciences*, No. 4, pp. 111–120.

Cavalieri, D. J. and Parkinson, C. L., 1981. Large-scale variations in observed Antarctic sea ice extent and associated atmospheric circulation. *Monthly Weather Review*, Vol. 109, pp. 2323–2336.

Cavalieri, D. J., Gloersen, P. and Campbell, W. J., 1984. Determination of sea ice parameters with Nimbus-7 SMMR. *J. Geophys. Res.*, Vol. 89(D4), pp. 5355–5369.

Cavalieri, D. J., Burns, B. A. and Onstott, R. G., 1990. Investigation of the effects of summer melt on the calculation of sea ice concentration using active and passive microwave data. *J. Geophys. Res.*, Vol. 95(C4), pp. 5339–5369.

Cavalieri, D. J., Parkinson, C. L., Gloersen, P. and Zwally, H. J., 1997. Arctic and Antarctic sea ice concentrations from multichannel passive-microwave satellite data sets: October 1978 to September 1995. *User's Guide. NASA Technical Memorandum 1045647*, 17 pp.

Chan, H. L., and Fung, A. K., 1977. A theory of sea scatter at large incident angles. *J. Geophys. Res.*, Vol. 82, pp. 3439–3444.

Comiso, J. C., 1994. Surface temperatures in the polar regions from Nimbus-7 temperature humidity infrared radiometer. *J. Geophys. Res.,* Vol. 99, No. C3, pp. 5181–5200.

Curlander, J. C. and McDonough, R. N., 1991. Synthetic aperture radar: systems and signal processing. New-York: John Wiley & Sons, Inc.

Elahci, C., 1987. Introduction to the physics and techniques of remote sensing. Wiley-Interscience Publications, John Wiley & Sons.

Fung, A. K., 1994. Microwave Scattering and Emission Models and Their Applications. Artech House, Inc., Norwood, MA, 573 p.

Fancy, S. C., Pank, L. F., Douglas, D. C., Curby, C. H., Garner, G. W., Amstrup, S. C., and Regelin, W. L., 1988. Satellite telemetry: a new tool for wildlife research and management. *U.S. Fish and Wildlife Serv. Resour. Publ.*, No 172, 55 pp.

Garner, G. W., Amstrup, S. C., Douglas, D. C., and Gardner, C. L., 1989. Performance and utility of satellite telemetry during field studies of free-ranging polar bears in Alaska. *Proc. Tenth Int. Symp. on Biotelemetry. Fayetteville: The Univ. Arkansas Press*, pp. 66–76.

Garner, G. W., Belikov, S. V., Stishov, M. S., Barnes, V. G., Jr., and Arthur, S. M., 1994. Dispersal patterns of maternal polar bears from the denning concentration on Wrangel Island. *Proc. Int. Conf. Bear Res. and Management*, No 9, pp. 410–410.

Garner, G. W. and Chadwick, V. J., 1997. Apical Predators in the Arctic Marine System and Potential Radioactive Contamination. *Arctic Research, National Science Foundation*, Vol. 11, pp. 30–32.

Gloerson, P., Zwally, H. J., Chang, A. T., Hall, D. K., Campbell, W. J. and Ramseier, R. O., 1978. Time dependence of sea ice concentration and multiyear ice fraction in the Arctic Basin. *Boundary Layer Meteorol.*, 13, pp. 339–359.

Gloersen, P. and Campbell, W. J., 1988. Variations in the Arctic, Antarctic, and global sea ice covers during 1978–1987 as observed with the Nimbus-7 SMMR. *J. Geophys. Res.*, Vol. 93, No. C9, pp. 10666–10674.

Harris, F. J., 1978. On the use of windows for harmonic analysis with the discrete Fourier transform, *Proceedings of the IEEE*, 66(1), pp. 51–83.

Habbs, R. I. and Mooney, H. A., 1990. Remote Sensing of Biospheric Functioning. Springer Verlag.

Hall, D. K., 1988. Assessment of polar climate change using satellite technology. *Reviews of Geophysics*, Vol.26, No. 1, pp. 26–39.

Kondratyev, K. Ya., 1992. The Global Climate. Nauka, St. Petersburg, 358 pp.

Kondratyev, K.Ya., 1995. Present Stage in the Study of Global Change Problems: The U.S. Program, Earth Observation and Remote Sensing, Hardwood Academic Publishers, Vol.13, pp. 295–306.

Krishen, K., 1973. Detection of oil spills using a 13.3 GHz radar scatterometer. *J. Geophys. Res.*, Vol. 78, pp. 1952–1963.

Lillesand, T. M. and Kiefer, R. W., 1979. Remote sensing and image interpretation. John Wiley & Sons, Inc., 612 pp.

Olmsted, C., 1993. Alaska SAR Facility SAR User's Guide. Geophysical Institute, University of Alaska, Fairbanks.

Onstott, R. G., Moore, R. K., Goginenni, S. and Delker, C., 1982. Four years of low-altitude sea ice broad-band backscatter measurements. IEEE. *Journal of Oceanic Engineering*, OE-7, 1, pp. 44–50.

Stirling, I., 1996. The importance of polynyas, ice edges, and leads to marine mammals and birds. Journal of marine systems, Elsevier Science, 298, pp. 1–13.

Swift, C. T., Fedor, L. S. and Ramseier, R. O., 1985. An Algorithm to Measure Sea Ice Concentration With Microwave Radiometers, *J. Geophys. Res.*, Vol. 90, No. C1, pp. 1087–1099.

Thomas, R. H., 1992. Polar Research from Satellites. Joint Oceanography Institute Inc.

Ulaby, F. T., Moore, R. K. and Fung, A. K., 1981. Microwave Remote Sensing: Active and Passive. Vol. 1, Artech House, Inc., Dedham, MA.

Wohl, M. G., 1995. Operational Sea Ice Classification from Synthetic Aperture Radar Imagery, *Photogram. Eng. & Remote Sens.*, Vol. 61, No. 12, pp. 1455–1466.

2

Russian Microwave Satellites: Main Missions, Characteristics and Applications

This chapter will discuss Russian microwave satellites that can collect both active (radar) and passive microwave (radiometer) data, and synthetic aperture radar (SAR) data. The satellite imaging radars and radiometers are the major focus of attention. To a lesser extent, we also treat the principal national and international microwave and multispectral optical satellite systems that are of particular interest to the Arctic research community and that were used in this book.

2.1 A BRIEF REVIEW OF THE MAIN NATIONAL AND INTERNATIONAL MICROWAVE SATELLITE SYSTEMS

Since 1972, three generations of space borne passive microwave imagers have been launched by the United States. These include the Electrical Scanning Microwave Radiometer (ESMR) on the National Aeronautics and Space Administration (NASA)'s Nimbus-5 satellite (1972), the Scanning Multichannel Microwave Radiometer (SMMR) on NASA's Nimbus-7 satellite (1978) and a series of the Defense Meteorological Satellite Program's (DMSP) Special-Sensor Microwave Imagers (SSM/I) (beginning from June 1987).

The ESMR was a single-channel (19 GHz) instrument with cross-track scanning and varying incidence angles. The Nimbus-5 ESMR data were used for studies of the large-scale characteristics of summer Arctic sea ice (Gloerson et al., 1978; Carsey, 1982). The SMMR was designed primarily for remote observations of oceanic, cryospheric and tropospheric moisture-related phenomena. This sensor measured vertically and horizontally polarized radiance at five wavelengths ranging from 0.81 to 4.5 cm, providing a total of ten radiance data with spatial resolution varied with wavelength and ranges from 30 to 150 km.

The Nimbus-7 SMMR was an experimental earth-imaging passive microwave sensor launched as a follow-on to earlier experimental NASA radiometers (ESMR, NEMS and SCAMS) launched on previous satellites in the Nimbus series (Nimbus-5, 6). The SMMR was a multi-channel, dual-polarized (vertical and horizontal), constant incidence-angle microwave instrument (five wavelengths ranging from 0.81 cm to 4.5 cm; the highest resolution 30 km was for 0.81-cm wavelength). These characteristics permitted a more accurate calculation of geophysical parameters (sea surface wind speed, sea ice concentration and type, surface temperature, snow extent and depth, soil moisture, vegetation extent). Satellite determinations of multiyear ice fraction and surface temperatures have been carried out on a routine basis since the Nimbus SSMR became operational (Cavalieri et al., 1984).

The main mission of DMSP is to provide global coverage of specialized near real-time meteorological, oceanographic and solar-geophysical data. The SSM/I instrument is a seven-channel, four-frequency, linear-polarized, passive microwave radiometric system that measures atmospheric, ocean and terrain microwave brightness temperatures at 19.35 GHz, 22.235 GHz, 37.0 GHz and 85.5 GHz. The 19.35 and 85.5 GHz channels have dual polarization, while the 22 GHz channel has vertical polarization. The SSM/I is a conical scanning instrument sweeping out an approximately 1400 km wide swath as it looks forward at an angle of about 45 degrees from vertical. The sampling resolution is 25 km for 19.35 GHz, 22.235 GHz and 37.0 GHz frequencies and 12 km for 85.5 GHz frequency. The sensor is noted for day/night, almost all-weather capability and daily coverage of the entire Arctic. The Global Hydrology Resource Center (GHRC) associated with Marshall Space Flight Center (MSFC) receives the SSM/I data from the Fleet Numerical Meteorology and Oceanography Center (FNMOC). GHRC processes SSM/I data within hours of its receptions. Each day, full-resolution "swath" brightness temperatures and reduced resolution "gridded" data sets are produced from data recorded by SSM/I instruments aboard

Table 2.1 Main Radiometer Parameters and Mission Characteristics

	Mission			
	DMSP F8	DMSP F10	DMSP F11	DMSP F13
Orbit				
Altitude (km)	838–882	740–853	841–878	840–875
Inclination (degrees)	98.8	98.8	98.8	98.8
Period (min)	101.8	100.7	101.9	101.9
Ascending equator-crossing time (UTC)	06:15	22:09	18:24	17:43
Orbit eccentricity	0.00145	0.00814	0.00129	0.00075
Other characteristics				
Launch date	19 June 1987	1 Dec. 1990	28 Nov. 1991	5 May 1995
Mission end	31 Dec. 1991	14 Nov. 1997	30 Sep. 1995	
Sensor characteristics				
	SSM/I characteristics			
Band	K	K	K_a	W
Frequency (GHz)	19.35	22.235	37.0	85.5
Wavelength (cm)	1.55	1.35	0.8	0.4
Polarization	V, H	V	V, H	V, H
Footprint size (km)				
Along-track	69	60	37	15
Along-scan	43	40	28	13
Sampling resolution (km)	25	25	25	12.5
Angular sector (degrees)	102.4			
Incidence angle (degrees)	53.1			
Swath width (km)	1400			

DMSP Block 5D-2 F8, F11 and F13 satellites. The NASA Goddard Space Flight Center sea ice concentration data set in the polar stereographic projection is derived from the SMMR and SSM/I sensors (Cavalieri et al., 1997). Main radiometer parameters and mission characteristics are presented in Table 2.1.

The Spaceborne Imaging Radar Program of the U.S. National Aeronautics and Space Administration has evolved through SEASAT and Shuttle Imaging Radar (SIR) to multi-frequency, multipolarization systems capable of providing measurements of biophysical and geophysical parameters of the earth with a variety of imaging geometry, resolutions and swaths. The first synthetic aperture radar (SAR) was launched in 1978 on board the SEASAT Spacecraft. Since then, the U.S. has carried out SIR-A, SIR-B and SIR-C/X SAR missions using the Space Shuttle. SIR missions have defined basic parameters for permanently orbiting SAR: three-frequency (L-, C- and X-band) multipolarization radar with variable resolution (25–250 m) and swath width (15–500 km) and selectable incidence angle (15–60°). This SAR is a part of the Earth Observing System (EOS) series. It will provide a 15-year data volume in the microwave band (Way, 1991). Main SAR parameters and mission characteristics are presented in Table 2.2.

The SEASAT SAR was launched into a 800 km high near-polar orbit. It operated at a wavelength of 23.5 cm with horizontal (HH) polarization, fixed viewing geometry and swath width of 100 km. The resolution of the SEASAT SAR imagery was 25 by 25 m at 4-looks. The primary objective of SEASAT experiment was the ocean and sea observations. A substantial volume of data was acquired over land areas situated within the coverage masks of various SEASAT receiving stations in Europe and North America. The SEASAT mission demonstrated the value of space-based SAR for sea ice observation. The images of sea ice far exceeded any other all-weather satellite system in resolution and overall usefulness. SEASAT data were used to determine the directional spectra of ocean waves, surface internal waves and mesoscale eddies, polar ice motion, geological structural parameters, soil moisture boundaries and vegetation characteristics.

The SIR-A SAR was launched in November 1981 into a 38-degree inclined orbit with altitude of 259 km. SAR operated at a wavelength of 23.5 cm with horizontal (HH) polarization, swath width of 50 km and spatial resolution of 40 by 40 m at 6-looks. The main results of the SIR-A mission included imaging of frequently cloud-covered tropical regions and successful use of SAR data for geological mapping of structural and geomorphological features. SIR-A provided better data for geological analysis, as they were relatively free of the layover distortion as SEASAT data.

The SIR-B SAR was launched in October 1984 into a 57-degree inclined orbit. The altitude varied between 354 km, 257 km and 224 km. Wavelength and polarization of the SAR were the same as on SEASAT and SIR-A. The new feature was an antenna steering mechanism that allowed imagery of the same site at various incidence angles during consecutive Shuttle overpasses to be obtained. The swath width ranged between 10 km and 60 km, the spatial resolution was 25 m in the azimuth direction and varied between 17 m and 58 m in range. The main results of the SIR-B mission included the demonstration of radar stereo imagery, repetitive observations of dynamic ocean phenomena and the derivation of "angular" signatures of terrain and vegetation using SAR data.

The SIR-C experiment began with launching two space shuttles in 1994. The radar system was transported into a 57-degree-inclined and approximately 225-km-high orbit. The IR experiment was based on multifrequency SAR with wavelengths of 23.9 cm, 5.7 cm, 3.1 cm and a multipolarization capability. The SIR-C allowed electronic steering of the antenna beam for data acquisition at varying incidence angles. The swath width ranged

Table 2.2 Missions and Main SAR Parameters

	Missions						
	SEASAT	SIR-A	SIR-B	ERS-1	JERS-1	SIR-C / X-SAR	RADARSAT
Orbit							
Altitude (km)	800	259	224–354	785	568	225	789
Inclination (degrees)	108	38	57	98.5	97.7	57.	98.6
Sun synchronous	no	no	no	yes	yes	no	yes
Exact repeat cycle (days)	3, 17			3, 35, 175	44		24
Approximate repeat cycle (days)	3	1	1			1	
Instrument							
Band	L-band	L-band	L-band	C-band	L-band	L/C/X-bands	C-band
Transmit frequency (GHz)	1.275	1.278	1.282	5.3	1.275	1.25/5.3/9.6	5.3
Wavelength (cm)	23.5	23.5	23.4	5.65646	23.5	23.9/5.7/3.1	5.7
Polarization	HH	HH	HH	VV	HH	quad / quad / VV	HH
Look (incidence) angle (degrees)	20 (23)	47 (50)	15–60 (15–64)	20 (23)	35 (38)	(15–55)	(20–59)
Right/left looking	right	right	right	right	right*	right**	right
Swath (km)	100	50	10–60	100	75	15–60	50–500
Peak power (kW)	1.2	1.0	1.2	4.8	1.3	3.5/2.2/3.35	5.0
Antenna type	corporate feed	corporate feed	corporate feed	corporate feed	corporate feed	phased array	phased array
Antenna size (m × m)	2.16 × 10.7	2.16 × 9.4	2.16 × 10.7	1.0 × 10.0	2.4 × 11.9	(2.95+0.75+0.4) × 12	1.5 × 15
Pulse duration (μs)	33.8		30.4	37.1	35	40	42
Pulse bandwidth (MHz)	19.0	6.0	12.0	15.55	15.0	20.0, 10.0	30.0, 17.3, 11.6
Pulse compression ratio	642:1		365:1	580:1	525:1	800, 400:1	1260, 727, 491:1
Pulse repetition frequency (Hz)	1463–1645			1640–1720	1505.8–1606	1240–1736	1270–1390
A to D sampling rate (MHz)	45.03		30.35	18.96	17.1		32.3, 18.5, 12.9
Signal quantization (bits)			6(3I,3Q)–12(6I,6Q)	10 (5I,5Q)	6 (3I,3Q)	8 (4I,4Q), 12 (6I,6Q)	8 (4I,4Q)
Noise equivalent σ_0 (dB)	–24	–32	–28	–18	–20.5	–48/–36/–28	–23
Footprint, Rg (km) × Az (km)	100 × 20	65 × 10	25–120 × 5–10	50 × 4.7	90 × 14	18–50 × 0.6–7	30–100 × 3–5
Doppler bandwidth (Hz)				±1260	±1157		±939
Coherent integer time (s)				0.6	1.7		0.46
Windowing function				hamming	hamming		hamming
Imaging							
Range resolution (m)	25	40	58–17	20	18	60–10	10, 100
Azimuth resolution (m) / looks	25/4	40/6	25/4	8/1, 25/3, 30/4	18/3, 30/4	25/4	8/1, 30/4, 100/8

from 15 km to 60 km, depending on the number of simultaneously used frequencies and polarizations. The resolution of the SIR-C was 25 m in the azimuth direction and varied between 10 m and 60 m in range. The SIR-C mission represented a very important step and provided the first opportunity to simultaneously acquire multifrequency SAR data, use multipolarization capability and get the first SAR imagery during different seasons for change detection. The system was radiometrically and geometrically calibrated. A number of SIR-C test sites were considered for experiments in agriculture, forestry and geology. Because of the orbit inclination, opportunities for sea ice investigations were very limited.

On July 1991 and on April 1995, accordingly, the European Space Agency's ERS-1, ERS-2 were launched into a near polar, sun-synchronous orbit with an altitude of 785 km. The primary goals were global and regional land surveys, monitoring of various resources and polar investigations (European Space Agency, 1993). ERS-1, 2 satellites carry the Active Microwave Instrument (AMI), which operates at C-band and combines the functions of a SAR and a wind scatterometer. The SAR mode can be operated for a maximum of 12 min per orbit. It is not possible to operate the scatterometer and the SAR simultaneously. ERS SAR is a single frequency SAR instrument with vertical (VV) polarization, swath width of 100 km and spatial resolution of 20 m in the range direction and 30 m in the azimuth direction at 4-looks. The SAR system is radiometrically and geometrically calibrated. ERS-1 SAR featured two different modes of data collection: a standard image mode and "sampled" wave mode, which is used to derive ocean wave direction and length from 5 km × 5 km small SAR scenes sampled every 200 to 300 km anywhere within the image swath. ERS-2 carries the radar together with a new instrument GOME (Global Ozone Monitoring Experiment) and an enhanced ATSR (Along Track Scanning Radiometer).

ERS-1, 2 SAR program concentrates on global use of active microwave techniques that enable a range of measurements to be made of ocean, sea and ice surfaces — monitoring of coastal zones, polar ice, geological features, vegetation, land superficial processes, hydrology and digital elevation models. ERS satellite data from receiving and processing stations around the world are being used for geophysical and operational applications as diverse as soil moisture monitoring and ocean wave spectra determination (European Space Agency, 1993). The ERS-1, 2 SAR data have been used in studies of ocean and sea ice circulation, climate processes, mesoscale processes near the sea ice edge, freshwater fluxes and convection, and have also proven useful in support of operations in ice-covered seas. ERS-1 SAR image products began to arrive at the Navy/NOAA Joint Ice Center (JIC) in March, 1992, providing an operational demonstration of the utility of SAR for routine ice analysis. ERS SAR operates at a different frequency from SEASAT's, which enhances the ERS imagery of sea ice (Wohl, 1995).

The simultaneous operation of ERS-1 and ERS-2 in so-called "tandem mode" for a period of several months allowed the collection of uniform data sets from the identical SARs on board the two satellites. Such unique data are especially useful for generating accurate digital elevation models, which are of high value for hydrology, cartography etc. (NASA, 1991; NASDA, 1990).

The Japanese Earth Resources Satellite (JERS) SAR program includes imaging radar JERS-1 SAR. Its primary goals are global and regional land surveys, monitoring of various resources and polar investigations. JERS-1 SAR satellite was launched into a sun-synchronous 568-km-high polar orbit in 1992. The payload includes both a SAR and an optical sensor. The SAR is an L-band sensor operating at a wavelength of 23.5 cm with horizontal (HH) polarization. The image swath is approximately 75 km. The spatial resolution is 18 m×18 m at 3-looks. The JERS-1 SAR Mission was designed primarily for land applications that include: resource exploitation, land surveying, forestry, environmental protection, disaster prevention and coastal monitoring (NASDA, 1990). JERS-1 carries onboard data recorders and can collect almost global coverage. Data already acquired from ERS-1, 2 and JERS-1, including limited amounts of JERS-1 data from its near-global coverage, are archived in the Alaska SAR Facility (ASF).

The Canadian RADARSAT SAR satellite program is a joint program between the Canadian Space Agency (CSA) and NASA. The RADARSAT-1 satellite was launched in November 1995. It is 3-axis stabilized and is designed to operate in a high-inclination (98.6°), sun-synchronous, dawn–dusk, low altitude (nominal equatorial height of 789 km) orbit with a repetition cycle of 343 revolutions in 24 days, or 14.291 revolutions per 24-hour period. The payload consists of a single SAR instrument, on-board storage facility for the SAR data, and a data transmission downlink. The SAR operates at 5.3 GHz (a wavelength of 5.7 cm., C-band) with horizontal transmit and horizontal receive polarizations. This frequency was selected as having the best overall utility to the variety of SAR imaging applications included in the mission objectives. The onboard recorders can collect data from any location in the world for up to 28 minutes per orbit when the satellite is in sunlight during the entire orbit. The 24-day repeat cycle gives sub-cycles of 7 days and 3 days. In the Antarctic Mode, the SAR instrument is able to operate for not less than 12 minutes, with a goal of 18 minutes per orbit.

The main RADARSAT objective is to generate data to assist in supporting the Mission to Planet Earth and a variety of applications. These applications include studies of the earth's land, ocean, and ice cover, and the protection of human life and property from natural disasters. Canadian national requirements consist of operational monitoring of ice areas, routine surveillance of coastal offshore areas, seasonal coverage of forested areas, periodic coverage of agricultural regions, global stereo-coverage and limited coverage for ship detection and regional wave statistics in the North Atlantic and East Pacific Oceans (Carsey, 1993). The SAR system has a sufficient flexibility. It is capable of imaging in seven different modes (Fine, Standard Wide, ScanSAR Narrow and Wide, Extended Low and Extended High) over an accessibility swath of over 900 km. The modes provide options to users in selecting spatial resolution, angle of incidence and swath width. Various levels of SAR products are being planned, with each level having different specifications as to resolution, geometric/positional corrections applied, format and turn-around time. The sensor has the ability to shape and steer the radar beam in one of several positions over a 500-km-wide accessibility swath that can be extended to include a 425-km-wide experimental swath in the far range. The incidence angle geometry may vary depending on the swath width. The user can select from a variety of imaging modes, ranging from the full 500-km swath to a fine-resolution beam. The standard beams are approximately 100 km wide. In addition, the SAR swath can be electronically altered in width and spatial resolution (10–100 m) to the specification of a particular application requirement. The different swath choices include:

1. Seven standard high-sensitivity, 4-look beams of 100 km width across the entire 500-km accessibility swath.
2. Two wide-swath, 4-look beams at a width of 150 km each (located in the near-range of the accessibility swath).
3. Five fine-resolution, single-look, steep-incidence angle swaths at a swath width of 50 km each (located in the far-range portion of the accessibility swath).

4. Two very-wide-swath, single-look beams at low resolution ScanSAR mode with swath width of 300 km and 500 km.
5. Six experimental beams with a swath width of 75 km each covering the 300-km-wide experimental region.

All mission characteristics are presented in Table 2.2.

The main SAR satellite data are accepted and processed by the Canadian Data Processing Facility (CDPF) and by the Alaska SAR Facility (ASF). Other countries have their own SAR data reception, data processing and archiving facilities. Antarctic data will be collected on the satellite tape recorder and then downlinked to ASF. The image products are fully calibrated.

Of particular interest to the Arctic research community are data from the ASF in Fairbanks, which was developed by NASA to support the international spaceborne missions. This facility was designed and implemented by the Jet Propulsion Laboratory and is located at and operated by the University of Alaska's Geophysical Institute, which receives, processes and distributes the data (Way, 1991; Olmsted, 1993). The ASF became operational in 1991 and presently receives ERS-1, 2, JERS-1 and RADARSAT data. The prime objective is applications of SAR products to land, ocean and ice research, particularly those associated with the U.S. Global Change Program and NASA's Mission to Planet Earth. SAR data used for sea ice analysis at the Navy/NOAA Joint Ice Center (JIC) are processed at the ASF and then digitally relayed to the JIC in Washington, D.C. Two high-resolution picture transmission (HRPT) stations became active in 1992 and 1993; they receive the advanced very-high-resolution radiometer (AVHRR) data, the DMSP data and SeaWiFS data. Advanced Earth Observing System (ADEOS) satellite data will be received at the ASF in the mid-1990s and will tremendously increase data availability by the end of this century and into the next. The ASF is also one of the Distributed Active Archive Centers (DAAC) of the NASA EOSDIS program. EOS will provide data at multiple wavelengths, spatial resolutions and temporal coverages. These data will play a very important role in assessing, monitoring and analyzing the polar environment. Data sets are normally processed at the ASF using the Geophysical Processor. Characteristics of ASF data products are presented in Table 2.3.

2.2 EVOLUTION OF RUSSIAN MICROWAVE SATELLITES

The Russian microwave satellite program has evolved primarily through the KOSMOS–OKEAN satellite series with three onboard instruments (real aperture radar, microwave radiometer, multispectral scanning system) and the KOSMOS-1870 synthetic aperture radar (SAR), ALMAZ-1 SAR to a multifrequency and multipolarization ALMAZ-1B, 2 SAR and RESURS–ARKTIKA SAR series. These systems were developed for various resources studies, all-weather earth observations and Arctic and Antarctic studies. The KOSMOS–OKEAN program began in February, 1979 with the launching of the KOSMOS-1076 and the KOSMOS-1151 satellites in January, 1980, and continued with launching the OKEAN-01 satellite series. The ALMAZ SAR satellite program began in 1973–1977, when three ALMAZ manned orbital stations (SALYUT-2, 3, 5) were launched with a visible telescope on board (resolution of 4 m). Since 1976, Russia has created the ALMAZ satellite series equipped with SAR imaging radars.

Table 2.3 List of ERS and JERS Data Products Available from the ASF

Level	Product name	Processing &	system	Data type (samples)	Bit quantization or units	# Lines	# Samples	Pixel spacing (meters)	Resolution (meters)	Extent	Distribution mode
Standard products:											
0	Computer compatible signal data	raw video signal	RGS	complex	5I × 5Q	NA	NA	NA	NA	12 sec	CCT, DOD
1A	Complex image	1-look processed	SPS	complex	16I × 16Q	6250	3750	8	10	30 × 50 km	CCT, DOD
1B	Full resolution image	4-looks processed	SPS	integer	8	8192	8192	12.5	30	100 × 100 km	CCT, DOD, film
1B	Low resolution image	256-looks (8 × 8 avg.) SAR processed	SPS	integer	8	1024	1024	100	240	100 × 100 km	CCT, DOD, film
Geocoded products:											
1B	Geocoded full resolution image	geolocated 4-looks	AOS	integer	8	8192	8192	12.5	30	100 × 100 km	CCT, DOD, film
1B	Geocoded low resolution image	geolocated 256-looks	AOS	integer	8	1024	1024	100	240	100 × 100 km	CCT, DOD, film
Geophysical products:											
2	Ice type classification	segmented into 3 or 4 classes	GPS	integer	4	1024	1024	100	240	100 × 100 km	CCT, DOD
2	Ocean wave spectra	2D PS contour plot wavelength & direction	GPS	real	meters, degrees	17	17	6000	6000	100 × 100 km	CCT, DOD
3	Ice type fraction	3 or 4 band concen-tration of ice type	GPS	real	%	20	20	5000	5000	100 × 100 km	CCT, DOD
3	Ice motion vectors	(Δx, Δy) displacement	GPS	real	km	20	20	5000	5000	100 × 100 km	CCT, DOD

CCT – computer compatible tape; DOD – 5.25″ digital optical disk; film – 8″ × 10″ format.

The KOSMOS-1076 and 1151 satellites were launched into nearpolar circular orbits with inclination of 82.5 degrees and average altitude of 650 km. These satellites contained a six-channel spectrometer (range, 0.455–0.675 μm; channel bandwidth, 3–8 nm; spatial resolution, 20 km), a ten-channel infrared radiometer (range, 9.04–18.4 μm, channel bandwidth, 0.135–0.325 μm, spatial resolution, 25 km) and a four-channel microwave radiometer (wavelengths 0.8, 1.35, 3.2, 8.5 cm; spatial resolution, from 18 km at 0.8 cm to 85 km at 8.5 cm). They had equipment for collecting and transmitting data from buoys and research vessels. Their main goal included improvement of synchronous remote measurements of the ocean and atmosphere parameters in various spectral bands (Nelepo et al., 1984).

The KOSMOS-1500 satellite was launched on September 28, 1983 into orbit with apogee of 679 km, perigee of 649 km and orbital inclination of 82.6 degrees. The satellite was equipped with a real aperture radar (wavelength of 3.15 cm), a microwave radiometer (wavelength of 0.8 cm), two multichannel scanning systems and an along-track three channel microwave radiometer (wavelengths of 0.8 cm, 1.35 cm, 8.5 cm) (Afanas'ev et al., 1989). The main goals were the development of spaceborne equipment, methods and tools for the remote sensing of the oceanographic systems; mapping Arctic and Antarctic sea ice fields and detection and study of dynamic formations in the open waters of the Pacific Ocean (Bushuev and Bychenkov, 1989). Other satellites of KOSMOS–OKEAN series were launched with the same orbital and instrument parameters: KOSMOS-1766 (July 29, 1986), KOSMOS-1869 (July 12, 1987), OKEAN-01 N3 (July 5, 1988), OKEAN-01 N5 (March 16, 1990), OKEAN-01 N6 (June 4, 1991), OKEAN-01 N7 (October 11, 1994), OKEAN-01 N8 (SICH-1) (August 31, 1995).

The ALMAZ SAR (KOSMOS-1870) satellite was launched on July 25, 1987 into orbit with inclination of 71.9 degrees and altitude of 250–300 km. Its mission was finished on July 30, 1989. This satellite was equipped with two synthetic aperture radars, one on each side of the satellite, providing 11–15 m resolution. The swath width was nearly constant at 2×20 km. The KOSMOS-1870 SAR surveys enabled investigation of wind waves (energy components) and ripples; hydrological fronts; dynamics of streams; surface manifestation of inner waves, whirls and rings of various scales; oil spills and underwater relief and circulating flows of a leeward type (Chelomey et al., 1990; Salganik et al., 1990).

The ALMAZ-1 SAR was launched on March 31, 1991 on a Proton booster into orbit with inclination of 72.7 degrees and altitude of 270–380 km. This mission was completed on October 17, 1992. The ALMAZ-1 SAR was equipped with advanced single-frequency (S-band) single-polarization SAR providing 10–15 m spatial resolution, depending on the incidence angle, and swath width equal to 2×20 km. There were several successful application demonstrations using ALMAZ-1 SAR data in the following areas: ocean research, geological mapping and mineral resources reconnaissance, study of the ecological state of coastal areas and inland waters; ecological monitoring; remote control of land usage; monitoring of ice situation, etc.

ALMAZ-1B SAR will be the third Russian satellite of ALMAZ satellite series. It was planned to be launched in 1999. ALMAZ-1B differs from the previous satellite since it is planned to carry the onboard complex of the remote sensing tools which include a set of various instruments providing a great volume of data on the features of earth's surface. This complex will include a radar system with three radars operating at the wavelengths of 3.6 cm, 9.6 cm and 70 cm (X-, S- and P-bands, respectively, spatial resolution of 5–7 m, 15–40 m, 20–40 m), two multispectral scanning systems (MSU-E, MSU-SK), optical electronic instruments for high-resolution stereo observations, a spectroradiometer and a lidar.

Further evolution of the ALMAZ program will be represented by the launch of the spaceborne multifunctional ALMAZ-2 satellite. This satellite was planned for 2000. It will provide observations for use in various fields of earth and atmosphere studies. Installation of about 15 instruments is planned onboard the satellite. Besides, ALMAZ-1B, ALMAZ-2 will be equipped with instruments for studying the composition and cleanliness of the atmosphere.

The RESURS-ARKTIKA SAR satellite was planned to be launched in 1999. This satellite will have a RAR (frequency of 9.52 GHz, wavelength of 3.15 cm, VV polarization), a five-channel microwave radiometer (40 km spatial resolution), a five-channel conical scanner (spatial resolution of 10 m), a three-channel scanner (spatial resolution of 5 m) and a new SAR (swath of 500 km, middle spatial resolution of 300 m, moderate spatial resolution of 130 m, high spatial resolution of 30 m, wavelength of 3.5 cm). Combined use of optical and radar equipment will allow gathering of better characteristics and details in sea ice recognition (Volkov et al., 1995).

2.3 KOSMOS–OKEAN SATELLITE SERIES

2.3.1 Orbit Characteristics

KOSMOS–OKEAN satellite series are launched into orbits with the following parameters: apogee of 679 km, perigee of 649 km, orbital inclination of 82.6 degrees and period of revolution of 97.7 min. The shift of the sub-satellite track per one period is 24.57 degrees. The orbit of the satellite in each period of revolution is virtually quasisynchronous, with the trajectory recurring in a 3-day cycle. Thus, for example, after 3 days (44 orbits) the satellite trajectory is displaced from its original position by 1.2 degrees. This cyclic recurrence of the trajectory is convenient: on the one hand for a fast full survey of the Arctic over a period of 3 days, and on the other hand for the regular observation of specific areas of the Pacific Ocean. KOSMOS–OKEAN satellite series main orbit characteristics are in Table 2.4.

2.3.2 Sensors

KOSMOS–OKEAN satellite series are launched with three remote sensing systems onboard: side-looking RAR, left-looking RAR, scanning microwave radiometer (RM-0.8) and multispectral scanner system (MSS or MSU-M). RAR, RM-0.8, two MSU-M and long-track microwave radiometer with three channels (0.8 cm, 1.35 cm, 8.5 cm) were installed only onboard the KOSMOS-1500 satellite.

Side-looking RAR is an active microwave imaging sensor that transmits pulses of energy and measures various characteristics of the reflected radiation. RAR operates at microwave frequency of 9.52 GHz (wavelength of 3.15 cm), VV polarization, and can observe the earth's surface in all weather conditions day and night. The image's resolution of side-looking RAR is determined by the radar pulse duration T, the signal wavelength λ, the antenna size L, the flight height H, and the look angle α. The azimuth resolution (resolution in the along-flight-track direction), Δ_a, is determined by $\Delta a = H\varphi_a/\cos\alpha$; $\varphi_a = \lambda/L$. The range resolution (in the direction perpendicular to the satellite's flight direction), Δ_y, is determined by $\Delta_y = CT/2$, where C is the speed of electromagnetic radiation (in air,

Table 2.4 Main Parameters of KOSMOS–OKEAN Satellite Series

Orbit	
Apogee (km)	679
Perigee (km)	649
Inclination (degrees)	82.6
Period (min)	97.7
Sub-satellite track shift per one period (degrees)	24.57
SLR Parameters	
Band	X-band
Transmit frequency (GHz)	9.52
Wavelength (cm)	3.15
Polarization	VV
Incidence angle (degrees)	20–47
Looking	Left
Swath width (km)	450–500
Range resolution (km)	0.6–0.9
Azimuth resolution (km)	2.1–2.8
Average spatial resolution (Rg. × Az.)	
for f = 466.5 MHz (km × km)	0.8 × 2.2
for f = 137.4 MHz (km × km)	2.8 × 2.5
Peak power (kW)	121
Antenna type	waveguide
Antenna size (m × m)	11 × 0.04
Antenna gain (dB)	34.8
Antenna beamwidth (–3 dB width)	
range (degrees)	42
azimuth (degrees)	0.15
Pulse duration (μsec)	3.1
Pulse repetition frequency (Hz)	100
Number of noncoher. accumulation	8
Kind of downlink	analog
Duration of record (min)	9
RM-08 Parameters	
Band	K_a-band
Frequency (–3 dB) (GHz)	36.62
Wavelength (cm)	0.8
Polarization	H
Looking angle (degrees)	20–50
Looking	left
Swath width (km)	550
Range resolution (km)	11.7–25.0
Azimuth resolution (km)	8.0–11.6
Sensitivity (degrees K)	3.37
Measured temperatures range	
mode 1 (degrees K)	110–330
mode 2 (degrees K)	150–250
Antenna beamwidth (–3 dB width)	
range (degrees)	0.895
azimuth (degrees)	0.650
Duration of record (min)	9

Table 2.4 (continued)

MSU-M parameters	
Spectral channels (μm)	
1	0.5–0.6
2	0.6–0.7
3	0.7–0.8
4	0.8–1.1
Looking angle (degrees)	–52.5 to 52.5
Swath width (km)	1900
Range resolution (km)	1.0–1.6
Azimuth resolution (km)	1.7–2.7
Momentary field of view	
range (rad)	0.00153
azimuth (rad)	0.00261
Scanning speed (lines/sec)	4
Duration of record (min)	6–12

3.0×10^8 μ/sec). For the OKEAN-01 radar parameters T = 3.0 μsec, $\lambda = 3.15$ cm, L = 11 m, the spatial resolution on the flight height of 660 km (the antenna beamwidth, φ_a, is 0.15 deg. at the level of –3 dB) varies from 2.1 km ground along-track, 0.6 km ground cross-track at $\alpha = 20$ deg. to 2.8 km ground along-track, 0.9 km ground cross-track at $\alpha = 47$ deg. The swath width is 450 km, providing almost global coverage every day and pixel size ranging about 1.5 km. The vertical polarization was chosen so as to maximize the reflections from the sea surface. The level of reflected returns from ice cover is almost independent of the polarization (Onstott et al., 1982).

Microwave Radiometer (RM-0.8) is a side-looking and scanning passive-microwave imaging sensor operating at a frequency of 36.62 GHz (wavelength — 0.8 cm) and H polarization. Data can be obtained day and night, and are only slightly affected by cloud cover. The spatial resolution of the radiometer along-track (Δ_a) and cross-track (Δ_y) is determined, respectively, by the along- and cross-track angular beamwidths, φ_a and φ_y, of the antenna. Let us define α — the scanning angle, and H — the flight height. Then, assuming the earth to be a plane: $\Delta_a = H\varphi_a/\cos\alpha$, $\Delta_y = H\varphi_y/\cos\alpha$. The spatial resolution on the flight height of 660 km (the antenna beamwidths are $\varphi_a = 0.650$ deg. and $\varphi_y = 0.895$ deg. at the level of –3 dB) varies from 8.0 km along-track, 11.7 km cross-track at $\alpha = 20$ deg. to 11.6 km along-track, 25 km cross-track at $\alpha = 50$ deg. The swath width is 550 km, providing almost global coverage every day and pixel size ranging about 15 km.

Multispectral Scanner System (MSU-M) is a passive scanning sensor operating at several wavelengths in the visible and near-infrared ranges. KOSMOS–OKEAN satellite series RAR, RM-08 and MSU-M parameters are given in Table 2.4.

2.3.3 Scanning Configuration

Side-Looking Real Aperture Radar (RAR) — To image terrain, the RAR is carried on a spacecraft platform moving at uniform speed and altitude. The forward motion provides scanning in the along track (azimuth) direction. The RAR beam is directed to the perpendicular to the track, and down toward the surface. The beam is wide in the vertical direction and so intersects the surface in an oval, with the long axis extended in the across-track (range) direction. The echo of a short pulse is received from surface points at increasing range. Thus, digitizing the signal in time provides scanning in the range direction. This direction is determined by the side to which the radar looks. Side-looking makes it unique as opposed to a nadir-looking beam, which would extend on either side of the nadir track. Then each travel time would correspond to a return from the left side.

Microwave Radiometer (RM-0.8) — A microwave radiometric image is constructed by means of an antenna system that includes a cutout parabolic reflector and linear scanning irradiator. The antenna system collects energy data of every scene. A microwave switch permits rapid alternate sampling between the antenna signal and a calibration temperature reference signal.

The low antenna signal is amplified and compared with the internal reference signal. The difference between the antenna signal and the reference signal is electronically detected and input to some mode of readout and recording. Because of the very low levels of radiation available to be passively sensed in the microwave region, a large antenna beamwidth is required to collect enough energy to yield a detectable signal. Consequently, passive microwave radiometers are characterized by low spatial resolution. A microwave radiometer's output is recorded on magnetic tape in an analog format. Data from a single scan (30 samples) are converted to 728 pixels and transmitted at 466.6 MHz frequency. Because of slow scanning speed (one line per second) and for superposition with other channels (RAR and MSU-M) one radiometer line is repeated and transmitted four times. An error of absolute calibration is about 4 K.

Multispectral Scanner System (MSU-M) — The line-scanning collection geometry of the MSU-M is next: a rotating mirror moves the field of view of the scanner along a scan line perpendicular to the direction of flight. The forward motion of the spacecraft platform advances the viewed strip between scans, causing a two-dimensional image data set to be recorded. In the MSU-M, the incoming energy is separated into several spectral components. By placing an array of detectors, the incoming beam is essentially "pulled apart" into multiple narrow bands, each of which is measured independently. The signals generated by each of the detectors are amplified by the system electronics and recorded by a multichannel magnetic tape recorder. The scanner output is normally recorded on an analog recorder and converted to images using a ground-base film recorder. In addition, it is possible to electronically convert the analog scanner output signal to a digital format by an onboard converter.

General scanning configuration for KOSMOS–OKEAN satellite series is illustrated in Figure 2.1.

2.3.4 Radiometric Calibration

2.3.4.1 Passive Microwave Image Calibration

The passive microwave radiometer RM-08 raw data include the set of the brightness calibration wedges and imagery. Every calibration wedge has eight temperature gradations in degrees Kelvin (1 — 110 K, 2 — 142 K, 3 — 174 K, 4 — 194 K, 5 — 226 K, 6 — 252 K, 7 — 280 K, 8 — 310 K). The calibration wedge is constructed by means of the reference generator and attenuator. Every gradation is presented on microwave image by set of plates. A plate is a rectangle with pixels that have the same intensity. It consists of 48 lines corresponding to the definite lines on the image and has 29–30 similar points in every line of plate. A full raw microwave image contains 3–4 calibration wedges with eight gradations, so one

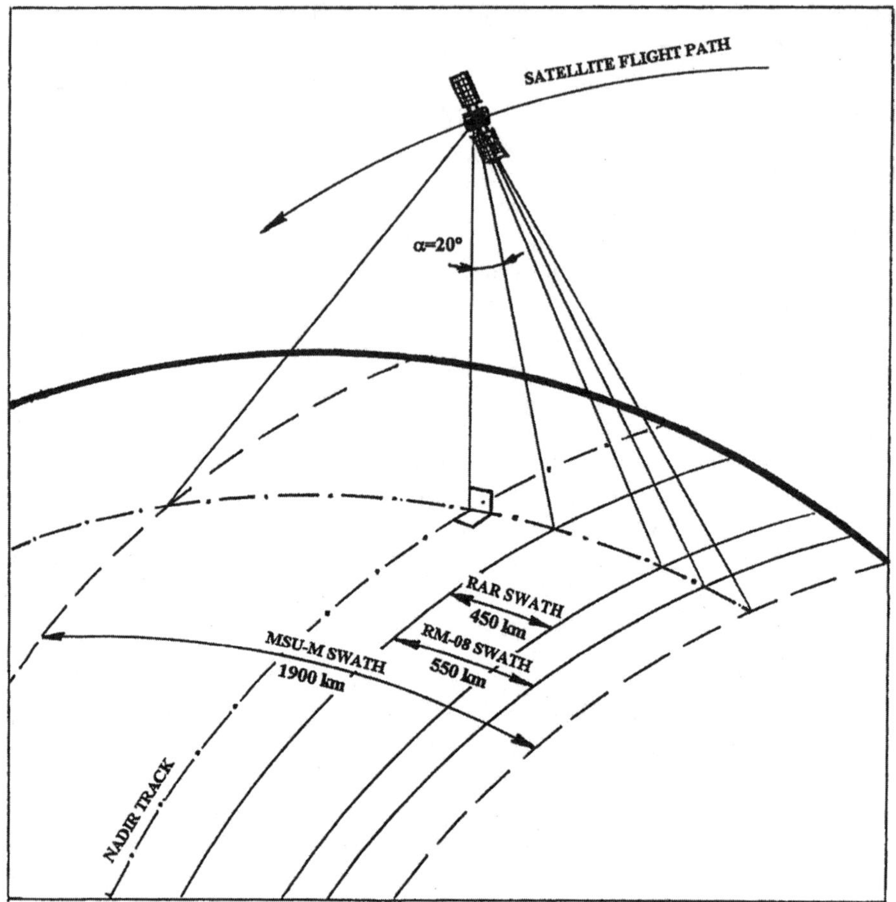

Figure 2.1 General scanning configuration for KOSMOS–OKEAN satellite series.

image has 24–32 plates plus k plates where $k < 8$. All plates provide calculating brightness temperature and correcting amplification coefficient.

Brightness temperature (T_b) calculation (Belchansky et al., 1997; Belchansky and Douglas, 2000) is performed after image georegistration. It includes estimating and correction of T_b across the image lines based on the calibration wedge and correction of T_b along image lines based on the reference objects on the image. Figure 2.2 shows a flow diagram creating a brightness image using raw RM-08 data.

Each byte of microwave data file describes one image or wedge pixel. The documentation file describes image and wedge parameters and is used for splitting data to the wedge and image and for definition of separate wedge gradation placement (for example if N is a number of gradation starting line, then $N + 48$ is a number of the next gradation starting line). The wedge data correction is performed by elimination of rough errors (noise), then averaging intensity values within wedge gradation. The table of correspondence between calibration wedge gradations and plates is developed based on partially linear representation. Dependence between the calibration wedge intensity $I(r, L)$ for every image line r and calibration wedge gradation L is derived as

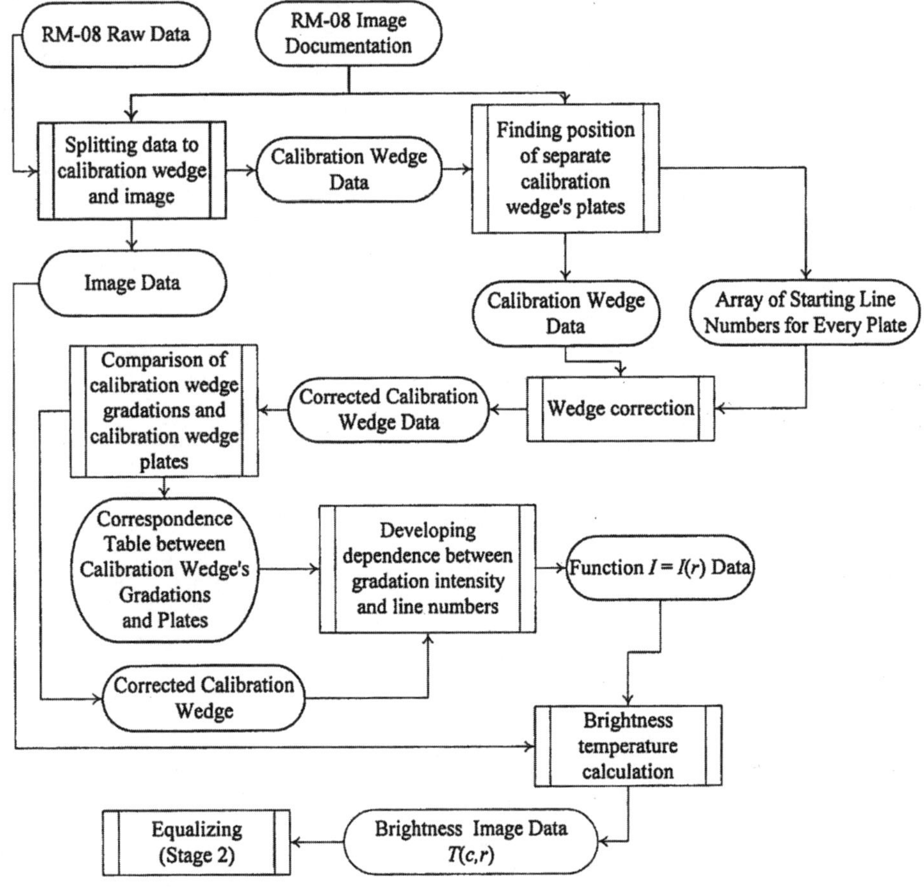

Figure 2.2 A flow chart of estimating and correction of T_b across the image lines based on the calibration wedge.

$$I(r, L) = I_{k1} + (I_{k2} - I_{k1})\frac{r - r_{k1}}{r_{k2} - r_{k1}}, \tag{2.1}$$

where $k1$, $k2$ are numbers of the calibration wedge plates, corresponding to the gradation number L, chosen to be situated as close as possible to line r (desirably before and after it); I_{ki} is the intensity of plate number ki ($i = \{1, 2\}$) and r_{ki} is the middle line of plate k_i ($i = \{1, 2\}$). There is a set of calibration wedges on every image. The plate is the intensity on every gradation calibration wedge. Every plate has a personal number for computing. An example of the function $I(r, L)$ for eight gradations is presented in Figure 2.3.

Brightness temperature $T_b(I, r)$ of pixel with intensity I in the image line r is calculated by the formula

$$T_b(I, r) = T^*(L) + (T^*(L+1) - T^*(L))\frac{I - I(r, L)}{I(r, L+1) - I(r, L)}, \tag{2.2}$$

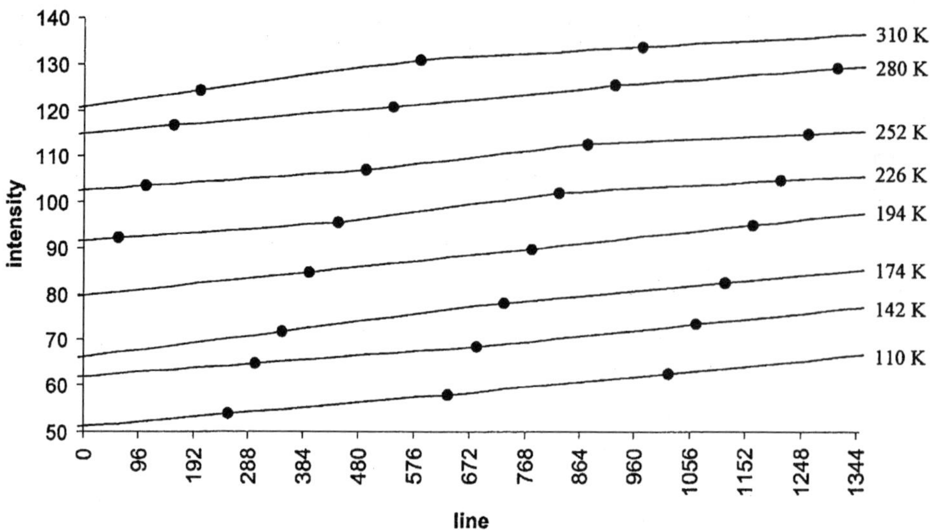

Figure 2.3 An example of the intensity function $I(r, L)$ for eight gradations of the calibration wedge.

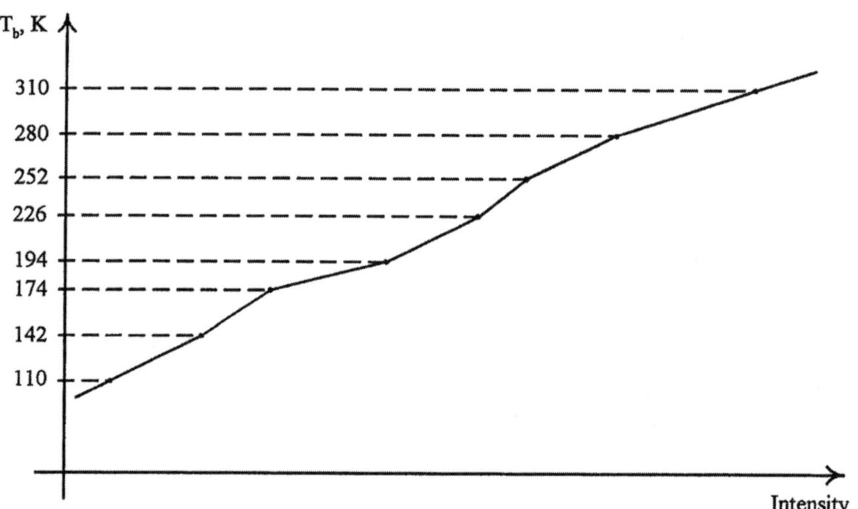

Figure 2.4 An example of possible brightness temperature $T_b(I, r)$ in an image line r.

where $T^*(L)$ is the brightness temperature corresponding to the calibration wedge gradation number L. Values of L are defined by inequality $I(r, L) \leq I < I(r, L + 1)$.

An example of possible brightness temperature $T_b(I, r)$ in an image line r is shown in Figure 2.4.

The correction of T_b along the image lines is performed using a reference object on the image. Figure 2.5 shows a flow chart of correction of T_b along image lines based on the reference objects on the image.

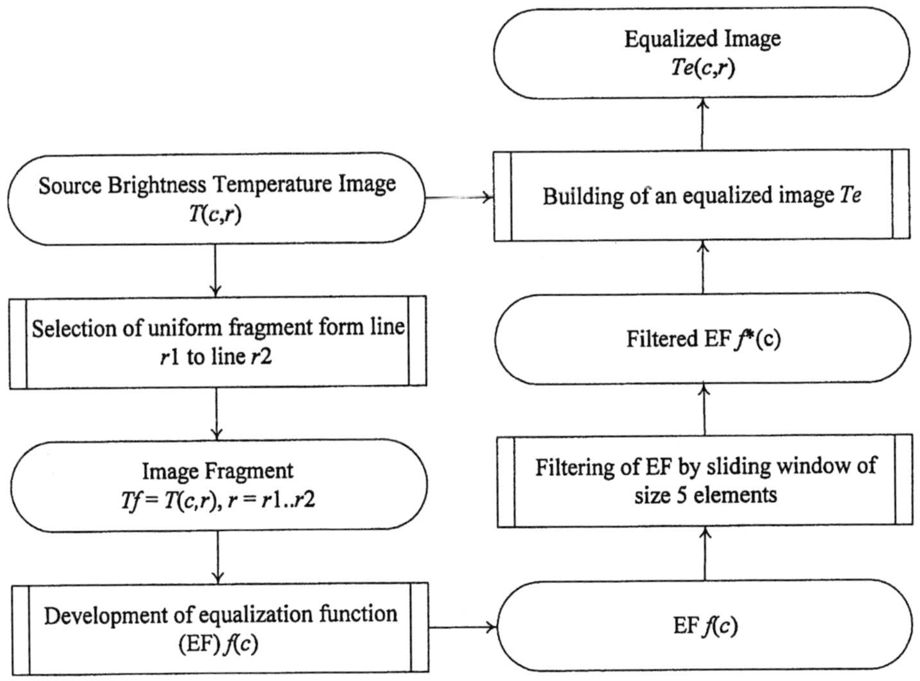

Figure 2.5 A flow chart of correction of T_b along the image lines based on the reference objects on the image.

The equalization function is presented as

$$f(c) = \frac{\sum_{r=r1}^{r2} T(c,r)}{\max_{c}\left(\sum_{r=r1}^{r2} T(c,r)\right)}, \tag{2.3}$$

where c is the image column number of line r, and $r1$, $r2$ are the first and last lines of the image fragment.

2.3.4.2 *Brightness Temperature Calibration Wedge Correction*

OKEAN passive microwave data can be adjusted to more closely emulate other microwave data, for example SSM/I results. Let us consider how the coincident brightness temperature (T_b) from another microwave sensor can be used for correction of RM-0.8 calibration wedge based on the scatterplot of the comparison values between T_b of RM-08 and T_b of another sensor (Belchansky et al., 1999; Belchansky and Douglas, 2000). The algorithm includes developing the scatterplot T_b (RM-08), T_b (another sensor), approximation of the scatterplot by the partially linear function using the least squares method (the nodes T_0,

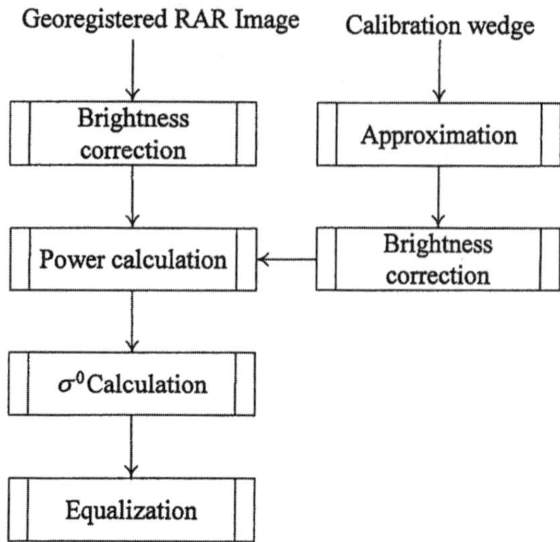

Figure 2.6 A flow chart of the RAR image brightness correction and estimating back-scatter coefficient σ^0 based on the calibration wedge.

$T_1, \ldots, T_7$ of this function correspond to the RM-08 wedge temperature gradations) and calculating the values of new RM-08 calibration wedge temperature gradation by substituting old values in the approximating function. The structure of approximation function and restrictions are presented as

$$\varphi(x) = \begin{cases} a_0 x + b_0, \ x \in S_0, \ \text{where } S_0 = \{T_0 \le x < T_1\} \\ a_1 x + b_1, \ x \in S_1, \ \text{where } S_1 = \{T_1 \le x < T_2\} \\ a_6 x + b_6, \ x \in S_6, \ \text{where } S_6 = \{T_6 \le x < T_7\} \end{cases}, \tag{2.4}$$

where the piecewise nodes $T_0, T_1, \ldots, T_7$ correspond to the RM-08 wedge temperature gradations; and x is brightness temperature, T_b (RM-08); $\varphi(x)$ is brightness temperature, T_b (another sensor).

The iteration wedge correction procedure is used to improve the correction accuracy. This procedure includes the repeated processing of raw RM-08 data to calculate a new brightness temperature image, new approximating function and new calibration wedge gradation values. The iteration process is finished when the square root of calibration wedge gradations is not greater than a given level.

2.3.4.3 *Active Microwave Image Calibration*

The RAR raw data include the set of the brightness calibration wedges and imagery. Every calibration wedge has eight power gradations in Watts (1:E_0 = 1.07×10^{-12}, 2:E_1 = 8.71×10^{-13}, 3:E_2 = 5.89×10^{-13}, 4:E_3 = 4.07×10^{-13}, 5:E_4 = 2.40×10^{-13}, 6:E_5 = 1.58×10^{-13}, 7:E_6 = 7.08×10^{-14}, 8:E_7 = 7.94×10^{-15}). The general principles of the radar calibration wedge construction are as follows: The delayed impulse from the transmitter passes through the whole receiver chain, the processing unit and the data transmission line and is recorded among the image data in terms of the brightness of a periodic recurrent wedge with eight levels of intensity, located at the start of the image. The radar calibration wedge is formed by alteration of the damping in the direct attenuator by passing to it a stepwise changing voltage, which is predetermined by the control plan. So it is possible to generate an absolute calibration signal for any well known natural formation with a known scattering coefficient (Kalmykov et al., 1989). The calibration wedge's structure is similar to the RM-08 wedge's structure. It consists of eight brightness gradations. Every gradation consists of 48 lines (every line = 30 pixels). This gradation is presented on the image by a set of plates. Every plate has a number and is referred to the concrete image lines on the RAR image. There are about 27–35 plates (three to four full calibration wedges plus some plates that present only part of the calibration wedge) on every RAR image. The radar calibration wedge is used for backscatter (σ^0) calculation based on the georegistered data. The calibration algorithms perform the approximation and correction of the calibration wedge brightness, the calculation of the returned power for each RAR image pixel, the calculation of the radar backscatter coefficient and the equalization of the backscatter coefficient based on the image reference objects. Figure 2.6 shows a flow chart of the RAR image brightness correction and estimating backscatter coefficient σ^0 based on the calibration wedge.

A set of the OKEAN-01 satellite radar parameters that is needed for RAR image calibration is presented in Table 2.5.

Table 2.5 A Set of the OKEAN-01 N7 sAtellite Radar Parameters Required for RAR Image Calibration

Parameter	Design.	Units of meas.	Value
Wavelength	l	cm	3.15
Peak power	P_0	kW	1.21
Antenna gain	G	dB	34.8
Antenna beamwidth (–3 dB width)			
range	V_d	degrees	42.0
azimuth	H_d	degrees	0.15
Pulse duration	t	μsec	3.1
Receiver noise power	E_n	W	1.26×10^{-15}
Loss in the transmitter-antenna circuit	L_T	dB	–1.70
Loss in the antenna-receiver circuit	L_R	dB	–1.70

Each byte of the radar data file describes one image or wedge pixel. The documentation file contains the image and wedge parameters. This file is used for splitting data to the wedge and image and for defining separate wedge gradation placement (for example, if N is the number of a gradation starting line, then $N + 48$ is the number of the next gradation starting line). The wedge correction is performed by averaging gradation intensity values and by eliminating rough (three standard deviations) errors in onboard data-processing systems. The table of correspondence between the calibration wedge gradations and the plates is performed based on the partially linear representation. Let us consider all main steps of the backscatter coefficient σ^0 calculation.

1. Approximation of the Calibration Wedge

   The dependence of the gradation (L) intensity, $I(r_i, L)$, from the image line number r_i is represented as

$$I(r_i, L) = I_{k1} + (I_{k2} - I_{k1})\frac{r_i - r_{k1}}{r_{k2} - r_{k2}}, \tag{2.5}$$

where $k1$, $k2$ are numbers of the calibration wedge plates, corresponding to the gradation number L; I_{ki} is intensity of the plate number k_i ($i = \{1, 2\}$), r_{ki} is number of the plate ki middle line ($i = \{1, 2\}$). $k1$ and $k2$ are chosen so that those plates were situated as close as possible to the line r_i and preferably before and after it. There is a set of calibration wedges on every RAR image. The plate is a calibration wedge gradation on every calibration wedge. Every plate has number value for computing the intensity of the calibration wedge as a function of the image line.

2. Correction of the Wedge Brightness

   The information signal has the constant offset, p, which is caused by the peculiarities of a RAR signal modulation in the radar system. The offset value, p, is calculated using the equation

$$p(r_i) = \frac{I(r_i, 7) \times b - I(r_i, 6)}{b - 1}.$$

   A priori known ratio, b, is defined as

$$b = \sqrt{\frac{E(6)}{E(7)}},$$

   where $E(6)$ and $E(7)$ are the powers in watts corresponding to sixth and seventh gradations of the radar calibration wedge. The value of b is set during the ground calibration and is constant during the entire flight time. The correction of the wedge gradation brightness $I^*(r_i, L)$ is performed using the following equation:

$$I^*(r_i, L) = I(r_i, L) - p(r_i). \tag{2.6}$$

3. Correction of the Image Brightness

   The temporal automatic amplification tuning (TAAT), δ, and the offset, p, are removed at this stage. The pixel brightness along the image line decreases because of the strong dependence of receiver input power on the incidence angle. For the convenience of the image visual interpretation, the brightness losses along the line are compensated by the TAAT, with a linear changing of the receiving gain from 1 at the first pixel to δ at the last one. Depending on the TAAT mode for OKEAN-01 N7, $\delta = \{2.99, 6.62, 1.02, 4.46\}$. The corrected brightness of the pixel c_i in the image line r_i is calculated by the following equation:

$$I^*(c_i, r_i) = \frac{I(c_i, r_i) - p(r_i)}{1 + \delta \times \frac{c_i}{c}}, \tag{2.7}$$

where c is the total number of elements in the line and $p(r_i)$ is just the same as for wedge brightness.

4. Calculation of the Image Power

   The received power corresponding to pixel with image coordinates (c_i, r_i) and corrected brightness I^* is calculated by the equation:

$$W(I^*, c_i, r_i) = E(L+1) + [E(L) - E(L+1)] \times \frac{[I^*(c_i, r_i)]^2 - [I^*(r_i, L+1)]^2}{[I^*(r_i, L)]^2 - [I^*(r_i, L+1)]^2} - E_n, \quad (2.8)$$

   where $E(L)$ is the power level of the wedge gradation number L. Values of L are defined by inequality $I(r_i, L+1) \le I(c_i, r_i) < I(r_i, L)$; $L = 0, 1, \ldots, 7$; E_n is the receiver noise power.

5. Calculation of the Image Backscatter Coefficient σ^0

   The basis for calculating the backscatter coefficient σ^0 is the main radar equation presenting the backscattered power, P_w, as follows (Belchansky et al., 1994):

$$P_w = \frac{P_o G^2 \lambda^2 S \sigma^0}{64\pi^3 D^4} A^2 L_T L_R F_d^2. \quad (2.9)$$

   The first parameter on the right of this equation is the transmitter power (P_o) delivered to the antenna during the time τ of a transmitted pulse. This power would result in a power density $P_o/4\pi D^2$ at range D from antenna to a target on the earth's surface. $4\pi D^2$ is an area of sphere with radius D. In the beam direction, the beam power is increased by G times, where G is an antenna gain, and power density becomes $P_o G/4\pi D^2$. Therefore, the power beam falling on the target with area S is $P_o G \times S/4\pi D^2$. The power is decreased by the atmosphere, the influence of which is given by coefficient A and by the transmitter–antenna circuit (the coefficient L_T). It is necessary also to involve the factor that takes into account the nonuniform power field distribution of the antenna (the coefficient F_d = the directivity multiplier). With these factors, the power falling on the target would be $P_o G A L_T F_d \times S/4\pi D^2$. If the target has a backscatter coefficient σ^0, then it scatters the σ^0 part of the falling power, i.e. $P_o G A L_T F_d \times S/4\pi D^2 \times \sigma^0$.

   The reflected radiation is affected by the following factors: a loss in atmosphere (A), a loss in the antenna–receiver circuit (L_R) and the antenna directional pattern (directivity multiplier F_d). The power beam is diffused in all spatial directions and power density near the antenna becomes $P_o G A L_T F_d \times S/4\pi D^2 \times \sigma^0 A L_R F_d \times 1/4\pi D^2$.

   The received power can be calculated from density by multiplying it by the antenna effective area S_a, which is $S_a = (\lambda^2/4\pi)G$ according to the reciprocity theorem, where G is an antenna gain and λ is a wavelength. Then $P_w = P_o G A L_T F_d \times S/4\pi D^2 \times \sigma^0 A_R F_d \times 1/4\pi D^2 \times (\lambda^2/4\pi) \times G$ and the backscatter coefficient σ^0 may be calculated using equation

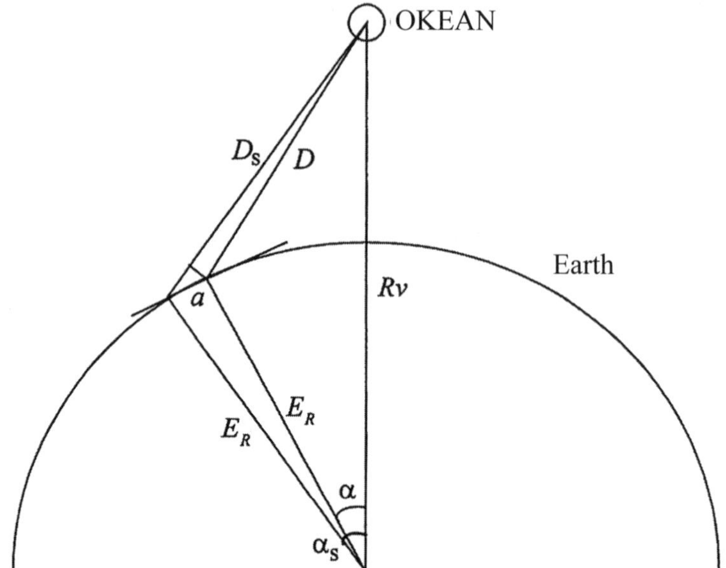

Figure 2.7 Definitions for the range plate.

$$\sigma^0 = \frac{64\pi^3}{P_0 G^2 \lambda^2 A^2 L_T L_R F_d^2} \frac{1}{S} \frac{D^4}{S} P_w. \tag{2.10}$$

The value P_w is a received power $W(I^*, c_i, r_i)$ corresponding to the pixel with the image coordinates (c_i, r_i) and the corrected brightness I^*, that is $P_w = W(I^*, c_i, r_i)$. $W(I, c_i, r_i)$. D is the slant range (Figure 2.7) to the surface element. It is defined by the time delay $T(c_i)$ depending on the pixel location c_i in the image line, and the light velocity C as $D(c_i) = T(c_i)C/2$. The antenna directivity multiplier, F_d, for RAR is calculated using the following equation $F_d(\varphi) = \cos(2\varphi\pi/3V_d)$, where φ is an angle between the target direction and the antenna electrical axis, and V_d is the antenna range. For RAR, the diagram multiplier changes from 0.7 at the angle with the antenna electrical axis of 15 degrees to 1.0 in the beam direction.

The surface element area is $S = ab$, where $b = DH_d$ (H_d is the antenna azimuth beamwidth) and $a = E_R\sqrt{2(1-\cos\Delta\alpha)}$, where α is an angle between satellite radius-vector R_v and the normal of the surface element (Figure 2.7), E_R is the Earth radius at the target location, $\Delta\alpha = \alpha_S - \alpha$, $\cos\alpha_S = \dfrac{R_v^2 + E_R^2 - D_S^2}{2R_v E_R}$, $D_S = D + \dfrac{C\tau}{2}$, τ is pulse duration. Figure 2.7 shows the elements location for a calculation.

6. Backscatter Coefficient σ^0 Image Equalization

The correction of the backscatter coefficient σ^0 along the image lines is performed using the reference object on the image. Figure 2.8 shows a flow chart

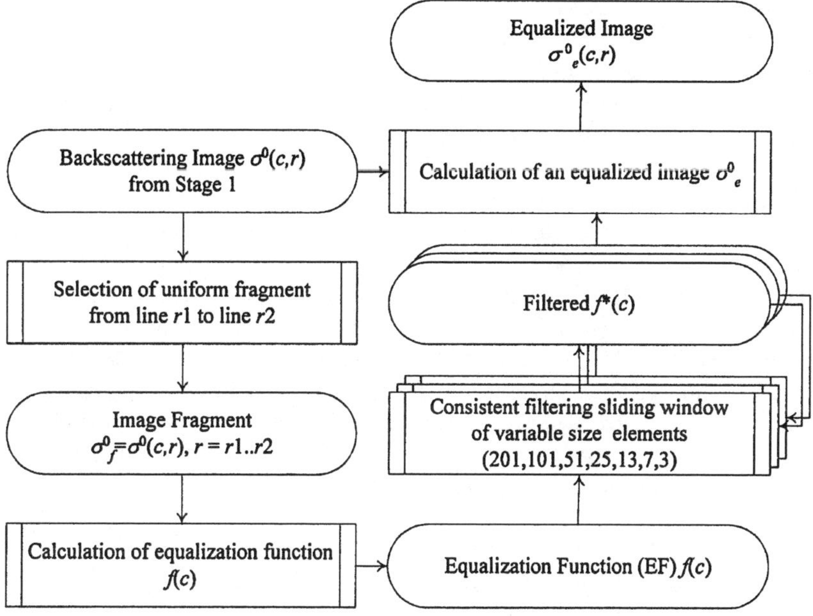

Figure 2.8 A flow chart of the correction of the backscatter coefficient σ^0 along image lines based on the reference objects on the image.

of the correction of the backscatter coefficient σ^0 along image the lines based on the reference objects on the image.

The equalization function is presented as

$$f(c) = \frac{\sum_{r=r1}^{r2} T(c,r)}{\max_{c}\left(\sum_{r=r1}^{r2} T(c,r)\right)}, \tag{2.11}$$

where c is a column number (pixel's number in the line), $r1$, $r2$ are first and last lines of image fragment.

The equalization function for RAR passes through several steps of filtering by variable window size. For RM-08, it has a fixed-size window.

2.3.5 Geometric Calibration and Registration

Geometric calibration of OKEAN-01 microwave data is the process by which the absolute location, image orientation errors and the relative image scale and skew errors are estimated. Registration, or geocoding, is the process of the geometrical correction where the image is resampled to some map projection. A significant difficulty lies in determining the overall locations of the image in geographic (latitude, longitude) coordinates (geolocation). The most effective method of geolocation for passive radiometric sensors is tiepointing, in which known geographic features are matched between the image and map data derived by conventional methods. This method can be used only for previously mapped regions. Arctic sea ice scenes are without distinguishing features. Therefore, Arctic satellite remote sensing data cannot be geolocated using tiepointing. Each OKEAN-01 microwave image is geometrically corrected and equalized to compensate for sensor look-angle. A set of geographic tiepoints are derived for each image using the satellite's orbital ephemeris data, a single user-defined ground-control point and iterations of user adjustments to the spacecraft's attitude (pitch, yaw, roll). The brightness temperature T_b and backscatter coefficient σ^0 images are converted to ARC/GRID, transformed to polar stereographic projection with 1-km pixel resolution. The geolocation of the OKEAN-01 microwave data is performed based on two main steps. On the first step, the one ground control point with known geographic coordinates is used for correction of the satellite orbital parameters (Belchansky et al., 1994). On the second step, the operator performs an image adjustment using some ground lines. The algorithm can be divided in the following stages:

- Identification of the image control point
- Calculating the geographic coordinates of the control point using given satellite's orbital ephemeris data (orbital parameters)
- Correcting orbital parameters so that the ground point coordinates be as match as close as possible to the reference point coordinates using the minimum distance criteria
- Calculating the geographic coordinates of all image points using orbital parameters corrected at the previous step
- Iterations of user adjustments to the spacecraft attitude (pitch, yaw, roll)

2.3.6 Structure of the Satellite Data

All satellite data are transmitted in the UHF and VHF bands. The data from the RAR and MSU-M having a spatial resolution of about 1 km are transmitted in the UHF (466.5 MHz) band to the ground station either directly or by readout from an onboard storage device. These stations are sided in regional centers in Moscow, Novosibirsk and Khabarovsk. The VHF (137.4 MHz) radio link is used for online transmission of data from one of the MSU-M spectral channels (by choice) and of lower-resolution data (about 2 km) from the RAR to a number of autonomous points for the reception of information on state hydrometry and other subjects. All satellite data can be transmitted in the UHF band in four distinct modes. For instance, microwave radiometer data + side-looking radar data + multispectral scanner are transmitted in mode 1. Special information can be placed in the beginning or in the end of line.

The structure of transmitted satellite data for every mode and an example of KOSMOS–OKEAN image (mode 1) are presented in Figures 2.9 and 2.10, respectively. Special information in Figure 2.9 is divided into the following intervals: interval 1, 15.2 + 1.0 ms = special board information; interval 2, 34.3 + 1.0 ms = the levels of black; interval 3, 12.0 + 1.0 ms = special board information. The complete length of line is about 3335 pixels. The complete length of the image information line is about 2825 pixels: MSU-M — about 1430 pixels, microwave radiometer — about 795 pixels, side-looking radar — about 625 pixels. The number of lines in the frame is 1500.

2.3.7 Data Acquisition, Processing and Distribution

Acquisition, processing and distribution of the KOSMOS–OKEAN satellite series data are carried out at the Scientific Research Centers for Natural Resources Studies (NITS IPR), Moscow Region, Dolgoprudny. NITS IPR includes the Main Space Data Collection and Processing Center (MSDCPC) and Region Space Data Collection and Processing Centers (RSDCPC). These centers are engaged in the research by consumers' demand and exploration of new kinds of information, new methods and technologies of space data collection and processing. Annually, MSDCPC and RSDCPC distribute information on orbit parameters, sensors characteristics, kinds of data and additional services to their potential consumers. The data are stored in the archives in the data-processing center, and the infrastructure of the users has been developed within the Russian Academy of Sciences and among the industrial organizations (Afanas'ev et al., 1989).

It is possible to receive directly transmitted information about a larger part of Russia and about the adjacent waters of the Arctic and Pacific Oceans. For each receiving station the radius of reception is almost 2500 km. One session of recorded information covers a region of approximate area 1930×2500 km in the case of the multiband MSU-M and of 470×2500 km in the case of the RAR.

Mode 1

Special Information		RM-08 Image		RAR Image	MSU-M Image
61.9 ± 0.1 ms		48.8 ± 0.1 ms		40.0 ± 0.1 ms	95.0 ± 0.9 ms
	↑	RM-08 Calibration Wedge, 1.8 ms	↑	RAR Calibration Wedge, 1.8 ms	

Mode 2

Special Information		RM-08 Image		RAR Image
61.9 ± 0.1 ms		97.6 ± 0.1 ms		79.7 ± 0.1 ms
	↑	RM-08 Calibration Wedge, 3.7 ms	↑	RAR Calibration Wedge, 3.9 ms

Mode 3

Special Information	Space		RAR Image
61.9 ± 0.1 ms	28.1 ms		120.3 ± 0.1 ms
		↑	RAR Calibration Wedge, 5.6 ms

Mode 4

Special Information	MSU-M Image
62.5 ms	187.5 ms

Figure 2.9 Structure of KOSMOS–OKEAN satellite series data.

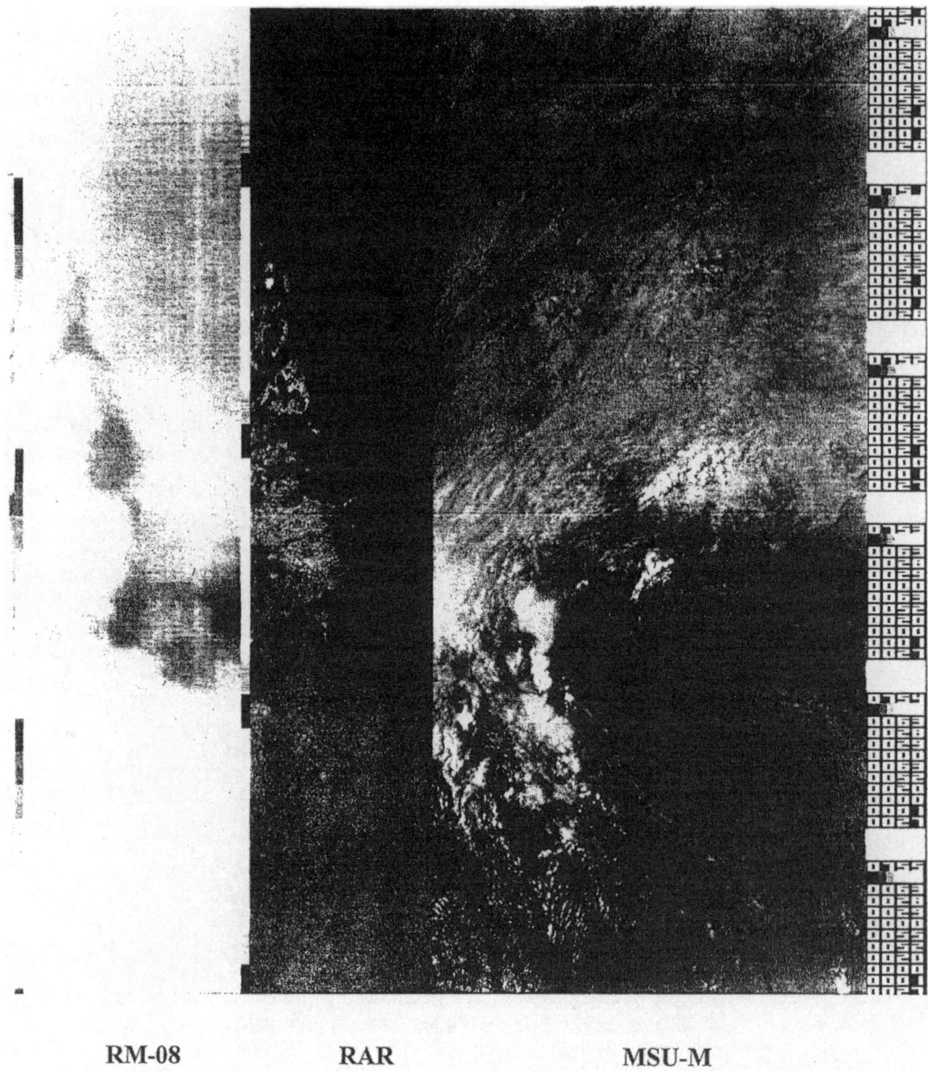

Figure 2.10 Example of KOSMOS–OKEAN satellite image (mode 1).

NITS IPR has a vast experience in using remote sensing, both Russian (METEOR-3, -3M; ELECTRO; RESURS-01; OKEAN-01; OKEAN-O) and foreign (SPOT, AVHRR, ERS and others), satellite spaceborne sensors for studying natural resources, providing hydrometeorological support, and estimating sea ice cover for ship navigation. Since 1974 NITS IPR has been responsible for carrying out maintenance of all operational satellite systems of the former Soviet Republic, and for the new Russian systems.

Let us consider the data acquisition, collection and processing system (DASPS) developed at the Institute of Ecology and Evolution, Russian Academy of Sciences, which supports the Arctic ecological research using KOSMOS–OKEAN satellite series data (Belchansky et al., 1994). The structure of this system is illustrated in Figure 2.11. It includes: Experiment Planning System (EPS); Data Acquisition System (DAS); Data Collection

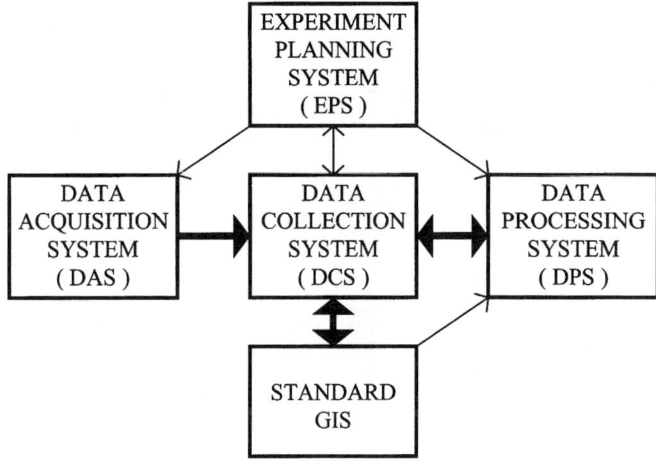

Figure 2.11 Data acquisition, collection and processing system (DACPS).

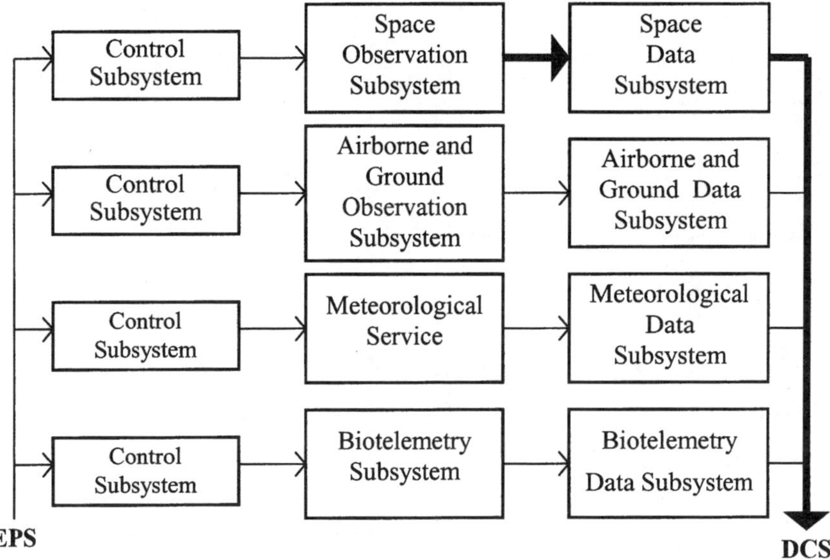

Figure 2.12 Data acquisition system (DAS).

System (DCS); Data Processing System (DPS); Intelligent GIS. The basic components of DAS, DCS, DPS and Intelligent GIS are shown in Figures 2.12, 2.13, 2.14 and 2.15.

The Experiment Planning System (EPS) is used for optimization of remote sensing observation and field measurements.

The Data Acquisition System (DAS) (Figure 2.12) includes:

- Control Subsystems
- Space Observation Subsystems (OKEAN, ALMAZ, AVHRR, ERS)

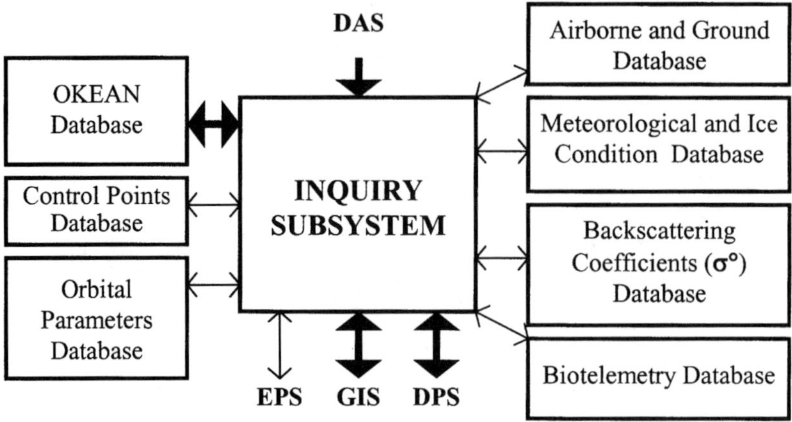

Figure 2.13 Data collection system (DCS).

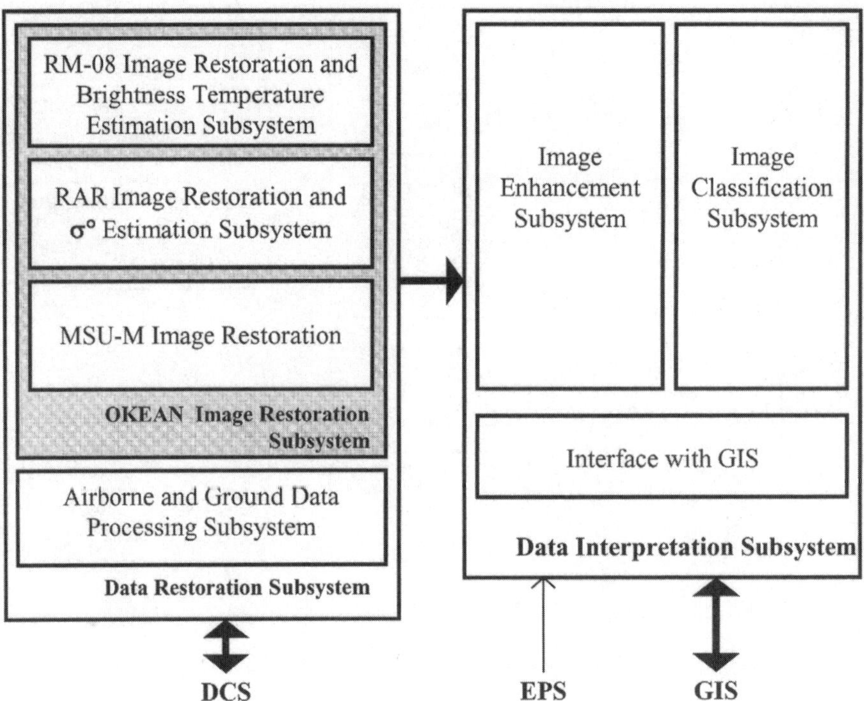

Figure 2.14 Data-processing system (DPS).

- Airborne and Ground Observation Subsystems
- Meteorological Service
- Biotelemetry Subsystem
- Space Data Subsystem

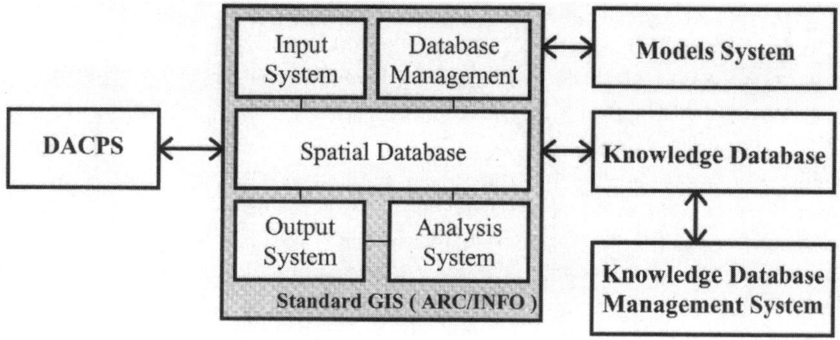

Figure 2.15 Structure of intelligent GIS.

- Airborne and Ground Data Subsystem
- Meteorological Data Subsystem
- Biotelemetry Data Subsystem

The Data Collection System (DCS) (Figure 2.13) includes:

- OKEAN Database (image data — RAR, RM-08, MSU-M, image attributes, board system parameters, orbital parameters measurements)
- Database of Air and Ground Data; Meteorological and Ice Condition Database
- Sigma Nought Ground Objects Database
- Control Points Database
- Inquiry Subsystem for connection with database
- Orbital Parameters Database

The Data Processing System (DPS) (Figure 2.14) consists of OKEAN Data Restoration Subsystem (DRS) and Data Interpretation Subsystem (DIS).

DRS includes:

- RM-0.8, RAR and MSU-M Image Restoration Subsystems
- Airborne and Ground Data Processing Subsystem
- Output File Subsystem
- Sets of Special Modules (first step processing, correcting for a variety of geometric distortion based on orbital parameters and reference points, Sigma Nought estimation based on calibration wedge, output file creation)
- Image Enhancement and Classification Subsystem

The structure of Intelligent GIS is shown in Figure 2.15. All systems are realized on a set of IBM PC/AT-compatible computers.

To date, this DASP is used for processing raw OKEAN-01 satellite images, including both digital calibration and automated registration to geographic map coordinates and for production of value-added ice products and charts.

The developed special software permits to radiometrically calibrate, equalize and georeference raw RM-08 and RAR imagery (Belchansky et al., 1993). Brightness temperatures in degrees Kelvin (T_b) are derived for every RM-08 pixel based on linear relationships with calibration data that are embedded in the raw imagery during data collection. The radar backscatter (σ^0) is derived for each RAR pixel as the ratio of the backscatter return power divided by the original output power (Fung, 1994). Each image is geometrically corrected and equalized to compensate for sensor look-angle. A set of geographic tie-points are derived for each image using the satellite's orbital ephemeris data, a single user-defined ground-control point, and iterations of user adjustments to the spacecraft attitude (pitch, yaw, roll). The T_b and σ^0 images are converted to ARC/GRID (Environmental Systems Research Institute, Redlands, CA, USA), transformed to polar stereographic projection with 1-km pixel resolution and then averaged to 3-km resolution using a low-pass filter. Each microwave satellite pass is presented by three types of files: (i) raw RM-08 and RAR imagery files (intensity value ranges 0–255, step 1, format — byte/pixel, pixel size — 1×1 km and documentation files (orbital parameters, time and conditions of survey, image size); (ii) geometrically and radiometrically corrected and calibrated imagery (pixel size 1×1 km, format — 2 bytes/pixel, polar stereographic projection; (iii) sea-ice type and concentration images (band 1 — water, band 2 — first-year ice, band 3 — multiyear ice, band 4 — deformed ice, band 5 — other ice, pixel size — 3×3 km, format-byte/pixel, sea-ice concentration value ranges from 0 to 100 with 0.1 increment, polar stereographic projection).

2.3.8 Main Applications

The main OKEAN-01 satellite data applications include the open water studies, the pollution observations, the sea ice cover studies (ice cover structure, ice age and ice thickness associated with it, sea-ice boundary and its variability in time) and the soil studies (Kalmykov et al., 1989; Efimov et al., 1985; Bushuev, 1989; Belchansky and Pichugin, 1991).

2.4 ALMAZ SATELLITE SERIES

2.4.1 ALMAZ SAR (KOSMOS-1870) Satellite

The first ALMAZ SAR satellite (KOSMOS-1870) was launched on July 25, 1987 into the orbit with inclination of 71.9 degrees and altitude of 250–300 km. The satellite turned round the Earth once each 89.3–90 minutes, daily westward shift in trajectory was 2–5 degrees, one-period sub-satellite track shift was 24 degrees. This satellite functioned on the orbit for more than two years and its mission was finished on July 30, 1989 (Figures 2.16 and 2.17).

KOSMOS-1870 was equipped with two synthetic aperture radars (SAR), one on each side of satellite, providing 11–15 m resolution. The spacecraft was capable of rolling around its axis, thereby extending the pointing range of SAR antennas in the cross-track direction to 250 km. Swath width was nearly constant at 2×20 km.

KOSMOS-1870 had a recording capability and subsequent data were transmitted to the data acquisition and processing center near Moscow via a radio channel at 90 Mbit/sec.

Data transmission antennas
Solar array
SAR Antennas

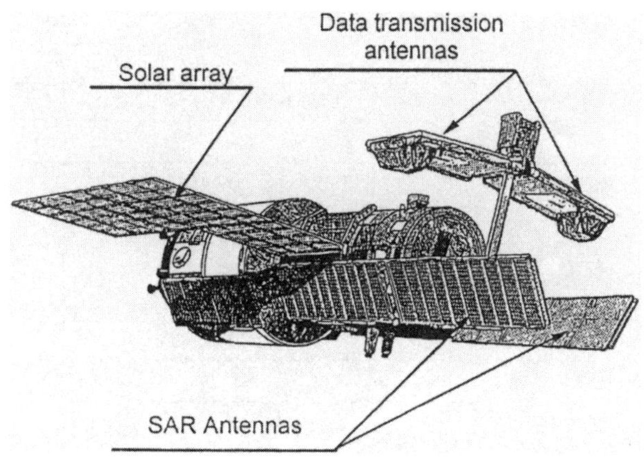

Figure 2.16 General configuration of ALMAZ-1 space station.

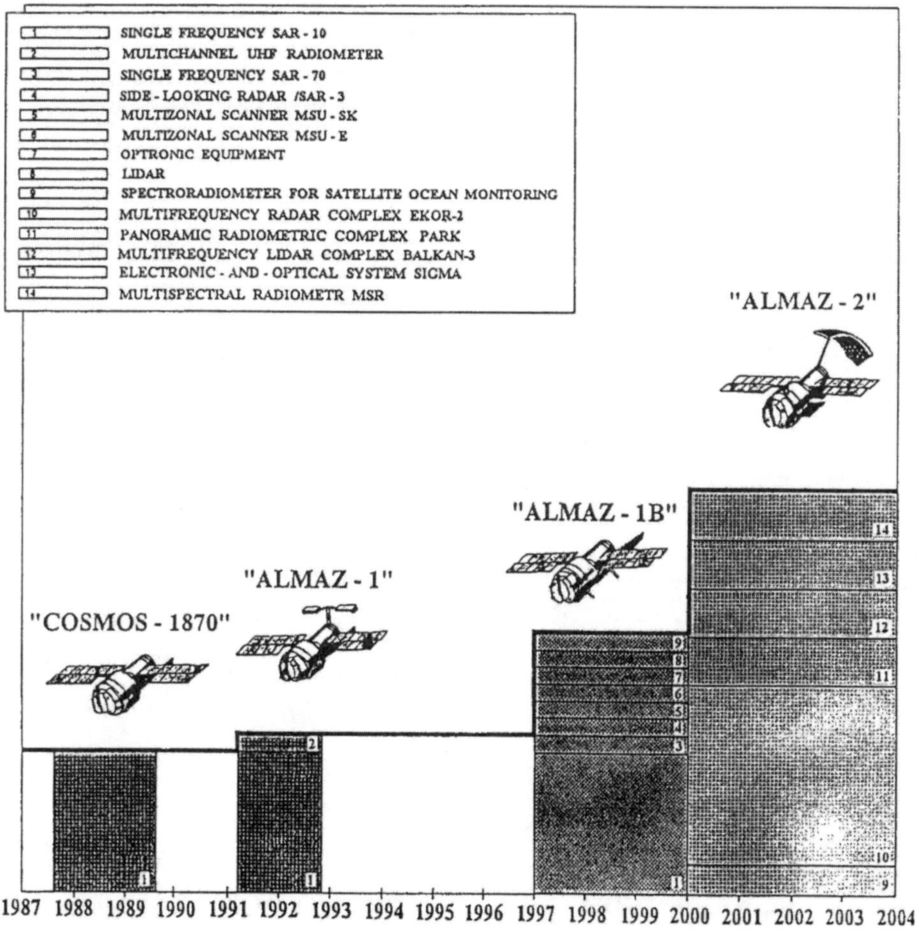

1 SINGLE FREQUENCY SAR - 10
2 MULTICHANNEL UHF RADIOMETER
3 SINGLE FREQUENCY SAR - 70
4 SIDE - LOOKING RADAR /SAR - 3
5 MULTIZONAL SCANNER MSU - SK
6 MULTIZONAL SCANNER MSU - E
7 OPTRONIC EQUIPMENT
8 LIDAR
9 SPECTRORADIOMETER FOR SATELLITE OCEAN MONITORING
10 MULTIFREQUENCY RADAR COMPLEX EKOR-2
11 PANORAMIC RADIOMETRIC COMPLEX PARK
12 MULTIFREQUENCY LIDAR COMPLEX BALKAN-3
13 ELECTRONIC - AND - OPTICAL SYSTEM SIGMA
14 MULTISPECTRAL RADIOMETR MSR
"ALMAZ - 2"
"ALMAZ - 1B"
"ALMAZ - 1"
"COSMOS - 1870"
1987 1988 1989 1990 1991 1992 1993 1994 1995 1996 1997 1998 1999 2000 2001 2002 2003 2004

Figure 2.17 Structure of ALMAZ SAR satellite program.

Table 2.6 Main Parameters of KOSMOS-1870 and ALMAZ-1 SAR Satellites

	Mission	
	KOSMOS-1870 (ALMAZ)	ALMAZ-1
Orbit		
Altitude (km)	250–300	270–380
Inclination (degrees)	71.9	72.7
Period (min)	89.3–90	89–92
24-h-sub-satellite track shift (degrees)	2–5	2–5
Attitude precision (s)	15–20	15–30
Stabilization precision (s)	4–6	4–6
Other Characteristics		
Initial mass on the operational orbit (kg)	18550	18550
Launch vehicle	Proton	Proton
Launch date	July 25, 1987	March 31, 1991
Mission end	July 30, 1989	October 17, 1992
Methods of image transmission to the ground data receipt station	via relay-satellite	via relay-satellite
Data transmission rate (Mb/s)	10	10
SAR Characteristics		
Band	S-band	S-band
Transmit frequency (GHz)	3.0	3.0
Wavelength (cm)	9.6	9.6
Polarization	HH	HH
Look (incidence) angle (degrees)	30–60	30–60
Right/left looking	right & left	right & left
Swath width (km)	2×(20–30)	2×(20–30)
Swath range km)	>200	350
Peak power (kW)	250	190
Middle power (kW)	80	80
Antenna type	waveguide	waveguide
Antenna size (m×m)	15×1.5	15×1.5
Pulse duration (μs)	0.1	0.07 or 0.1
Pulse repetition frequency (Hz)	3000	3000
Radiometric resolution (dB)	5	3–5
A to D sampling rate (MHz)	28.8 or 20.16	28.8 or 20.16
Signal quantization (bits)	10 (5I, 5Q)	10 (5I, 5Q)
Noise equivalent σ_0 (dB)	3	3
SAR Imaging		
Range resolution (m)	15–8	16–6 (0.07 μs), 20–8 (0.1 μs)
Azimuth resolution (m)/looks	11–15/1	11–15/1
UHF-Radiometer Characteristics		
Frequencies: 1 (GHz)	—	6.0
2 (GHz)	—	37.5
3 (GHz)	—	2500.0
4 (GHz)	—	2720.0
Swath width (km)	—	10–30
Swath range (km)	—	500
Temperature resolution (degrees K)	—	0.1–0.3
Spatial resolution (km)	—	5

The KOSMOS-1870 surveys proved the possibility to investigate wind waves (energy components) and ripples; hydrological fronts, dynamics of streams, surface manifestation of inner waves, whirls and rings of various scales, oil spills and underwater relief, circumvent flows of a leeward type (Chelomey et al., 1990; Salganik et al., 1990). Main KOSMOS-1870 SAR satellite parameters are presented in Table 2.6.

2.4.2 ALMAZ-1 SAR Satellite

The ALMAZ-1 SAR satellite was launched on a Proton booster into the orbit with inclination of 72.7 degrees and altitude of 270–380 km. The satellite turned round the Earth once each 89.0–92.0 minutes. After a year and a half of operation the satellite mission was completed.

The ALMAZ-1 SAR was equipped with advanced single frequency (S-band) single polarization SAR providing 10–15 m spatial resolution, depending on the incidence angle and double swath width equal to 40 km. Incidence angles were selectable between 30 and 60 degrees. There were two slotted waveguide scanning antennas each 15 by 1.5 m in size, one on each side of the satellite providing image data on either side of nadir. The ALMAZ-1 SAR satellite had the multispectral UHF radiometer and more efficient data transmission system. Digital satellite data could be stored on onboard recorders with 150 seconds of recording time and down-linked from the ALMAZ-1 to a ground acquisition station via a relay-satellite. ALMAZ-1 SAR observation geometry, earth coverage and example of image are shown in Figures 2.18, 2.19 and 2.20, respectively. The main ALMAZ-1 SAR satellite parameters are presented in Table 2.6.

The structure of ALMAZ-1 SAR satellite (Figure 2.16) includes: propulsion system, solar batteries, SAR antennas, data reception and transmission units, navigation equipment, part of space exploration equipment, etc. ALMAZ-1 SAR data collection, processing and distribution were carried out by the NPO Mashinostroenie (the scientific engineering center for planning and data processing and distribution mission). The user's infrastructure has been developed within the Academy of Sciences and among industrial organizations. A series of international programs on remote sensing have been implemented with the participation of the U.S. and France.

The ALMAZ-1 SAR was used in the Russian Ocean Scientific Program aimed at the development of radiophysical studies of the ocean, of the generation and development of surface gravitation and gravity capillary waves; evaluation of radar data content for fundamental oceanology and ecology studying (Viter et al., 1994). The estimating of imaging radar efficiency has shown its high data content in the context of Global Change investigations: sea ice, atmosphere–sea-ice interaction, oceanography, ice sheets and glaciology, hydrology, land cover, geomorphology and geology, volcanology and interferometry.

There were several successful application demonstrations using ALMAZ SAR data in the following areas: ocean research, geological mapping and mineral resources reconnaissance, study of the ecological state of coastal areas and inland waters, ecological monitoring, remote control of land usage, monitoring of ice situation, etc.

2.4.3 ALMAZ-1B and ALMAZ-2 SAR Satellites

ALMAZ-1B SAR will be the third Russian satellite of ALMAZ satellite series. It was to be launched in 1999 into an orbit with an altitude of 350–400 km and inclination of 73 degrees.

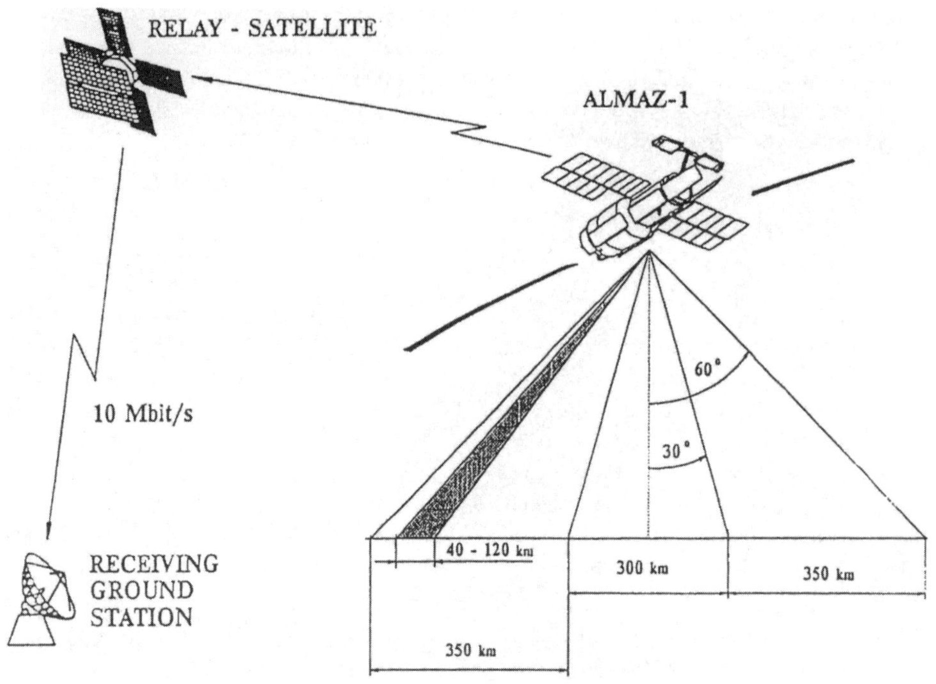

Figure 2.18 ALMAZ-1 SAR observation geometry.

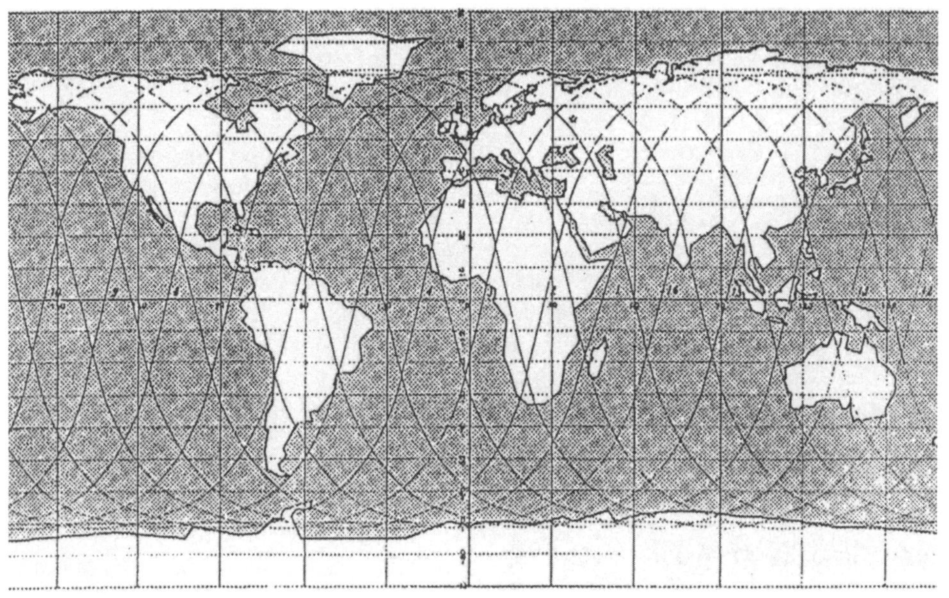

Figure 2.19 ALMAZ-1 SAR earth coverage.

Figure 2.20 Example of ALMAZ-1 SAR image.

ALMAZ-1B differs from the previous satellite since it is planned to carry the onboard complex of the remote sensing tools which includes a set of various instruments providing great data volume of the earth surface features. This complex will include a radar system with three radars operating at the wavelengths of 3.6 cm, 9.6 cm and 70 cm (X-, S- and P-bands, respectively, spatial resolution of 5–7 m, 15–40 m and 20–40 m), two multispectral scanning systems (MSU-E, MSU-SK), optical electronic instruments for the high resolution stereo observations, a spectroradiometer and a lidar.

Data transmission modes will include: direct transmission to the acquisition stations in Russia and in other countries (data transmission rate of 122 Mb/sec); onboard data accumulation and down-link to the ground stations when flying above them (data transmission rate of 3 or 0.96 Mb/sec), onboard data accumulation and down-link to the central acquisition station using a relay-satellite (data transmission rate 10 Mb/sec). Main ALMAZ-1B SAR satellite parameters and characteristics of other instruments are presented in Tables 2.7 and 2.8, respectively.

The further evolution of the ALMAZ program will be represented by the launch of the spaceborne multifunctional ALMAZ-2 satellite. The satellite will provide the observations for use in various fields of the earth and atmosphere studies. It is planned to install about 15 instruments onboard the satellite. Besides, ALMAZ-1B and ALMAZ-2 will be equipped with instruments for studying the composition and cleanliness of the atmosphere. Many of these instruments because of their mass and size can only be placed on an ALMAZ type satellite. ALMAZ-2 will secure a place for itself in the system of heavy universal space platforms. These platforms study the earth in a way which is not possible for the light class spacecrafts or manned complexes. The ALMAZ satellite series, beginning with ALMAZ-1B, could take part in the international EOS system. Main ALMAZ-2 SAR satellite parameters and instruments are presented in Tables 2.7 and 2.8.

2.5 CONCLUSION

Multispectral optical satellite systems are widely used for Arctic ecological studies. However, periods of prolonged darkness or persistent cloud cover often preclude systematic data acquisition. Active and passive microwave sensors are not limited by illumination, or restricted by clouds and fog since the radio brightness temperature and the radar backscatter measured in the microwave band for fixed sensor parameters (emission frequency, polarization) and angles of observation depend on the physical parameters of surface.

Since 1972, three generations of space borne passive microwave imagers have been launched by the United States, including the ESMR (1972), the SMMR (1978) and a series of the DMSP SSM/I sensors (beginning from 1987). The U.S. space borne SAR systems have evolved through SEASAT and Shuttle Imaging Radar to a multifrequency, multipolarization systems. The European (ERS-1, 2 SAR), Japanese (JERS-1 SAR) and Canadian RADARSAT-1 SAR satellites carry the one frequency and one polarization imaging radar system with different parameters.

The Russian microwave satellite program includes the KOSMOS–OKEAN, RESURS–ARKTIKA SAR, KOSMOS-1870 SAR and ALMAZ SAR satellite series. The ALMAZ-1B will be the third Russian satellite of ALMAZ satellite series. This series differs from the previous satellites since it is planned to carry three SAR operating, respectively, at X-, S- and P-bands, two multispectral scanning systems, optical electronic

Table 2.7 Main ALMAZ-1B, ALMAZ-2 SAR satellite parameters.

	Mission				
	ALMAZ-1B			ALMAZ-2	
Orbit					
Altitude (km)	350–400			600	
Inclination (degrees)	73			73	
Period (min)	92				
Observed regions	78° N–78° S				
Other Characteristics					
Initial mass on the operational orbit (kg)	18550			21000	
Mass of playload for remote sensing (kg)	4500			6500	
Launch vehicle	Proton			Proton	
Active lifetime (years)	up to 3			5	
Power supply (kW)	2.3–3.3			2.5	
Methods of image transmission to the ground data receipt station	via relay-satellite or direct			via relay-satellite or direct	
Data transmission rate (Mb/s)	10 (relay-satellite) 122 (direct)			10 (relay-satellite) 122 (direct)	
SAR Characteristics					
Instrument	SAR-10	SAR-70	SAR-3	SAR	SAR
Band	S-band	P-band	X-band	S/C/X-bands	
Wavelength (cm)	9.6	70	3.6	10, 5.0, 3.0	120
Viewing angle off the nadir (degrees)	25–51	25–46			
Resight swath (km)	330	330	330		
Observing mode					
Polarization	VV	H/VH,V/VH			
Spatial resolution (m)	15–40	20–40	5–7	10–15, 50	20–200
Swath width (km)	120–170	120–170	20–35	75–100, 500	200–300
Data rate (Mb/s)	104–288	172–488	116–370		
Side looking	right	right			
Intermediate mode					
Polarization	V/HV,H/HV				
Spatial resolution (m)	15				
Swath width (km)	60–70				
Data rate (Mb/s)	354–740				
Side looking	right				
Detailed mode					
Polarization	HH				
Spatial resolution (m)	5–7				
Swath width (km)	30–50				
Data rate (Mb/s)	172–582				
Side looking	left				

instruments for stereo observations, a spectroradiometer and a lidar. ALMAZ SAR data are used in ocean research, geological mapping, ecological monitoring, land usage, etc. Today OKEAN-01 satellites are used for sea ice monitoring, Arctic ecological research, mapping of the Arctic and Antarctic ice fields.

Special Sensor Microwave/Imager (SSM/I) of Defense Meteorological Satellite Program (DMSP) is reliably used for Arctic monitoring but because of its relatively coarse resolution (near 25 km) application has been limited to broad-scale investigations.

Table 2.8 ALMAZ-1B, ALMAZ-2 Instruments Parameters

ALMAZ-1B	
Multispectral Scanning System MSU-E2	
Waveband (μm)	0.5–0.6, 0.6–0.7, 0.8–0.9
Spatial resolution (m)	10
Number of sets on board	2
Swath width (km)	2×24
Data rate (Mb/s)	11.5 (3 channels)
Multispectral Scanning System MSU-SK	
Waveband (μm)	0.5–0.6, 0.6–0.7, 0.7–0.8, 0.8–1.1, 10.3–11.8
Spatial resolution (m)	170 (1–4 channels), 600 (5 channels)
Swath width (km)	600
Data rate (Mb/s)	11.5 (5 channels)
Spectroradiometer for Sea Surface Monitoring SROSM	
Waveband (μm)	0.4–12.5
Spatial resolution (m)	600
Swath width (km)	2×100
Data rate (Mb/s)	2.4 (11 channels)
Temperature resolution (degrees K)	0.1
Optical & Electronic Instrumentation for Stereo Survey SILVA	
Waveband (μm)	0.5–0.6, 0.6–0.7, 0.7–0.8, 0.58–0.8
Spatial resolution (m)	2.5–4
Swath width (km)	80
Resight swath (km)	±300
Data rate (Mb/s)	560 (1 channel)
Lidar BALKAN-2	
Waveband (nm)	532
Vertical resolution (m)	
lidar mode	3–10
range-finding mode	0.5–1.0
Beam width (arc sec)	40
Swath width (degrees)	±10 from nadir
Sensing frequency (Hz)	1
ALMAZ-2	
Microwave Radiometer	
Waveband (cm)	0.3–21
Spatial resolution (m)	3000–20000
Number of channels	7
Swath width (km)	1200
Rain Radar System	
Waveband (cm)	1–2
Spatial resolution (m)	250
Number of channels	2
Swath width (km)	100
Multispectral Scanning System MSS	
Waveband (μm)	0.4–1.1
Spatial resolution (m)	20
Number of channels	11

Table 2.8 (continued)

Swath width (km)	1000
Lidar	
Waveband (μm) Number of channels	0.266–1.064 8
Infrared Spectrometer	
Waveband (μm)	2.5–10.5
Sidereal Photometer	
Waveband (μm) Number of channels	0.45–0.63 2
Solar Spectrometer	
Waveband (μm)	0.1–4.5

Compared to passive microwave, synthetic aperture radar (SAR) satellite systems (Russian ALMAZ, European ERS, Japanese JERS) collect high resolution data (spatial resolution near 20 m) over much smaller geographic areas. The potential for using SAR data to address Arctic ecological questions warrants thorough investigation. Therefore more studies are required to evaluate the diversity of methods for processing and analyzing these data in the context of specific ecological applications and specific system characteristics.

The polar-orbiting satellite series KOSMOS–OKEAN and RESURS–ARKTIKA SAR carries intermediate-resolution multisensor instruments and has the unique capability of simultaneously collecting passive microwave, real aperture radar and optical imagery. These satellites can record data for the entire Arctic over a 3-day period. Acquiring both passive and active data provides more distinct, integrated signatures.

However, all microwave instruments have identical problems with an inability to effectively distinguish ice cover during the presence of meltponding, wet snow or flooding (Cavalieri et al., 1990; Comiso, 1991). To this end, the multisensor system has advantage of differing sensitivities to different surface types, enhanced temporal resolution and multi-scale spatial resolutions that can be used to assess the precision of broad-scale or long-term interpretations.

REFERENCES

Afanas'ev, Yu. A., Nelepo, B. A. et al., 1989. Programme of experiments on the KOSMOS-1500 satellite. *Soviet Journal of Remote Sensing*, Vol. 5 (3), pp. 347–358.

Aleksandrov, V. Yu. and Loshchilov, V. S., 1989. Quantitative interpretation of satellite radar images of sea ice using a priori data. *Soviet Journal of Remote Sensing*, Vol. 5(3), pp. 391–396.

Bass, F. G. et al., 1968. Very high frequency radiowave scattering by a disturbed sea surface. *IEEE Trans.*, AP-16, pp. 554–559.

Belchansky, G. I. and Douglas, D. C., 2000. Classification methods for monitoring Arctic sea-ice using OKEAN passive/active two-channel microwave data. *J. Remote Sensing Environment*, 73(3), pp. 307–322.

Belchansky, G. I. and Pichugin, A. P., 1991. Radar Sensing of Polar Regions. Proceedings of the International Conference on the Role of the Polar Regions in Global Change, Edited by G. Weller, C. L. Wilson, and B. A. Severin, Geophysical Institute, University of Alaska Fairbanks, and Center for Global Change and Arctic System Research, University of Alaska Fairbanks, pp. 47–57.

Belchansky, G. I., Douglas, D. C. and Ovchinnikov, G. K., 1994. Processing of Space-monitoring Data to Document Parameters of the Habitat of Arctic Mammals. *Soviet Journal of Remote Sensing*, Vol.11, No. 4, pp. 623–636.

Belchansky, G. I., Douglas, D. C. and Kozlenko, N. N., 1997. Determination of the types and concentration of sea ice from satellite data from two-channel microwave active and passive observation systems. *Earth Observation and Remote Sensing*, Harwood Academic Publishers. Vol. 14, No 6, pp. 891–905.

Belchansky, G. I., Platonov, N. G. and Eremeev, V. A., 1999. Correction of OKEAN RM-08 radiometer calibration wedge. *Earth Research from Space*, No. 1, pp. 75–81.

Bushuev, A. V. and Bychenkov, Yu. D., 1989. Use of KOSMOS-1500 satellite radar images to study the distribution and the dynamics of sea ice. *Soviet Journal of Remote Sensing*, Vol. 5(3), pp. 382–390.

Carsey, F., 1982. Arctic sea ice distribution at end of summer 1973–1976 from satellite microwave data. *J. Geophys. Res.*, 87(C8), pp. 5809–5835.

Carsey, F., 1993. Science Requirements for RADARSAT/ERS-2 Modifications for the Alaska SAR Facility. JPL D-10443, Jet Propulsion Laboratory, Pasadena, 22 p.

Cavalieri, D. J., Burns, B. A. and Onstott, R.G., 1990. Investigation of the effects of summer melt on the calculation of sea ice concentration using active and passive microwave data. *J. Geophys. Res.*, 95(C4), pp. 5339–5369.

Cavalieri, D. J., Gloersen, P. and Campbell, W. J., 1984. Determination of sea ice parameters with Nimbus-7 SMMR. *J. Geophys. Res.*, 89(D4), pp. 5355–5369.

Cavalieri, D. J. and Parkinson, C. L., 1981. Large-scale variations in observed Antarctic sea ice extent and associated atmospheric circulation. *Monthly Weather Review*, Vol. 109, pp. 2323–2336.

Cavalieri, D. J., Parkinson, C. L., Gloersen, P. and Zwally, H. J., 1997. Arctic and Antarctic sea ice concentrations from multichannel passive-microwave satellite data sets: October 1978 to September 1995, User's Guide. NASA Technical Memorandum 1045647, 17 p.

Comiso, J. C., 1991. Top/bottom multisensor remote sensing of Arctic sea ice. *J. Geophys. Res.*, 96(C2), pp. 2693–2709.

Comiso, J. C., 1994. Surface temperatures in the polar regions from Nimbus-7 temperature humidity infrared radiometer. *J. Geophys. Res.*, Vol. 99, No C3, pp. 5181–5200.

Coordination for the Next Decade. 1995. Committee on Earth Observation Satellites, ESA, 133 pp.

Chan, H. L. and Fung, A. K., 1977. A theory of sea scatter at large incident angles. J. Geophys. Res., 82, pp. 3439–3444.

Chelomey, V. N., Efremov, G. A. et al., 1990. High resolution radar sensing of the sea surface from SAR KOSMOS-1870. *Earth Research from Space*, 2, pp. 80–90.

Curlander, J. C. and McDonough, R. N., 1991. Synthetic aperture radar: systems and signal processing. New-York: John Wiley & Sons, Inc.

Elahci, C., 1987. Introduction to the physics and techniques of remote sensing. Wiley-Interscience Publications, John Wiley & Sons, Inc.

ERS-1 User Handbook, 1992. Editors: Lois Proud and Bruce Battrick, ESA Publications Division, ESA SP-1148, 159 pp.

European Space Agency, 1993. Proceedings of the First ERS-1 Symposium, Space at the Service of our Environment, ESA SP-359.

Efimov, V. B., et al., 1985. The ice cover research by mean of spaceborne and airborne radiophysical systems. *Bulletin of USSR AS. Atmosphere and Ocean Physics*, 21, pp. 512–519.

Fung, K. A., 1994. Microwave Scattering and Emission Models and Their Applications. Artech House, Inc., 573 pp.

Gloersen, P. and Campbell, W. J., 1988. Variations in the Arctic, Antarctic, and global sea-ice covers during 1978–1987 as observed with the Nimbus-7 SMMR. *J. Geophys. Res.*, Vol. 93, No. C9, pp. 10666–10674.

Gloerson, P., Zwally, H. J., Chang, A. T., Hall, D. K., Campbell, W. J. and Ramseier, R. O., 1978. Time dependence of sea ice concentration and multiyear ice fraction in the Arctic Basin. *Boundary Layer Meteorol.*, 13, pp. 339–359.

Harris, F. J., 1978. On the use of windows for harmonic analysis with the discrete Fourier transform. *Proceedings of the IEEE*, 66(1), pp. 51–83.

JPL, Alaska SAR Facility Archive and Catalog Subsystem: User's Guide, JPL D-5496, March 22,

145 p., 1991.

JPL, Alaska SAR Facility: Geophysical Processor System Data User's Handbook, Version 2.0, JPL D-9526, 125 pp., 1993.

Kalmykov, A. I., Pichugin, A. P. and Tsymbal, V. N., 1985. The water surface wind field determination by the side-looking radar system on the KOSMOS-1500 satellite. *Earth Research from Space*, 4, pp. 65–77.

Kalmykov, A. I., Kurekin, A. S., Efimov, V. V. et al., 1989. KOSMOS-1500 Satellite Side-Looking Radar. *Soviet Journal of Remote Sensing*, V. 5, No. 3, pp. 471–485.

Krasnogorskiy, M. G. and Shirokov, P. A., 1994. The Meeting of the Section "Natural Resources and the Environmental Space-Borne Studies" of the Space Council of the Russian Academy of Sciences. *Soviet Journal of Remote Sensing*, Vol. 11(5), pp. 911–914.

Krishen, K., 1973. Detection of oil spills using a 13.3 GHz radar scatterometer. *J. Geophys. Res.*, 78, pp. 1952–1963.

Mitnik, L. M., Desyatova, G. I. and Kovbasyuk, V. V., 1989. Use of radar data to determine ice cover characteristics in the Sea of Okhotsk, winter 1983–1984. *Soviet Journal of Remote Sensing*, Vol. 5(3), pp. 371–381.

NASA, 1991. EOS Reference Handbook, NASA Headquarters, NP-202, 145 pp.

NASDA, 1990. ERS-1 Earth Resources Satellite-1, NASDA 1990 3/6 T, 30 pp.

Nelepo, B. A., Armand, N. A. and Khmyrov, B. E., 1984. The Oceanographic Experiment Involving KOSMOS-1076 and KOSMOS-1151. *Soviet Journal of Remote Sensing*, Vol.2 (3), pp. 383–392.

Olmsted, C., 1993. Alaska SAR Facility, Scientific SAR User's Guide. Geophysical Institute, University of Alaska, Fairbanks, ASF-SD-003, 53 pp.

Onstott, R.G., Moore, R. K., Goginenni, S. and Delker, C., 1982. Four years of low-altitude sea ice broad-band backscatter measurements. *IEEE. Journal of Oceanic Engineering,* OE-7, 1, pp. 44–50.

Parashar, S. et al., 1993. RADARSAT Mission Requirements and Concept. *Canadian Journal of Remote Sensing*, Vol. 19, No. 4, pp. 280–288.

Radar Data Development Programme, 1992. Brochure published by the Canada Center for Remote Sensing, Ottawa, Ontario.

Salganik, P. O., Efremov, G. A., Neronskii, L. B. et al., 1990. Radar sensing of the Earth from SAR KOSMOS-1870. *Earth Research from Space*, No. 2, pp. 70–79.

Volkov, A. M., Grischenko, V. D., Kurevleva, T. G. et al., 1995. Principles of new generation space system development for sea-ice observations. *Earth Research from Space,* No. 1, pp. 63–72.

Viter, V. V., Efremov, G. A., Ivanov, A. Yu., Etkin, V. S. et al., 1994. Programme "OKEAN-1" of SAR "ALMAZ-1": Preliminary Results of High Resolution Radar Surveys of Oceans. *Soviet Journal of Remote Sensing*, Vol. 11(6), pp. 990–1004.

Way, J., 1991. The Evolution of Synthetic Aperture Radar Systems and their Progression to the EOS SAR. *IEEE Transactions on geoscience and remote sensing*, Vol. 29, No. 6.

Wohl, M. G., 1995. Operational Sea Ice Classification from Synthetic Aperture Radar Imagery. *Photogrammetric Engineering & Remote Sensing*, Vol. 61, No. 12, pp. 1455–1466.

3

Arctic Marine Mammal Sea Ice Habitat Studies Using Active and Passive Microwave Satellite Data

3.1 INTRODUCTION

Satellite telemetry and remote sensing are used for Arctic sea ice habitat studies. Systematic monitoring of Arctic sea ice habitat parameters helps to assess the species' biological requirements, predict effects of habitat changes, justify protection of key areas and test hypotheses concerning underlying ecological processes (Holbrook and Schmitt, 1988; Andren, 1990; Carey et al., 1992; Lubin et al., 1993; Arthur et al., 1996). The AVHRR satellite data are effective for characterizing sea ice types, concentration and surface temperature except when clouds or darkness obscured the view (Massom et al., 1994). Many sea ice features such as types and concentrations, ridges and roughness, lead and polynya formations can be observed using passive and active microwave measurements (Chan et al., 1977; Cavalieri et al., 1981; Gray et al., 1982; Onstott et al., 1982).

The most important sea ice habitat parameters derived currently from passive microwave satellite observations are the position of the ice edge, the concentration and the surface temperatures. The utility of microwave measurements results from the strong contrast in thermal microwave emission between areas of ice-free ocean and ice-covered water (Fung, 1994). The SMMR and SSM/I passive microwave data are used for a more accurate calculation of total sea ice concentration and determination of both multiyear sea ice concentrations and the physical temperatures of the sea ice pack (Cavalieri et al., 1984; Gloersen and Cavalieri, 1986; Eppler et al., 1986; Grody and Basist, 1997). A sea ice surface temperature derived from passive microwave satellite observations is more robust than AVHRR data because microwaves pass easily through the polar atmosphere (Germain and Cavalieri, 1997).

The structure, age and thickness associated with the sea ice as well as its boundary can be efficiently derived from active microwave satellite observations (Onstott and Grenfell, 1987; Gloersen and Campbell, 1988; Zwally et al., 1991). For example, the ASF sea ice classification algorithm identifies four various types of sea ice based on measured radiometric brightness and lookup tables (LUT) containing parametric description of the backscatter statistics of different sea ice types (multi-year ice, first-year rough ice, first-year smooth ice, and new ice or open water). There is a lookup table for each season, since backscatter is affected by seasonal environmental conditions (Fetterer et al., 1994).

Active and passive microwave satellite sensors offer the potential of obtaining synoptic data of large expanses of remote ice-covered oceans under all weather conditions irrespective of the amount of solar illumination. This is of particular importance for Arctic sea ice habitat study where much of the polar ice canopy is under clouds or in darkness. Optimum use of these systems depends on knowledge of the magnitude and variability of radar backscatter and emissivity for various sea ice types and concentration as functions of different sea ice conditions (melt, temperature, snow cover, roughness etc.) and variation in the basic radar parameters (frequency, incidence angle, polarization). Combining emissivity with backscatter data improves the ability to classify sea ice habitats (Curlander and McDonough, 1991; Fung, 1994).

This chapter focuses on Arctic sea ice habitat studies using remotely sensed and animal-tracking satellite data. The sea ice types, concentrations and surface temperatures are analyzed based on single-channel passive and single-channel active microwave data. The principal goals are to evaluate contemporary information technologies and the efficiency of Russian OKEAN satellite data for marine mammal ecology studies.

3.2 STUDY AREA

The study area (Figure 3.1) covers the Barents, Kara and Laptev Seas and adjacent parts of the Arctic Ocean (70–85 degrees north latitude; 10–120 degrees east longitude). It is bordered by the northern coast of the continent, by the southern edge of the floating ice distribution and by the northern boundary of warm oceanic streams. The climate is continental, with frequent strong winds, fogs and storms, long periods of polar night and very low winter temperatures. During the winter, the prevailing wind directions determine the conditions of snow accumulation and thaw. From March or April till August, a period of steady windless weather is observed. The peculiarities of the ice drifting conditions are determined mostly by seasonal variations. The speed of drifting sea ice is estimated at 30 km or more over each 24-four period. There are zones of different drift intensity. The direct interaction with the Arctic Ocean determines some peculiarities (possibility of low-pressure systems moving farther north than anywhere else without being affected by continents, high level of meteorological factor variability etc.). The heat budget of the area is dominated by the world's most concentrated meridional exchanges of water and ice; the variability in both oceanic and atmospheric fluxes is mirrored in the high variability of sea ice extent in the region.

3.3 DATA

Satellite data includes the digital images (Figure 3.2) acquired by the polar-orbiting OKEAN-01 N7, N8 satellites (passive microwave radiometer, side-looking real aperture radar) and the polar bear satellite tracking data acquired by the NOAA satellites for regions within the study area during 1995–1998. OKEAN-01 satellite instruments (passive

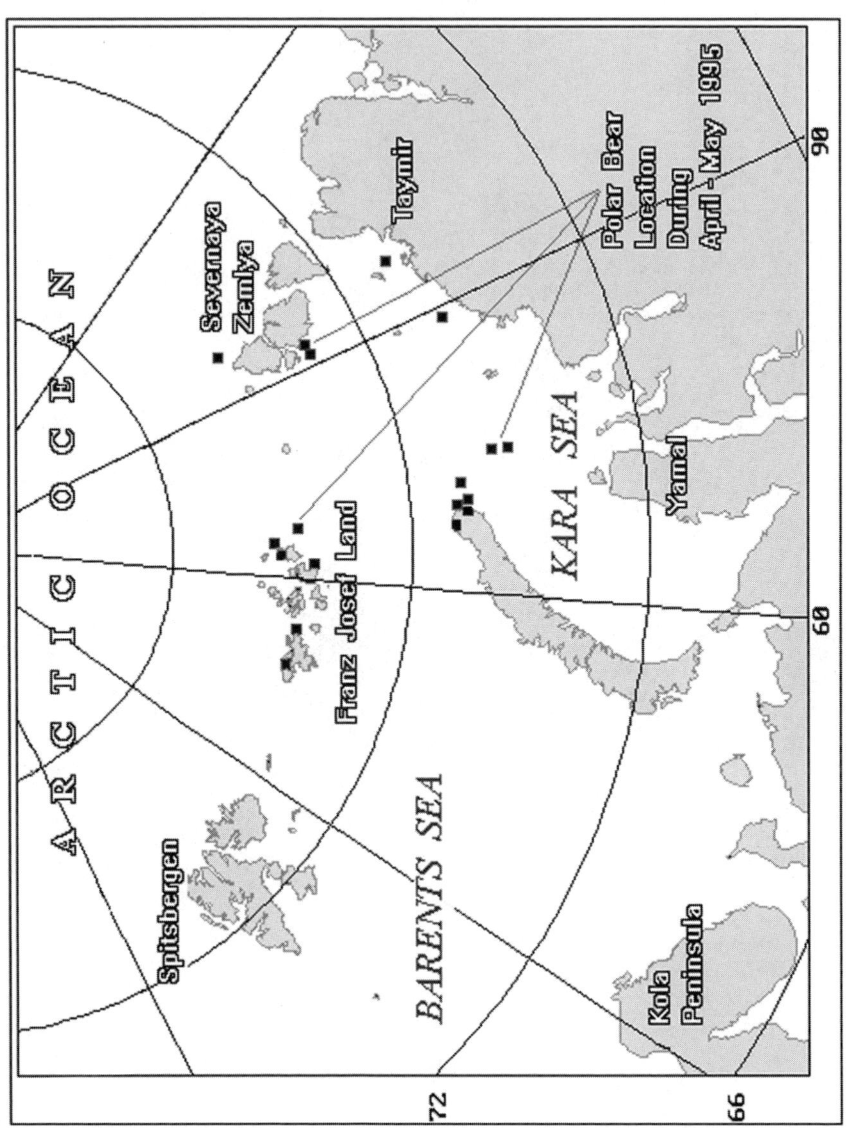

Figure 3.1 Location of study area.

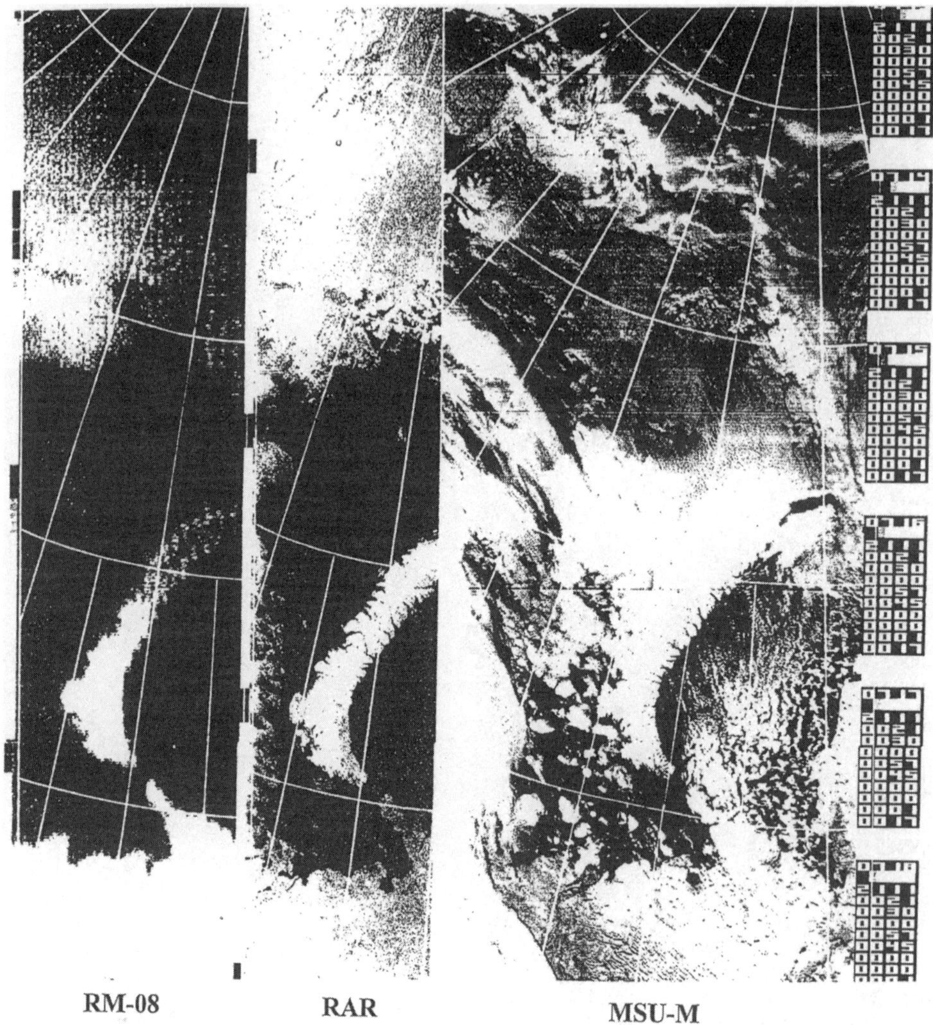

Figure 3.2 Examples of raw OKEAN-01 satellite data of simultaneously acquired passive microwave (RM-08), side-looking real aperture radar (RAR) and optical (MSU-M) imagery.

microwave radiometer — RM-08, and side-looking real aperture radar — RAR) and imagery characteristics are presented in Table 2.4.

The Arctic polar bear tracking satellite data have been collected by Dr. G. Garner (Alaska Science Center, U.S.A.) based on the real transmitter's duty cycle and where and when the polar bear transmitters were working on the study area.

OKEAN-01 radar and passive microwave radiometer data takes for the Barents–Kara and Laptev Seas region synchronous (within 24 hours) with polar bear satellite telemetry (geographic coordinates) measurements have been collected during 1995–1998 for subsequent analysis. Sea ice maps were acquired through the U.S. Navy/NOAA Joint Ice Center, 1995–1998.

3.4 STRUCTURE OF DATA PROCESSING SYSTEM

The problem-oriented Data Acquisition, Collection and Processing System (DASPS) has been developed for studying the sea ice extent, surface characteristics and marine mammal sea ice habitats. The structure of DASPS includes: Experiment Planning System (EPS); Data Acquisition System (DAS); Data Collection System (DCS) and Data Processing System (DPS) (Figure 3.3).

EPS was used for optimization of remote sensing observation and satellite telemetry measurements. DAS includes: the control subsystem, space observation subsystem, space data subsystem, meteorological data subsystem and telemetry data subsystem.

DCS includes the OKEAN-01 satellite database (image data — RAR, RM-08, MSU-M, image attributes, space board system parameters, orbital measurements); satellite tracking database, multithematic geographic databases, meteorological sea ice condition database; Sigma Nought ground objects database; control points database; inquiry subsystem for connection with database and orbital parameters database.

DPS consists of an orbital parameter estimation and OKEAN-01 data restoration subsystem and data interpretation subsystem. DRS includes RM-0.8, RAR and MSU-M image restoration subsystems; output file subsystem and sets of special modules (first step processing, correcting for a variety of geometric distortion, Sigma Nought (σ^0) estimation: the data values converting to backscatter coefficients using a calibration wedge, output file creation; image enhancement and classification subsystem).

Polar bear sea ice habitats database includes satellite images and polar bear tracking data and sea ice habitat data. All subsystems have been installed on a set of IBM PC/AT compatible computers.

The computer algorithms to process "raw" OKEAN-01 satellite images include both digital calibration and equalization (Estimating Sigma Nought), automated registration to geographic map coordinates (based on orbital parameters data and reference points), ice type classification and assessing the accuracy of the results.

3.5 ESTIMATING SEA ICE HABITAT PARAMETERS

The main documented sea ice habitat parameters include backscatter coefficients and upwelling brightness temperatures, sea ice types and concentrations, sea ice physical surface temperatures, sea ice habitat entropy and other integrated parameters. These parameters were estimated for an area defined by a circle with radius R around polar bear location (X, Y).

Backscatter coefficient (σ^0) — The backscatter coefficient σ^0 was determined as a result of raw OKEAN-01 satellite radar data processing (Section 2, Belchansky and Douglas, 2000).

Brightness temperature (T_b) — The respective composite upwelling brightness temperature was determined as measured from space by radiometer.

Sea ice physical parameters. Two-channel active and passive microwave measurements were used to quantify the fractional areas occupied by open water (OW), young (Y) sea ice, first-year (FY) sea ice, multiyear (MY) sea ice, and to estimate the sea ice surface's physical temperature.

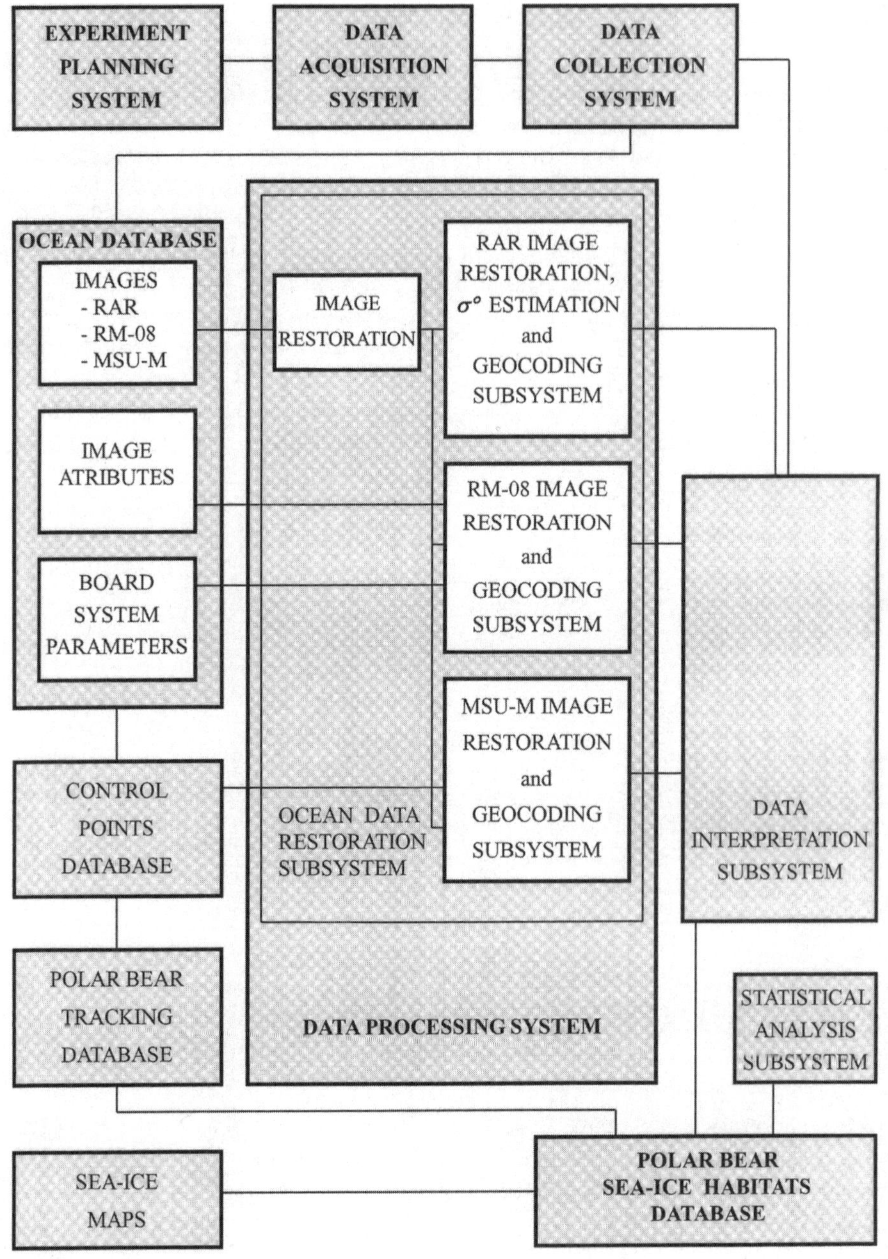

Figure 3.3 Data collection and processing system.

3.5.1 Estimating the Area Fractions of Open Water (OW), Consolidated Young/First-year Sea Ice (Y/FY) and Consolidated Multiyear Sea Ice (MY)

For this goal the following algorithm is used. If there is no land within the antenna footprint, the following constrain must hold:

$$C(OW) + C(Y/FY) + C(MY) = 1, \tag{3.1}$$

where $C(OW)$, $C(Y/FY)$, $C(MY)$ are the respective area fractions of open water (OW), consolidated young/first-year sea ice (Y/FY), and consolidated multiyear sea ice (MY).

The composite upwelling brightness temperature, T_b, in K as measured from space by radiometer, and backscatter coefficient, σ^0, determined from space radar data, is assumed to be given by

$$T_b = [1 - C(Y/FY) - C(MY)] \times T_b(OW) + C(Y/FY) \times T_b(Y/FY) + C(MY) \times T_b(MY), \tag{3.2}$$

$$\sigma^0 = [1 - C(Y/FY) - C(MY)] \times \sigma(OW) + C(Y/FY) \times \sigma(Y/FY) + C(MY) \times \sigma(MY), \tag{3.3}$$

where $T_b(OW)$, $T_b(Y/FY)$, $T_b(MY)$, $\sigma(OW)$, $\sigma(Y/FY)$, and $\sigma(MY)$ are the respective brightness temperatures and the backscatter coefficients of open water, consolidated young/first-year sea ice, and consolidated multiyear sea ice. These parameters can be received and used as *a priori* data from published information (Onstott et al., 1982, 1987; Eppler et al., 1986).

For a more accurate estimation of parameters, the two-dimensional histogram $\{F = F(T_b, \sigma^0)\}$ was constructed and analyzed based on published information about $T_b(OW)$, $T_b(Y/FY)$, $T_b(MY)$, $\sigma(OW)$, $\sigma(Y/FY)$ and $\sigma(MY)$. Using these parameters and two-channel microwave measurements [T_b and σ^0] it is possible to estimate the respective area fractions of open water, consolidated *Y/FY* sea ice, and consolidated *MY* sea ice ($C(OW)$, $C(Y/FY)$, $C(MY)$) from Equations (3.1)–(3.3).

3.5.2 Estimating the Sea Ice Surface Physical Temperature

If there is no land within the antenna footprint, the following constrain must hold:

$$\varepsilon(X_i,Y_j) = C[OW](X_i,Y_j) \times \varepsilon[OW] + C[Y/FY](X_i,Y_j) \times \varepsilon[Y/FY] + C[MY](X_i,Y_j) \times \varepsilon[MY], \tag{3.4}$$

where $C(OW)$, $C(Y/FY)$, $C(MY)$ are the respective area fractions of open water (OW), consolidated young/first-year sea ice (Y/FY), consolidated multiyear sea ice (MY) and $\varepsilon(X_i,Y_j)$, $\varepsilon[OW]$, $\varepsilon[Y/FY]$, $\varepsilon[MY]$ are the respective emissivities for passive microwave radiometer frequency, polarization considered.

The sea ice surface physical temperature, T, in K is assumed to be given by

$$T(X_i,Y_j) = [T_b(X_i,Y_j) - 26]/\varepsilon(X_i,Y_j) + 26, \tag{3.5}$$

where $T_b(X_i,Y_j)$ and $\varepsilon(X_i,Y_j)$ are the respective brightness temperatures in K as measured from space by radiometer and the emissivity in point with coordinates X_i, Y_j determined from (3.4) (Swift et al., 1985). Using these parameters, it is possible to estimate the sea ice surface physical temperature.

An estimation of sea ice parameters can be made based on the results of analysis of coregistered RAR and RM data by taking into account the data by Onstott and Eppler (Onstott et al., 1982; Onstott and Grenfell, 1987; Eppler et al., 1986).

3.5.3 Algorithm for Estimating Parameters of Sea Ice Habitat

Processing the passive and active microwave data having different spatial resolutions creates an optimum working pixel-size-choice problem. An optimum working pixel size is usually chosen by compromising between accuracy of estimating geophysical surface parameters and possible loss of spatial resolution.

In this study a choice of pixel size (3 km×3 km) is defined by the necessity of minimizing the loss of spatial resolution and georegistration mistake correction. A bilinear interpolation is more efficient for estimation of sea ice concentration and surface temperature than the more commonly used piece-linear interpolation. Depending on imagery type (RM and RAR together, or RAR and only RM) and scene type (open water is present — OW, open water is not present — NOOW) all algorithm variants are as follows:

A)	RM + RAR,	OW	—	*C*[*OW*]	:	*C*[*Y/FY*]	:	*C*[*MY*];
B)	RM + RAR,	NOOW	—	0	:	*C*[*Y/FY*]	:	*C*[*MY*];
C)	RAR,	OW	—	*C*[*OW*/*Y/FY*]			:	*C*[*MY*];
D)	RAR,	NOOW	—	0	:	*C*[*Y/FY*]	:	*C*[*MY*];
E)	RM,	OW	—	*C*[*OW*]	:	*C*[*OW*/*Y/FY*];		
F)	RM,	NOOW	—	0	:	*C*[*Y/FY*]	:	*C*[*MY*].

Corresponding area fractions can be determined for every variants by solving by simultaneous set of equations as derived from (3.1)–(3.3). For example, a set of equations for variants A, B, C, E looks like the following sets:

A. RM + RAR, OW: ESTIMATING *C*(*OW*), *C*(*Y/FY*), *C*(*MY*)

$$\left|\begin{array}{l} T_b = C(OW) \times T_b(OW) + C(Y/FY) \times T_b(Y/FY) + C(MY) \times T_b(MY) \\ \sigma^0 = C(OW) \times \sigma(OW) + C(Y/FY) \times \sigma(Y/FY) + C(MY) \times \sigma(MY) \\ 1 = C(OW) + C(Y/FY) + C(MY) \end{array}\right. ,$$

where: $C(OW) \to [0,1]$, $C(Y/FY) \to [0,1]$, $(MY) \to [0,1]$.

Marginal concentration processing:

a. $C(OW) < 0$; $C(Y/FY) > 0$; $C(MY) > 0$

$C''(OW) = 0$; $C''(Y/FY) = C(Y/FY)/[1 - C(OW)]$; $C''(MY) = C(MY)/[1 - C(OW)]$.

b. $C(OW) > 0$; $C(Y/FY) < 0$; $C(MY) > 0$

$C''(Y/FY) = 0$; $C''(OW) = C(OW)/[1 - C(Y/FY)]$; $C''(MY) = C(MY)/[1 - C(Y/FY)]$.

c. $C(OW) > 0$; $C(Y/FY) > 0$; $C(MY) < 0$

$C''(OW) = C(OW)/[1 - C(MY)]$; $C''(MY) = 0$; $C''(Y/FY)= C(Y/FY)/[1 - C(MY)]$.

d. $C(OW) < 0$; $C(Y/FY) < 0$; $C(MY) > 0$

$C''(OW) = 0$; $C''(Y/FY) = 0$; $C''(MY) = 1$.

e. $C(OW) < 0$; $C(Y/FY) > 1$; $C(MY) < 0$

$C''(OW) = 0$; $C''(Y/FY) = 1$; $C''(MY) = 0$.

f. $C(OW) > 0$; $C(Y/FY) < 0$; $C(MY) < 0$

$C''(Y/FY) = 0$; $C''(OW) = 1$; $C''(MY) = 0$.

B. RM+RAR, NOOW: ESTIMATING *C*(*Y/FY*), *C*(*MY*), *C*(*OW*) = 0

$$\begin{cases} T_b = C(Y/FY) \times T_b(Y/FY) + C(MY) \times T_b(MY) \\ \sigma^0 = C(Y/FY) \times \sigma(Y/FY) + C(MY) \times \sigma(MY) \\ 1 = C(Y/FY) + C(MY) \end{cases},$$

where: $C(Y/FY) \to [0,1]$, $C(MY) \to [0,1]$.

Parameters $C(Y/FY)$, $C(MY)$ were estimated using the method of least squares.

Marginal concentration processing:

a. $C(Y/FY) < 0$; $C(MY) > 1$

$C''(Y/FY) = 0$; $C''(MY) = 1$.

b. $C(Y/FY) > 1$, $C(MY) < 0$

$C''(Y/FY) = 1$; $C''(MY) = 0$.

C. RAR, OW: ESTIMATING $C[OW/Y/FY]$, $C[MY]$

$$\begin{cases} \sigma^0 = C(OW/Y/FY) \times \sigma(OW/Y/FY) + C(MY) \times \sigma(MY) \\ 1 = C(OW/Y/FY) + C(MY) \\ \sigma(OW/Y/FY) = [\sigma(OW) + \sigma(Y/FY)]/2 \end{cases},$$

where $C(OW/Y/FY) \to [0,1]$, $C(MY) \to [0,1]$.

Marginal concentration processing:

a. $C(OW/Y/FY) < 0$; $C(MY) > 1$

$C''(OW/Y/FY) = 0$; $C''(MY) = 1$.

b. $C(OW/Y/FY) > 0$; $C(MY) < 0$

$C''(OW/Y/FY) = 1$; $C''(MY) = 0$.

E. RM, OW: ESTIMATING $C(OW)$, $C(Y/FY/MY)$

$$\begin{cases} T_b = C(OW) \times T_b(OW) + C(Y/FY/MY) \times T_b Y/FY/MY \\ 1 = C(OW) + C(Y/FY/MY) \\ T_b(Y/FY/MY) = [T_b(Y/FY) + T_b(MY)]/2 \end{cases},$$

where $C(OW) \to [0,1]$, $C(Y/FY/MY) \to [0,1]$.

Marginal concentration processing:

a. $C(Y/FY/MY) < 0$; $C(OW) > 1$

$C''(Y/FY/MY) = 0$; $C''(OW) = 1$.

b. $C(Y/FY/MY) > 1$; $C(OW) < 0$

$C''(Y/FY/MY) = 1$; $C''(OW) = 0$.

A flow chart showing the methodology and algorithm used for sea ice parameter habitat estimation is presented in Figure 3.4.

The main result of the first step is developing the two-dimensional histogram $\{F = F(T_b, \sigma^0)\}$ and determining the two-dimensional clusters corresponding to the respective consolidated area fractions of OW, Y/FY and MY based on information from the following publications (Onstott et al., 1982, 1987; Eppler et al., 1986). The results ($T_b(OW)$, $T_b(Y/FY)$, $T_b(MY)$, $\sigma(OW)$, $\sigma(Y/FY)$, $\sigma(MY)$) from the two-dimensional histogram $\{F = F(T_b, \sigma^0)\}$ were used to measure the following parameters: $C(OW)$, $C(Y/FY)$, $C(MY)$.

Analysis of the two-dimensional histogram showed that most of the two-dimensional (backscatter coefficient, σ^0, and upwelling brightness temperature, T_b) clusters are inside

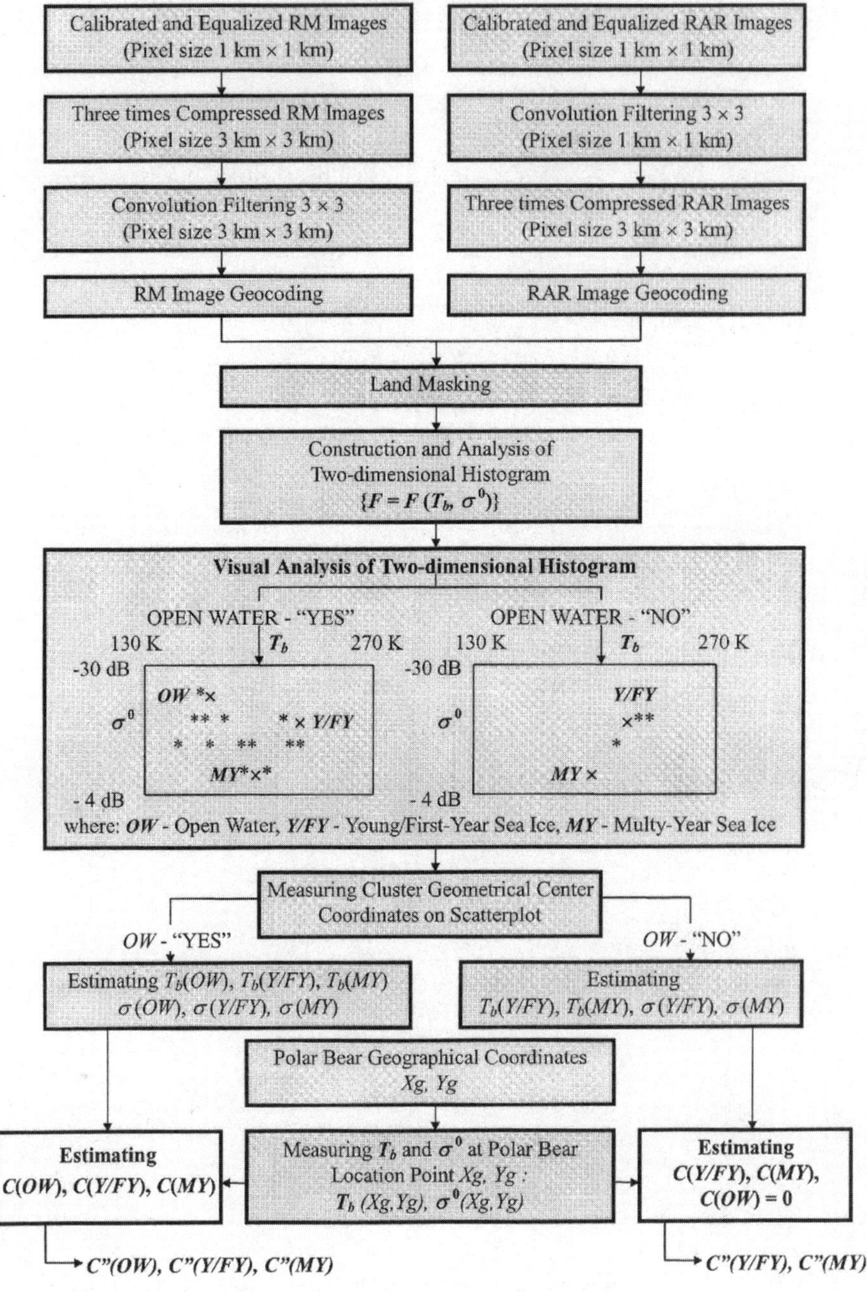

Figure 3.4 A flow chart showing the methodology and algorithm of estimating sea ice habitat parameters.

of the triangle formed by the centers of three base clusters (*OW*, *Y/FY*, *MY*). This fact is an indirect indicator of the correct use of the linear model for estimating the basic sea ice types.

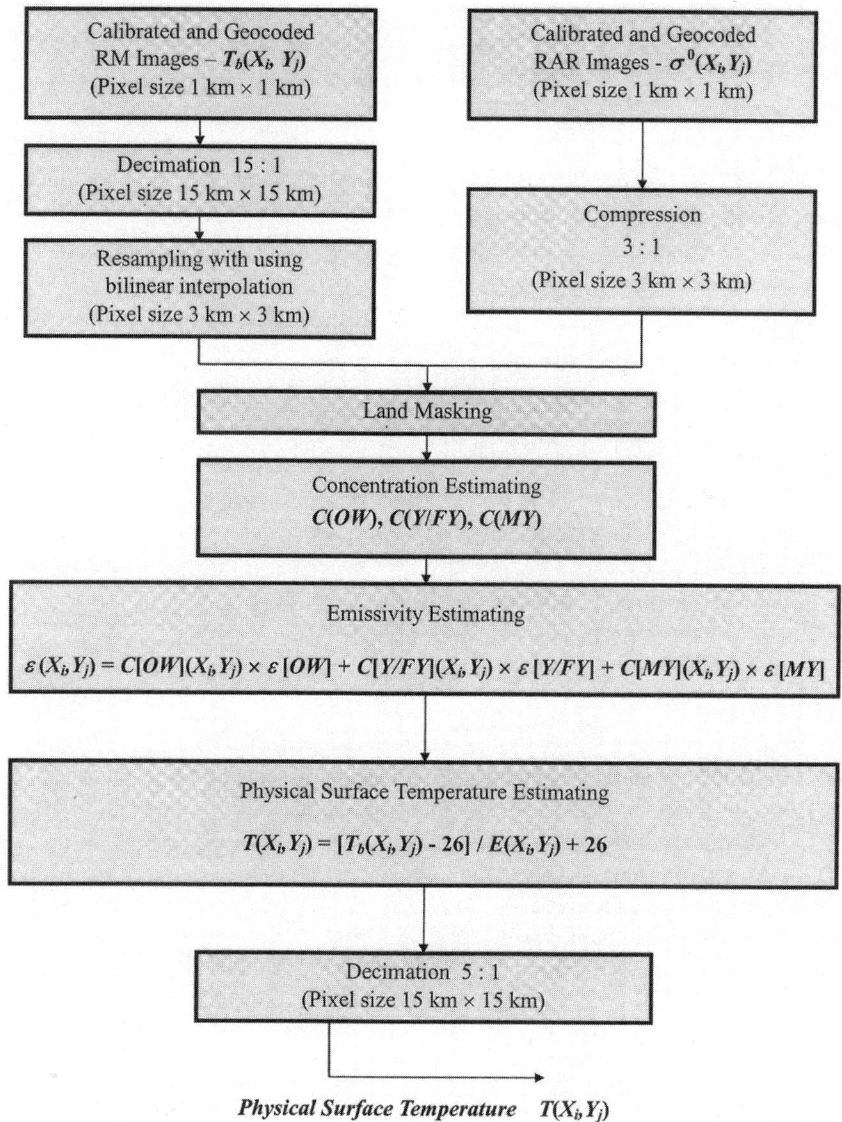

Figure 3.5a A flow chart showing the methodology and algorithm of estimating sea ice surface temperature.

As follows from Figure 3.4, the algorithm can be used for estimating two or three concentrations at animal location points when data from OKEAN satellites and polar bear transmitters are synchronous. For estimating the concentrations when OKEAN data and animal transmitter data are not synchronous, it would be useful to interpolate the concentration data using concentrations estimated for every point where images overlap longer than one day.

A flow chart showing the methodology and algorithm used for polar bear sea ice surface temperature studies is presented in Figure 3.5*a*.

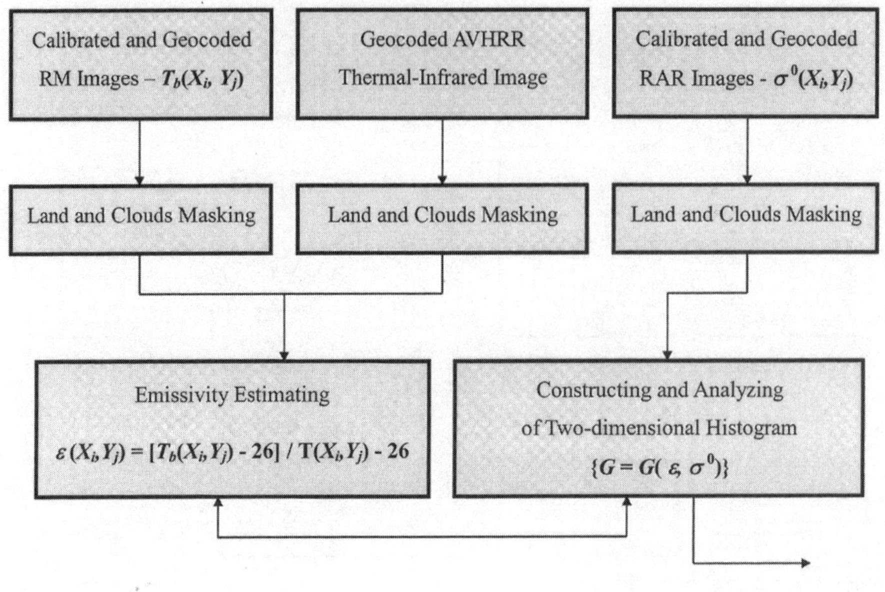

Figure 3.5b A flow chart showing the methodology of estimating emissivity based on AVHRR data.

As follow from Equation (3.5) the accuracy of sea ice surface temperature estimation (dT) depends on the accuracy of concentration (dC), emissivity ($d\varepsilon$) and radio brightness temperature estimation for main sea ice types.

The accuracy of sea ice surface temperature estimation can be better if spatial and temporal averaging are used. In this case, a loss of spatial resolution naturally takes place.

For polar bear sea ice habitat studies, a window of 7×7 was used for spatial averaging filtering of sea ice surface temperature $T(X_i,Y_j)$ estimation and a bilinear interpolation for receiving working pixel size. The predicted parameter accuracy estimation (error st. dev.) is characterized thus: Sea ice type concentration of 5%, Spatial resolution of (3:15) km×(3:15) km, Pixel size of 3 km×3 km, Bilinear (Interpolation); Sea Ice Surface temperature of 4.6 K for pure type and of 25.4 K for mixture of types, Spatial resolution of 105 km×105 km, Pixel size of 3 km×3 km, Bilinear Interpolation.

As follow from Equation (3.4), the accuracy of emissivity $\varepsilon(X_i,Y_j)$ estimation depends on the accuracy of the respective emissivity knowledge ($\varepsilon[OW]$, $\varepsilon[Y/FY]$ and $\varepsilon[MY]$: open water, consolidated Y/FY sea ice and consolidated MY sea ice) for passive microwave radiometer frequency, polarization and temporal and space parameters of study area point.

The AVHRR thermal-infrared data (the sea ice surface temperature estimation) and respective OKEAN passive microwave data (T_b) can be used for more accurate emissivity ($\varepsilon[OW]$, $\varepsilon[Y/FY]$ and $\varepsilon[MY]$) estimation in periods when some part of the study area is without clouds. To this end, the emissivity $\varepsilon(X_i,Y_j)$ can be estimated from Equation (3.5) for the temporal and space AVHRR and OKEAN data [$T_b(X_i,Y_j)$ – the respective brightness temperature in the point with coordinates X_i,Y_j from the microwave radiometer; $T(X_i,Y_j)$ – the respective surface temperature in the point with coordinates X_i,Y_j as a result of AVHRR data analysis]. Constructing and analyzing the two-dimensional histogram $\{G = G(\varepsilon, \sigma^0)\}$, it is possible to estimate the respective emissivity $\varepsilon[OW]$, $\varepsilon[Y/FY]$ and $\varepsilon[MY]$ and then to estimate the respective surface temperature based on the methodology and algorithm of sea ice surface temperature estimation for all OKEAN satellite data for the study area and for periods when the study area is masked by clouds.

A flow chart showing the methodology of estimating emissivity $\varepsilon[OW]$ of open water, $\varepsilon[Y/FY]$ of consolidated Y/FY sea ice, and $\varepsilon[MY]$ of consolidated MY sea ice based on AVHRR thermal-infrared data (the sea ice surface temperature estimate) and respective OKEAN-01 passive microwave data (T_b) is presented in Figure 3.5b.

3.6 ESTIMATING THE SEA ICE HABITAT'S INTEGRATED INFORMATION CONTENT

The integrated information content of sea ice habitats can be studied based on Shannon's (1948), Simpson's (1949), Sheldon's and McIntosh's indices (*H*, *ISM*, *ISL* and *IM*, respectively).

3.6.1 The Shannon's (Entropy) Analysis of Raw Satellite Data

The Shannon's analysis includes studying the relationship between polar bear sea ice habitat entropy, in terms of remote sensing data, and every animal movement in an area defined by a circle with a radius (R) around the animal's location for all satellite tracking data. The entropy ($H1$) can be estimated using only raw satellite data and the following equation:

$$H1 = -\sum_{i=1}^{N} P_i \log P_i, \tag{3.6}$$

where $P_i = N_i/N$, N_i is the number of pixels that have each value, which is called the frequency of each value on an image histogram (a. backscatter coefficient image histogram; b. upwelling brightness temperature image histogram; c. backscatter × upwelling brightness temperature two dimension image histogram); N are data file values ($m \times n$). The sea ice habitat Entropy 1 derived from the satellite image characterizes a diversity of sea ice surface.

3.6.2 The Shannon's (Entropy) Analysis of Satellite Data Classification Results

Entropy 2 ($H2$) can be estimated using the results of satellite data processing and the following equation:

$$H2 = -[C(OW) \times \log C(OW) + C(Y/FY) \times \log C(Y/FY) + C(MY) \times \log C(MY) + C(DI) \times \log C(DI) + C(SI) \times \log(SI)], \tag{3.7}$$

where $C(OW)$, $C(Y/FY)$, $C(MY)$, $C(DI)$, $C(SI)$ are the respective area fractions of OW, consolidated Y/FY sea ice, consolidated MY sea ice, deformed Y/FY sea ice, other sea ice, $C(OW) \rightarrow [0,1]$, $C(Y/FY) \rightarrow [0,1]$, $(MY) \rightarrow [0,1]$, $C(DI) \rightarrow [0,1]$, $C(SI) \rightarrow [0,1]$.

In the framework of our studies the sea ice habitat parameters can be estimated by a number of different pixel types in an area defined by a circle with the radius (R) around an animal's location for every satellite tracking data. These parameters may be such as:

- *Nland* — number of land-type pixels
- *Now* — number of water-type pixels
- *Ny/fy* — number of Y/FY sea ice-type pixels
- *Nmy* — number of MY sea ice-type pixels
- *Ndef* — number of deformed sea ice-type pixels
- *Ni_w_edg* — number of pixels on the ice–water border
- *Nf_w_edg* — number of pixels on the FY sea ice–water border
- *Nm_w_edg* — number of pixels on the MY sea ice–water border
- *Nm_f_edg* — number of pixels on the MY ice–FY ice border.

In this case the integrated information content of animal sea ice habitats can be estimated and studied based on *H, ISM, ISL, IM* and combined indices (Zahl, 1977; Belchansky et al., 1995).

$$H = \sum \frac{n_i}{N} \ln \frac{n_i}{N}, \tag{3.8}$$

where: n_i ($i = 1,...,9$) is: *Nland, Now, Ny/fy, Nmy, Ndef, Ni_w_edg, Nf_w_edg, Nm_w _edg, Nm_f_edg*
and N = *Nland* + *Now* + *Ny/fy* + *Nmy* + *Ndef* + *Ni_w_edg* + *Nf_w_edg* + *Nm_w_edg* + *Nm_f_edg*.

$$ISL = e^H, \tag{3.9}$$

where H — Shannon's index.

$$ISM = \frac{1}{D}; \quad D = \sum \frac{n_i(n_i - 1)}{N}(N - 1), \tag{3.10}$$

$$IM = \sqrt{\sum n_i^2}. \tag{3.11}$$

For the combined assessment of polar bear sea ice habitats, let us define $Y_j(X_j)$ as the means of index (*H, ISM, ISL*, or *IM*) that can be received using image ij for month number $j(1 \le i \le I, 1 \le j \le J)$. According to the so-called jackknife method applied to the estimation of the diversity index the so-called pseudo values Y_{ji} and new corrected estimates $E_j(Y_{ji})$ of the diversity indices are defined as

$$Y_{ji} = I_j Y_j(X_j) - (I_j - 1) Y_j(X_{ji}),\ 1 \leq i \leq I_j, \tag{3.12}$$

$$E_j(Y_{ji}) = \sum \frac{Y_{ji}}{I_j}, \tag{3.13}$$

where: Y_{ji} — pseudo values of index for j (the month when image number i is excluded); $Y_j(X_{ji})$ — means of the index for month number j without image number i; $E_j(Y_{ji})$ — index obtained using so-called jackknife method.

3.7 SCHEDULING OF ACQUISITION OF SATELLITE IMAGES

One problem connected with marine mammal sea ice habitat studies using satellite data is the discrete character of data collected from remote sensing platforms and the discrete chance character of satellite telemetry data. Under these conditions, the optimum scheduling of satellite time-series images that coincide spatially and temporally with polar bear satellite tracking locations data is needed. The first priority of the data-collection strategy was to collect satellite images for areas and dates where and when the polar bear transmitters were expected to be working (based on the real transmitter's duty-cycle, general transmitter location as determined from the real transmitter's data for every 4- to 6-day period). The second priority was to develop 3- to 4-day sea ice maps for the entire study area four times for every month. This strategy took into consideration a cost-efficiency criterion and maximized analytical possibilities of OKEAN-01 N7, N8 satellite/polar-bear transmitter data for analysis of regional and seasonal ice–bear distributions and movements.

The optimum scheme of OKEAN-01 N7, N8 data collection was developed. This scheme was based on the analysis of daily polar bear transmitter-signal histograms for every 6-day period on intervals from the beginning of monitoring to the day of data acquisition by the two satellites, and scheduling over a period of 2 weeks using a linear prediction. The OKEAN data-collection strategy included making completed sea ice maps of the entire study area — over 3 to 4-day periods, and collecting OKEAN scenes for the study area and dates where and when the polar bear transmitters were expected to be working (based on the real transmitter's duty-cycle, and general transmitter location).

3.8 RESULTS

The raw active and passive microwave OKEAN satellite data were converted into back-scatter coefficient and reflectance parameters (brightness temperature) (Figure 3.6).

These parameters and *a priori* data were processed using computer algorithms developed to estimate and map sea ice concentration, types, surface temperature (Figures 3.4, 3.5a) and information on sea ice habitat indices (*H*, *ISM*, *ISL*, *IM*, respectively). Figure 3.4 shows that calibration, compression, filtering and image geocoding are carried out in terms of backscatter coefficient and brightness temperature parameters.

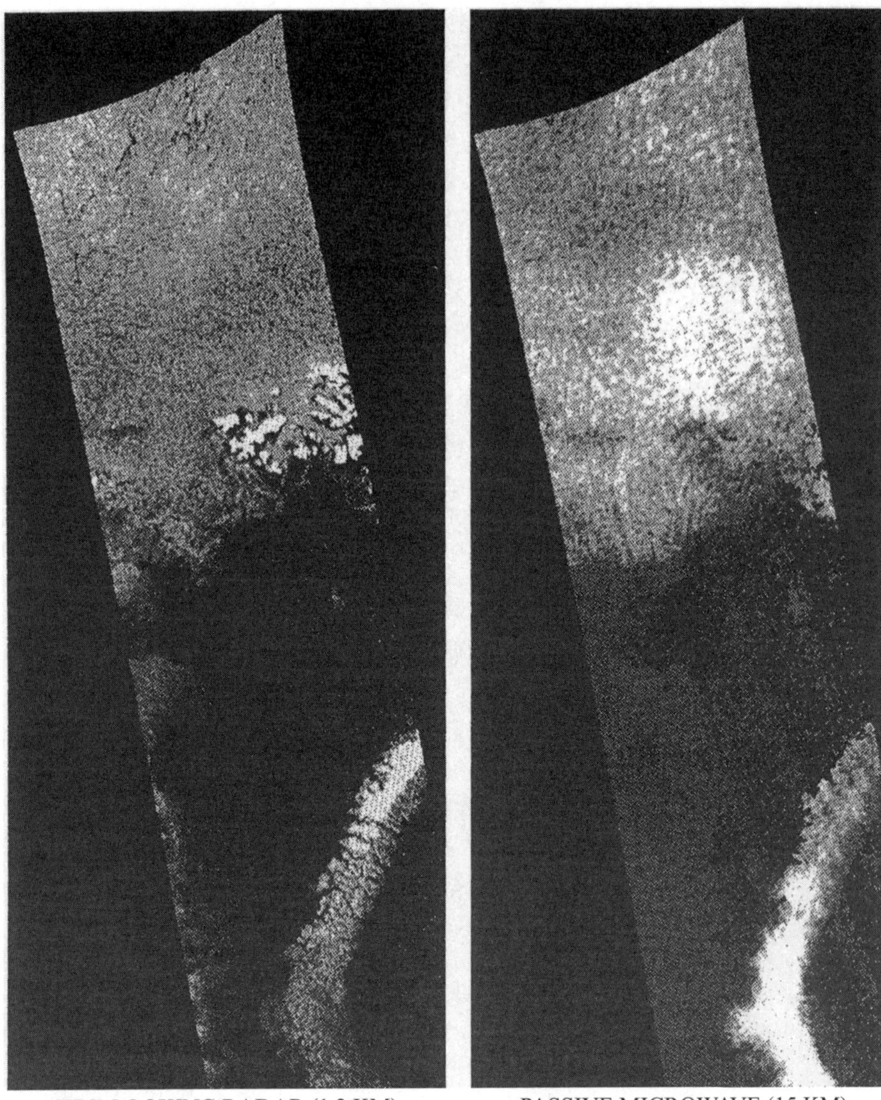

SIDE-LOOKING RADAR (1.2 KM) PASSIVE MICROWAVE (15 KM)

Figure 3.6 Examples of georegistered OKEAN-01 satellite data of simultaneously acquired passive microwave (RM-08) and side-looking real aperture radar (RAR).

An analysis of the two-dimensional histogram shows that most of the two-dimensional (the backscatter coefficient (σ^0) and upwelling brightness temperature (T_b)) clusters are inside the triangle formed by the centers of three base clusters (*OW*, *Y/FY*, *MY*).

Estimation of a physical surface temperature included the calibrating and geocoding, decimating, resampling and compression scheme (Figure 3.5a) differing from the scheme described on Figure 3.4. This difference is explained only by the wish to obtain greater accuracy of surface temperature estimation.

An example of the polar bear sea ice habitat parameters (brightness temperature (T_b), backscatter coefficient (σ^0), main sea ice type concentration, dominant sea ice types and

Table 3.1 An Example of Polar Bear Sea Ice Habitat Parameter Estimations in the Barents and Kara Seas

Biotele-metry data	Satell. orbit number	Satell. orbit date	Polar bear coordinates (lat.) (long.)	(T_b) (Ka-H) (K)	σ^0 (X-VV) (dB)	Concentration (*OW*:*Y*/*FY*:*MY*) (% : % : %)	Domin. S.I.T./*T*(C)
Polar Bear #06341							
07.11.95	#0987	06.11.95	78.715N 78.590E	241.3	−16.3	0.0:100.0:0.0	*Y*/*FY*/−9.2
13.11.95	#1090	13.11.95	78.492N 68.193E	212.4	−19.7	0.0:100.0:0.0	*Y*/*FY*/−21.5
19.11.95	#1207	21.11.95	79.691N 65.075E	241.1	−17.2	0.0:100.0:0.0	*Y*/*FY*/−9.5
25.11.95	#1251	24.11.95	81.503N 75.075E	269.2	−18.2	0.0:100.0:0.0	*Y*/*FY*/−17.2
01.12.95	#1347	30.11.95	81.127N 82.678E	222.1	−17.5	26.3:73.7:0.0	*Y*/*FY*/−5.4
01.12.95	#6130	01.12.95	81.127N 82.678E	244.0	−19.7	0.0:100.0:0.0	*Y*/*FY*/−8.2
07.12.95	#6232	08.12.95	80.981N 90.442E	225.6	−13.4	0.0:83.7:16.3	*Y*/*FY*/−11.3
31.12.95	#6555	30.12.95	81.024N 90.775E	230.7	−12.6	0.0:87.7:12.3	*Y*/*FY*/−21.1

surface temperature (*T*) estimations) are presented in Table 3.1 for a limited sample of polar bear locations in the Barents and Kara Seas.

An example of a spatial mosaic of the OKEAN images in terms of sea ice types is presented in Figure 3.7.

The content analysis of the polar bear sea ice habitats includes estimating the dynamics of change of indices *H*, *ISM*, *ISL*, *IM* as a function of a circle of area around the polar bear's location, estimating the dynamics of change of the indices for an unvarying circle of area around the polar bear's location as a function of the polar bear's movement (spatial and temporal sea ice habitat dynamics); studying a relationship between the indices of sea ice habitat and polar bear movement, behavior etc.

An example of the mean assessment of the dynamics of change of Shannon's indices of polar bear sea ice habitats with a 95% confidence interval for every month of the year is presented in Figure 3.8. The assessment was made using only satellite radar and telemetry data for four polar bears for the period of October 1995–December 1996 for an invariable circle of area around polar bear location (R = 50 km).

An example of the mean assessment of dynamics of change of the combined indices (obtained using *H*, *ISM*, *ISL* and *IM* indexes and the so-called jackknife method) of polar bear sea ice habitats with a 95% confidence interval for every month of the year is presented in Figure 3.9. The assessment was made using the sea-ice habitat parameters estimated by the number of different pixel types in an area defined by invariable circle around the polar bear's location (R = 50 km) (*Nland*, *Now*, *Ny/fy*, *Nmy*, *Ndef*, *Ni_w_edg*, *Nf_w_edg*, *Nm_w _edg*, *Nm_f_edg*) and telemetry data for four polar bears for the period of October 1995–December 1996.

The sea ice distribution–ice type temporal database has been developed for marine mammal ecology studies. The database can be used for studies of temporal and spatial migration activity, latitude and longitude distributions; for estimating the dynamics of the natural habitat boundaries; for comparative analysis of sea ice habitats based on the sea ice parameters (sea ice types, concentration, surface temperature, backscatter, brightness temperature entropy), etc.

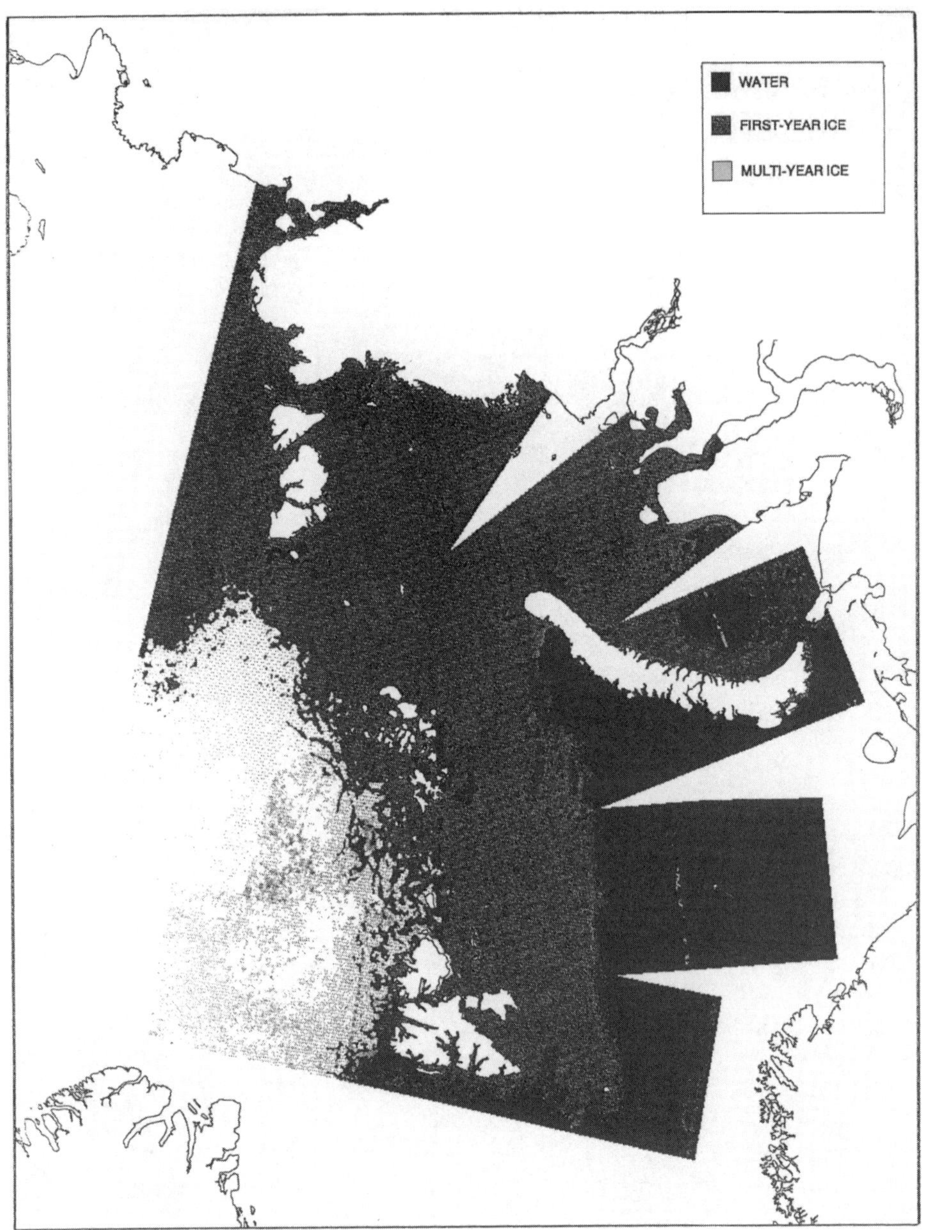

Figure 3.7 Spatial mosaic of the OKEAN-01 images.

3.9 DISCUSSION

This chapter demonstrates how active (X-VV) and passive (Ka-H) microwave satellite sensors are used for estimating and documenting marine mammal sea ice habitat parameters (sea ice types, concentration, physical surface temperature, entropy, etc.). It should be

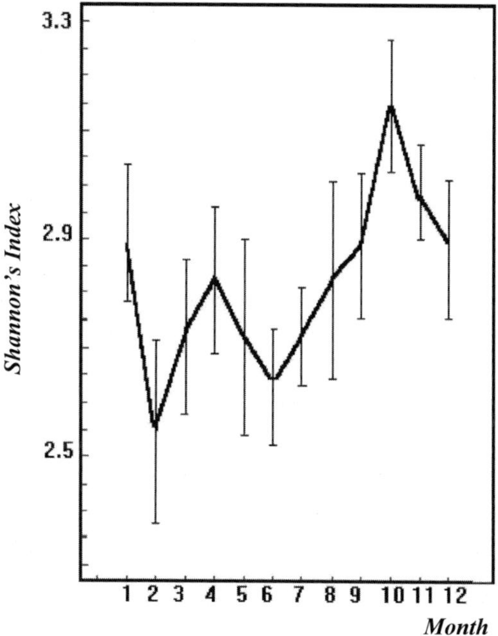

Figure 3.8 An example of the dynamics of mean change of Shannon's indices of polar bear sea ice habitats.

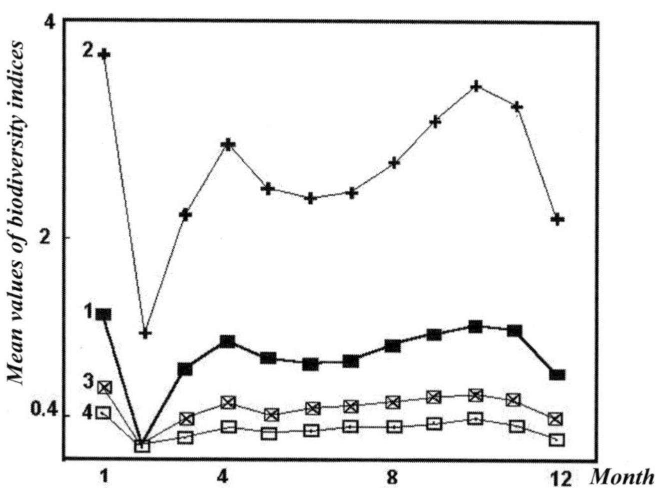

Figure 3.9 An example of the dynamics of change of combined indices of polar bear sea ice habitats (*1* – Shannon's, *2* – Sheldon's, *3* – McIntosh's, *4* – Simpson's).

emphasized that this is an intensity-based classification. The use of texture, higher-order moments and other features for sea ice classification can be easily incorporated into the sea ice classification process. The capability of the OKEAN satellites to concurrently acquire active and passive microwave data across extensive areas demonstrates the platform's utility for monitoring marine mammal sea ice habitats especially during cloudy periods.

Multispectral NOAA advanced very-high-resolution radiometer satellite images are used effectively to characterize marine mammal sea ice habitats both in terms of both surface type and physical temperature, except when clouds or darkness obscured the view (Massom et al., 1994). The SMMR and SSM/I data (brightness temperature and ice concentration maps) are archived at and distributed by the National Snow and Ice Data Center (NSIDAC) in Boulder, Colorado (Gloersen et al., 1993). These data provide important information for Arctic ecology studies, but both active and passive microwave systems are required to provide information on sea ice types, concentration, physical surface temperature, ridges and roughness, lead and polynya formations, movement, etc. Combining emissivity with backscatter data improves the ability to classify sea ice types (Curlander et al., 1991; Fung, 1994).

However, all microwave remote sensing systems have identical problems with the poor ability to distinguish ice cover during the presence of meltponding, wet snow, or flooding, which are often associated with seasonal temperature changes or diurnal freeze–thaw cycles (Winebrenner et al., 1993, 1994). The most sensitive detection of sea ice habitat parameters must include routine satellite monitoring with adequate spatial and temporal resolutions for compatibility with ecology parameters and models. An integrated multisensor remote sensing approach could take advantage of differing sensitivities to different surface types, enhanced temporal resolution and multiscale spatial resolution that could be used to assess the precision of broad-scale or long-term interpretations. In this context, integrated data processing of SMMR, SSM/I, OKEAN-01 and AVHRR data can open new possibilities for Arctic marine mammal studies.

REFERENCES

Andren, H., 1990. Despotic distribution, unequal reproduction success, and population regulation in the jay *Garrulus gradarius* L. *Ecology*, Vol. 71, pp. 17962–1803.

Arthur, S. M., Manly, B. F. J., McDonald, L. L. and Garner, W. G., 1996. Assessing habitat selection when availability changes. *Ecology*, Vol. 77(1), pp. 215–227.

Belchansky, G. I. and Douglas, D.C., 2000. Classification methods for monitoring Arctic sea-ice using OKEAN passive/active two-channel microwave data. *J. Remote Sens. Environm.*, 73(3): 307–322.

Belchansky, G. I., Douglas, D.C., Ovchinnikov, G. K., Pank, L. F. and Petrosyan, V. G., 1992. Processing of space monitoring data for studying large mammals in Arctic environment. *Earth Research from Space, Russian Academy of Sciences*, No. 2, pp. 75–81.

Belchansky, G. I. and Petrosyan, V. G., 1995. Methodological aspects of estimating changes in structure of small mammals. *J. Progress of Current Biology*, Russian Academy of Sciences, Moscow, No. 5, Vol. 115, pp. 573–585.

Belchansky, G. I., Ovchinnikov, G. K., Mordvintsev, I. N. and Douglas, D. C., 1995. Assessing trends in Arctic sea ice distribution using the KOSMOS–OKEAN satellite series. *Polar Record* 31(177), pp. 129–134. Scott Polar Research Institute and Cambridge University Press.

Belchansky, G. I., Douglas, D.C. and Kozlenko, N. N., 1996. The Estimation of Sea Ice Types and Concentration Using Two-channels Microwave Active and Passive Satellite Sensor Data. *Earth Research from Space, Russian Academy of Sciences*, No. 6, pp. 28–39.

Carey, A. B., Horton, S. P. and Biswell, B. L., 1992. Northern Spotted Ows: influence of prey base and landscape character. *Ecological Monographs* 62, pp. 223–250.

Cavalieri, D. J. and Germain, K. M., 1995. Arctic sea ice research with satellite passive microwave radiometers. Geoscience and remote sensing society, *IEEE NEWSLETTER*, December, pp. 6–12.

Cavalieri, D. J., Martin, S. and Gloersen, P., 1979. Nimbus-7 SMMR observation of Bering Sea Ice cover during March. *J. Geophys. Res.*, Vol. 88, pp. 2743–2754.

Cavalieri, D. J. and Parkinson, C. L., 1981. Large-scale variations in observed Antarctic sea ice extent and associated atmospheric circulation. *Monthly Weather Review*, Vol.109, pp. 2323–2336.

Cavalieri, D. J., Gloersen, P. and Campbell, W. J., 1984. Determination of sea ice parameters with Nimbus-7 SMMR. *J. Geophys. Res.*, Vol. 89, pp. 5355–5369.

Chan, H. L. and Fung, A. K., 1977. A theory of sea scatter at large incident angles. *J. Geophys. Res.*, Vol. 82, pp. 3439–3444.

Comiso, J. C., 1983. Sea ice effective microwave emissivities from satellite passive microwave and infrared observations. *J. Geophys. Res.*, Vol. 88, No. C 12, pp. 7686–7704.

Curlander, J. C. and McDonough, R. N., 1991. Synthetic aperture radar: systems and signal processing. New-York: John Wiley & Sons, Inc.

Eppler, D. T., Farmer, L. D. and Lohanick, A. W., 1986. Classification of sea ice types with single-band (33.6 GHz) airborne passive microwave imagery. *J. Geophys. Res.*, Vol. 91, No. C9, pp. 10661–10695.

Fetterer, F. M., Gineris, D. and Kwok, R., 1994. Sea ice type maps from Alaska Synthetic Aperture Radar Facility imagery: An assessment. *J. Geophys. Res.*, Vol. 99, No. C11, pp. 22443–22458.

Fung, A. K., 1994. Microwave Scattering and Emission Models and Their Applications. Artech House, Norwood, MA, 573 p.

Garner, G. W., Knick, S. T. and Douglas, D. C., 1990. Seasonal movements of adult female polar bears in the Bering and Chukchi Seas. *Int. Conf. Bear Res. and Manage*. 8 pp. 219–226.

Germain, K. M. and Cavalieri, D. J., 1997. A microwave technique for mapping ice temperature in the Arctic seasonal sea ice zone. *IEEE Transaction on Geoscience and Remote Sensing*, Vol. 35, No. 4, pp. 946–953.

Gloersen, P. and Barath, F. T., 1977. A scanning multichannel microwave radiometer for Nimbus-G and Seasat-A. IEEE, *J. Oceanic Eng.*, OE-2, pp. 172–178.

Gloersen, P. and Campbell, W. J., 1988. Variations in the Arctic, Antarctic, and global sea-ice covers during 1978–1987 as observed with the Nimbus-7 SMMR. *J. Geophys. Res.*, Vol. 22. 93, No. C9, pp. 10666–10674.

Gloersen, P. and Cavalieri, D. J., 1986. Reduction of weather effects in the calculation of Sea Ice concentration from microwave radiances. *J. Geophys. Res.*, Vol. 91, pp. 3913–3919.

Gloersen, P. and Campbell, W. J., 1991. Recent variations in Arctic and Antarctic sea-ice covers. *Nature*, 352 (6330), pp. 33–36.

Gloersen, P., Campbell, W. J., Cavalieri, D. J., Comiso, J. C., Parkinson, G. L. and Zwally, H. J., 1993. Satellite passive microwave observations and analysis of Arctic and Antarctic sea-ice, 1978–1987. *Annals of Glaciology*, Vol.17, pp. 149–154.

Gray, A. L., Hawkins, R. K. and Livingstone, C. E., 1982. Simultaneous Scatterometer and Radiometer Measurements of Sea-Ice Microwave Signatures. *IEEE Journal of Oceanic Engineering*, Vol. 0, No. 1, pp. 20–32.

Grody, N. C. and Basist, A. N., 1997. Interpretation of SSM/I measurements over Greenland. IEEE transaction. *Geoscience and Remote Sensing*, Vol. 35, No. 2, March, pp. 360–365.

Holbrook, S. J. and Schmitt, R. J., 1988. The combined effects of predation risk and food reward on patch selection. *Ecology*, Vol. 69, pp. 125–134.

Lubin, Y., Ellner, S. and Kotzman, M., 1993. Web relocation and habitat selection in a desert widow spider. *Ecology*, Vol. 74, pp. 1915–1928.

Massom, R. and Comiso, J. C., 1994., The classification of Arctic sea ice types and the determination of surface temperature using advanced very high resolution radiometer data. *J. Geophys. Res.*, Vol. 99, No. C3, pp. 5201–5218.

Onstott, R. G., Moore, R. K. and Gogineni, S., 1982. Four years of low-altitude sea ice broad-band backscatter measurements. *IEEE*, OE-7, pp. 44–50.

Onstott, R.G. and Grenfell, T. C., 1987. Evolution of microwave sea ice signatures during early summer and midsummer in the marginal ice zone. *J. Geophys. Res.*, Vol. 92, No. C7, pp. 6825–6835.

Parkinson, G. L. and Cavalieri, D. J., 1989. Arctic Sea Ice 1973–1987: Seasonal, Regional, and Interannual Variability. *J. Geophys. Res.*, Vol. 94, No. C10, pp. 14499–14523.

Swift, C. T., Fedor, L. S. and Ramseier, R. O., 1985. An Algorithm to Measure Sea Ice Concentration With Microwave Radiometers. *J. Geophys. Res.*, Vol. 90, No. C1, pp. 1087–1099.

Wickland, D. E., 1991. Mission to planet earth: the ecological perspective. *Ecology*, Vol. 72, pp. 1923–1933.

Winebrenner, D., Key, J., Schweiger, A., Nelson, E. D., Colony, R., Baebet, D. and LeDrew, E., 1993. On links between microwave and shortwave signatures of multiyear sea ice during the onset of melt. *Proceedings of the Topical Symposium on Combined Optical-Microwave Earth and Atmosphere Sensing*, IEEE, Albuquerque, N. M., March 22–25.

Winebrenner, D. P., Nelson, E. D. and Colony, R., 1994. Observation of melt onset on multiyear Arctic sea ice using the ERS-1 synthetic aperture radar. *J. Geophys. Res.*, Vol.99, No. C11, pp. 22425–22441.

Zahl, S., 1977. Jackknifing and index diversity. *Ecology*, Vol. 58, pp. 907–913.

Zwally, H. J., Comiso, J. C. and Walsh, J. E., 1991. Variability of Antarctic and Arctic sea ice. *Proc. Int. Conf. Role of the Polar Regions in Global Change*, G.Weller, C. L. Wilson and B. A. Severin, Eds., Geophysical Institute, University of Alaska Fairbanks and Center for Global Change and Arctic System Research, UAF, Vol. 1, p. 22.

4

Investigating Variability in Arctic Sea Ice Distribution Using OKEAN and ALMAZ SAR Satellite Data

4.1 INTRODUCTION

Satellite monitoring of sea ice cover in the Arctic is essential for understanding the climate and ecological processes. Results of numerical investigations show that Arctic sea ice affects the climate by regulating the exchange of heat, moisture and momentum between the ocean and atmosphere and is a potential early indicator of global climate change. Regional yearly variations in the seasonal distribution and abundance of sea ice have been shown to have significant effects on Arctic marine ecosystems and on the reproduction and survival of Arctic marine mammals and birds. For example, a decrease in sea ice cover may stimulate an initial increase in biological productivity. However, it is likely that polar bear, seal and other populations will decline wherever the quality and availability of breeding habitat are reduced (Stirling, 1980; Smith et al., 1991; Stirling et al., 1993; Stirling, 1996). A number of studies have suggested that changes in the global average air temperature might be detectable by observing changes in the extents of the Arctic sea ice cover. Minimum sea ice extent and concentration during the summer provide indirect information required for evaluation of heat and humidity fluxes among the ocean, atmosphere and perennial ice pack (Parkinson, 1991A; Gloersen and Campbell, 1991). One sensitive region in the context of global change is the Barents and Kara Seas and adjacent parts of the Arctic Ocean. Documenting variations in the annual minimum ice extent and concentration in this region is important for understanding historical periodicity, and for investigating long-term trends. However, long-term trends derived from ice maps are fixed estimates without variance.

ESMR, SMMR and SSM/I satellite measurements are the prime data used to derive sea ice concentration, distinguish FY ice from MY ice and determine and map the interannual variability of spatial distribution of North Polar sea ice (Cavalieri et al., 1984; Parkinson, 1991B; Gloersen and Campbell, 1993). Comparative analysis of the ice maps using Nimbus-7 SMMR data and the Navy/NOAA ice maps shows significant quantitative differences and the need to use a consistently calibrated and analyzed data set (Zwally et al., 1991). Analysis of spatial and temporal variations in global sea ice coverage from

October 1978 through August 1987 has been carried out based on Nimbus-7 SMMR data, and published in a series of publications (Cavalieri et al., 1984; Gloersen and Campbell, 1991). These works have demonstrated that interannual variability of the sea ice extents is much larger for the individual regions than for the Arctic as a whole, with some regions exhibiting decreasing trends and others increasing. Negative trends have been defined in the Kara and Barents Seas for this period.

This chapter provides a brief description of investigating spatial and temporal variations in sea ice distribution in the Barents and Kara Seas and adjacent parts of the Arctic Ocean using microwave satellite data. The main objectives of these studies included quantifying long-term trends and annual variations in the extent of seasonal sea ice cover based on KOSMOS–OKEAN satellite series and ALMAZ SAR satellite data and Monte-Carlo analysis. The objectives also included developing the regional-scale integrated remote sensing database to address historical variations and current trends of sea ice cover in the Arctic region, and assessing the capability of integrating ALMAZ, KOSMOS and OKEAN-01 satellite data for sea ice mapping. The sea ice extent was studied for a set of sea ice classification schemes produced using the regional-scale remote sensing database, a multithematic geographical database and a problem-oriented data processing system (Belchansky et al., 1992, 1994). KOSMOS and OKEAN-01 satellite image data were used to estimate confidence intervals for the 1974–1994 map-derived ice trends in the Barents and Kara Seas. Results from the 1974–1994 period were combined with ice extension data reported by Vinje (1991) to examine longer-term trends over a 28-year period between 1966 and 1994.

4.2 STUDY AREA

The study area included the Barents Sea, the Kara Sea and adjacent parts of the Arctic Ocean (Figure 4.1). Two subregions encompassing the Barents Sea and the western Kara Sea were analyzed separately for comparability, with the results reported in Vinje (1991). The more detailed study area parameters and characteristics are presented in Chapter 3.

4.3 DATA

The main data included the sea ice maps and images collected by KOSMOS-1500, KOSMOS-1766, KOSMOS-1869 and OKEAN-01 N3, N5, N6 polar-orbiting satellites (RM-08, RAR and MSU-M) and ALMAZ SAR satellite. Historical digital images, 1984–1994, were acquired for regions within the study area (Table 4.1).

The KOSMOS–OKEAN polar-orbiting satellites are capable of recording RM-08, RAR and MSU-M imagery either simultaneously or individually. The RM-08 radiometer (0.8 cm wavelength) has 15 km average resolution and a 550 km wide swath. The RAR (3.15 cm wavelength) has 1.2 km average resolution and a 450 km swath width, and the MSU-M scanner records in one of four visible or near-infrared bandwidths with 1.5 km average resolution and a 900 km swath (Figure 4.2).

ALMAZ SAR (9.6 cm wavelength, resolution 11 m) (Figure 4.3) data were acquired for the Kara Sea and included four images (orbit #01762; 19.07.91): (1) 70.445°N,

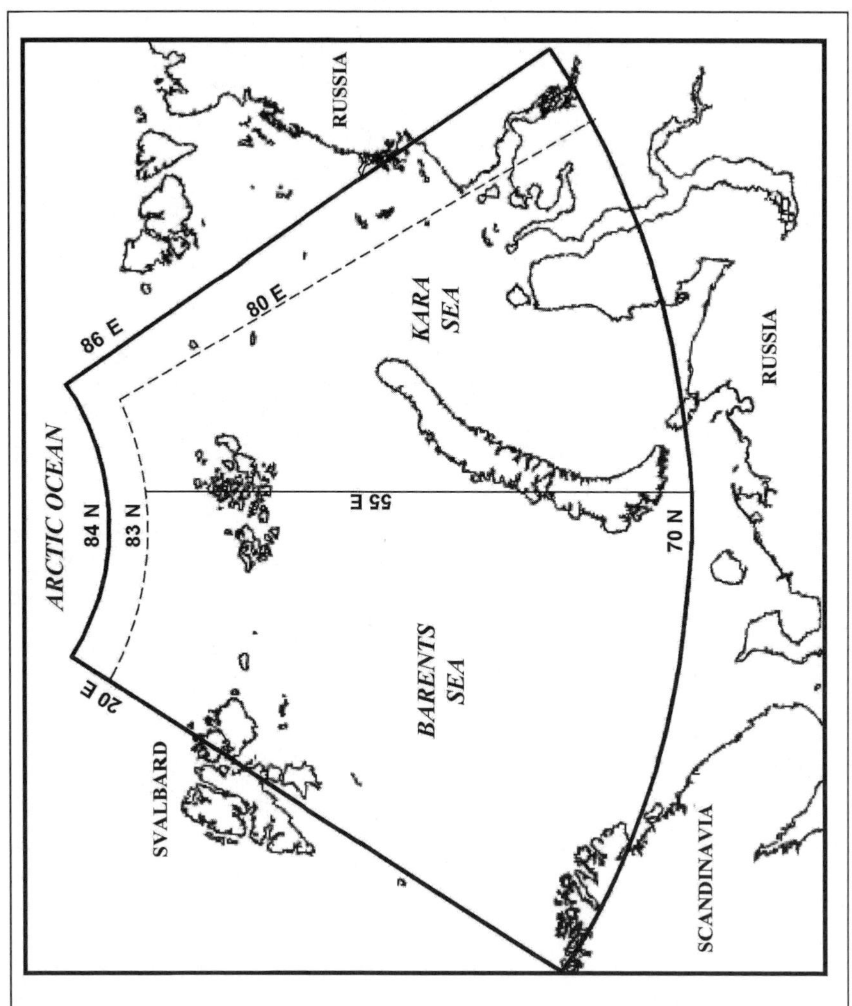

Figure 4.1 Location of study area.

Table 4.1 Acquisition Dates, Sensors, and Geographic Coverage of the KOSMOS and OKEAN-01 Satellite Images

Satellite	Date	Time GMT	Orbit	Sensors	MSU-M channel (μm)	Satellite track	
						Begin	End
KOSMOS-1500	25.07.84	02.38–02.45	4431	RAR+MSU-M	0.8–1.1	81.5N 13.8E	62.8N 58.1E
KOSMOS-1500	04.08.84	02.08–02.14	4578	RAR+MSU-M	0.8–1.1	82.3N 11.6W	64.9N 45.1E
KOSMOS-1500	11.08.84	13.52–13.58	4689	RAR+MSU-M	0.8–1.1	64.4N 53.2E	82.1N 106.4E
KOSMOS-1500	28.08.84	20.50–20.56	4943	RAR		82.4N 19.4E	65.1N 77.3E
KOSMOS-1766	28.07.87	23.30–23.35	5392	RAR		64.3N 28.8E	80.3N 61.3E
KOSMOS-1869	09.08.87	10.29–10.35	58	MSU-M	0.7–0.8	79.6N 24.4E	59.8N 55.0E
KOSMOS-1766	04.09.87	04.17–04.23	5925	RAR+MSU-M	0.8–1.1	81.8N 13.1W	63.5N 35.6E
KOSMOS-1766	06.09.87	01.54–02.00	5953	RAR+MSU-M	0.8–1.1	82.6N 06.3W	66.9N 65.1E
KOSMOS-1766	31.07.88	07.59–08.05	10802	MSU-M	0.7–0.8	81.6N 23.3E	63.0N 68.8E
				RAR+MSU-M	0.8–1.1		
KOSMOS-1766	02.08.88	08.52–08.58	10832	MSU-M	0.7–0.8	81.6N 06.2E	63.0N 51.6E
OKEAN-01 N3	15.08.88	15.22–15.27	608	RM-08+RAR+MSU-M	0.8–1.1	81.3N 13.1W	65.9N 26.8E
OKEAN-01 N3	20.08.88	12.41–12.46	680	RM-08+RAR+MSU-M	0.8–1.1	82.4N 26.8W	71.6N 50.4E
OKEAN-01 N3	22.08.88	13.36–13.42	710	RM-08+RAR+MSU-M	0.8–1.1	82.3N 17.6W	64.8N 39.5E
OKEAN-01 N3	28.08.88	11.25–11.30	797	RM-08+RAR+MSU-M	0.8–1.1	81.7N 14.4E	66.8N 59.5E
OKEAN-01 N5	19.07.90	16.17–16.23	2086	RAR+MSU-M	0.8–1.1	82.5N 29.0W	67.4N 46.4E
OKEAN-01 N5	05.08.90	14.09–14.15	2335	RAR+MSU-M	0.8–1.1	82.3N 38.1E	68.6N 45.2E
OKEAN-01 N5	12.08.90	02.33–02.38	2431	RAR+MSU-M	0.8–1.1	69.5N 68.0E	82.5N 130.5E
OKEAN-01 N5	13.08.90	04.36–04.42	2447	RAR+MSU-M	0.8–1.1	63.3N 29.9E	81.8N 77.5E
OKEAN-01 N5	20.08.90	04.30–04.38	2550	RAR+MSU-M	0.8–1.1	66.0N 20.0E	82.5N 85.6E
OKEAN-01 N5	24.08.90	03.03–03.08	2608	RAR+MSU-M	0.8–1.1	68.7N 36.8E	82.4N 93.9E
OKEAN-01 N6	17.07.91	12.45–12.51	637	RM-08+RAR+MSU-M	0.8–1.1	81.2N 01.5W	62.2N 39.8E
OKEAN-01 N6	31.07.91	23.41–23.47	850	RM-08+RAR+MSU-M	0.8–1.1	62.1N 58.3E	81.1N 99.2E
OKEAN-01 N6	04.08.91	09.18–09.24	900	RM-08+RAR+MSU-M	0.8–1.1	81.0N 18.0E	61.8N 57.7E
OKEAN-01 N6	11.08.91	22.02–22.08	1011	RM-08+RAR+MSU-M	0.8–1.1	65.5N 64.6E	82.5N 126.5E
OKEAN-01 N6	26.08.91	05.58–06.03	1222	RM-08+RAR+MSU-M	0.8–1.1	79.1N 39.9E	62.8N 66.7E

Sensors: RM-08 – Passive Microwave Radiometer; MSU-M – Multispectral Scanning System; RAR – Side-Looking Real Aperture Radar.

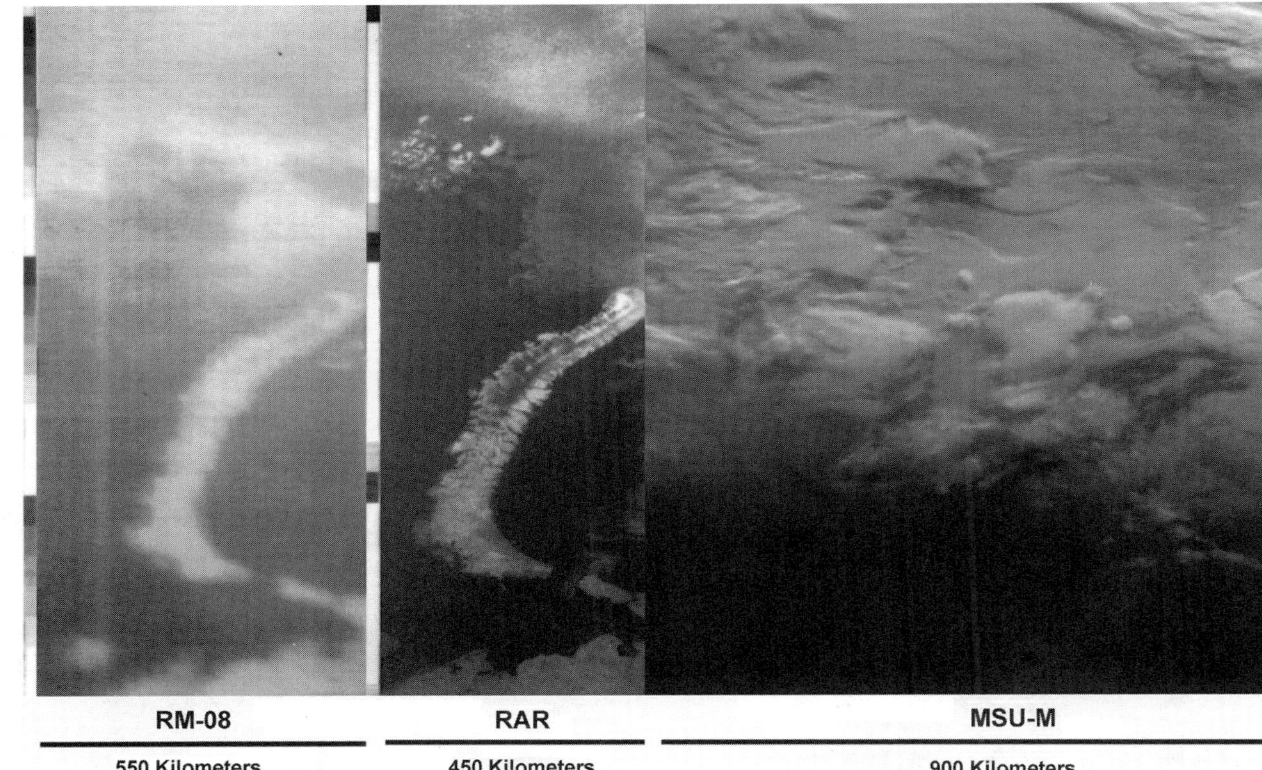

Figure 4.2 Examples of raw OKEAN satellite data of simultaneously acquired passive microwave (RM-08), side-looking real aperture radar (RAR) and optical scanner (MSU-M) imagery.

Figure 4.3 Example of ALMAZ SAR satellite data sea ice classification.

56.102°E; (2) 70.463°N, 57.064°E; (3) 70.475°N, 58.027°E; (4) 70.481°N, 58.992°E (center points coordinates).

The more detailed characteristics of the KOSMOS–OKEAN and ALMAZ SAR satellite series instruments are presented in Chapter 2.

Sea ice maps (1:5 000 000) were acquired from the Russian Hydrometeorological Center for 10-day intervals, July through October, 1974–1994. The Russian ice maps were manually derived by specialists at the Hydrometeorological Center from a variety of data sources, depending on availability, that included shipboard observations, aircraft reconnaissance, Scandinavian and American ice maps, and Russian and American satellite data.

Minimum ice-extent data in the Barents and western Kara Seas for the period 1966–1983 were extracted from Vinje (1991).

4.4 METHODS

The sea ice extent, surface characteristics and trends were studied using constructed multithematic geographical and regional-scale remote sensing databases. For this task, a problem-oriented data collection and processing system was developed.

The data collection subsystem includes the satellite database (RAR, RM-08, MSU-M, ALMAZ SAR image data; space board system parameters; orbital measurements) and the orbital parameter estimation subsystem. The data processing subsystem includes the data restoration system and the data interpretation system.

The remote sensing and multithematic geographic databases include, respectively, a set of satellite images for the period 1984–1994 (July–October) and a sea ice database developed using sea ice maps for the period 1974–1994 (July–October). Data structure of the sea ice database is characterized by the attributes shown in Table 4.2.

The Hydrometeorological Center sea ice maps were digitized into an ARC/INFO (ESRI, Redlands, California) geographical information system database.

For each year between 1974 and 1994, the total area covered by sea ice was quantified for the absolute annual minimum, the minimum monthly mean, and the sea ice extent at the end of August. Four definitions of ice extent, based on ice concentration, were considered for each analysis: E1 → 90%, E2 → 70%, E3 → 40% and E4 → 10% (Figure 4.5).

KOSMOS–OKEAN image analysis began with radiometric calibration and georegistration (Figure 4.6). Derivative bands that included an RAR/RM-08 ratio and the first principal component (PC1) of RM-08, RAR and MSU-M were calculated for OKEAN-01 N3 and N6. Supervised ice classification procedures were developed using randomly selected training polygons extracted from corresponding ice maps.

An algorithm was developed with minimum loss function criteria (Aronoff, 1984) to ascertain the most efficient classification procedure for discriminating five ice concentration classes: <10%, 10–40%, 40–70%, 70–90% and >90%. Differences between the satellite-derived classifications and the hydrometeorological ice maps were calculated (Figures 4.6–4.8), and the results were used in a Monte-Carlo analysis (Figure 4.4) to statistically estimate confidence limits for the ice map trends.

Ice extent trends during the 1966–1993 period were calculated by including data published in Vinje (1991).

Let us consider the scheme of a Monte-Carlo analysis (Figure 4.4).

Table 4.2 Data Structure of Sea Ice Database

COVER TYPE	CONCENTRATION	SURFACE FORMS	
Water	Compact pack ice	Floating ice:	Pack ice motion process:
Ground	Very close pack ice	Pancake ice	Drift dividing zone
Ice of land origin	Close pack ice	Giant floe	Shearing ice zone
Lake ice	Open pack ice	Vast floe	Diverging ice zone
River ice	Very open pack ice	Big floe	Faint pres. ice zone
Sea ice	Open water	Medium floe	Consid.pres. ice zone
	Bergy water	Small floe	Strong pres. ice zone
	Ice free	Ice cake	Hummocking forms:
		Small ice cake	Hummock
		Floeberg	Standing floe
		Ice breccia	Ridged ice zone
		Brash ice	Consolidated ridge
		Iceberg	Weathered ridge
		Ice island	Openings in the ice:
		Fast ice:	Crack
		Fast ice	Fracture
		Coastal ice	Lead
		Icefoot	Polynya
		Anchor ice	
		Grounded ice	
		Stranded ice	
		Grounded hummock	
STAGES OF DEVELOPMENT	ICE-SURFACE FEATURES		
	HUMMOCKED ICE	M-YEAR ICE HILLING	
New ice:	Ice plane surface	Plane MYI	
Frazil ice	Hummocked ice 0–20%	Smoothed MYI	
Grease ice	Hummocked ice 20–40%	Temperate hilled MYI	
Slush	Hummocked ice 40–60%	Strong hilled MYI	
Shuga	Hummocked ice 60–80%		
Nilas:	Hummocked ice 80–100%		
Ice rind			
Dark nilas	STAGES OF MELTING	SNOW-COVERED ICE	
Light nilas			
Pancake ice	Melting absent	Snow absent	
Young ice:	Puddle	Snow-covered 0–5 cm	
Grey ice	Patch of soggy snow	Snow-covered 5–20 cm	
Grey-white ice	Small lakes	Snow-covered >20 cm	
First year ice (F-Y):	Thaw hole	Sastrugi	
Thin F-Y ice	Gully	Snow-blow	
Medium F-Y ice	Dried ice	Snow-drift	
Thick F-Y ice	Rotten ice		
Old ice:	Flooded ice	ICE POLLUTION	
Residual f-y ice	Ram		
Second-yearice		Ice pure	
Multiyearice		Ice pollution 10–40%	
Old ice		Ice pollution 40–60%	
Unknown age		Ice pollution >70%	

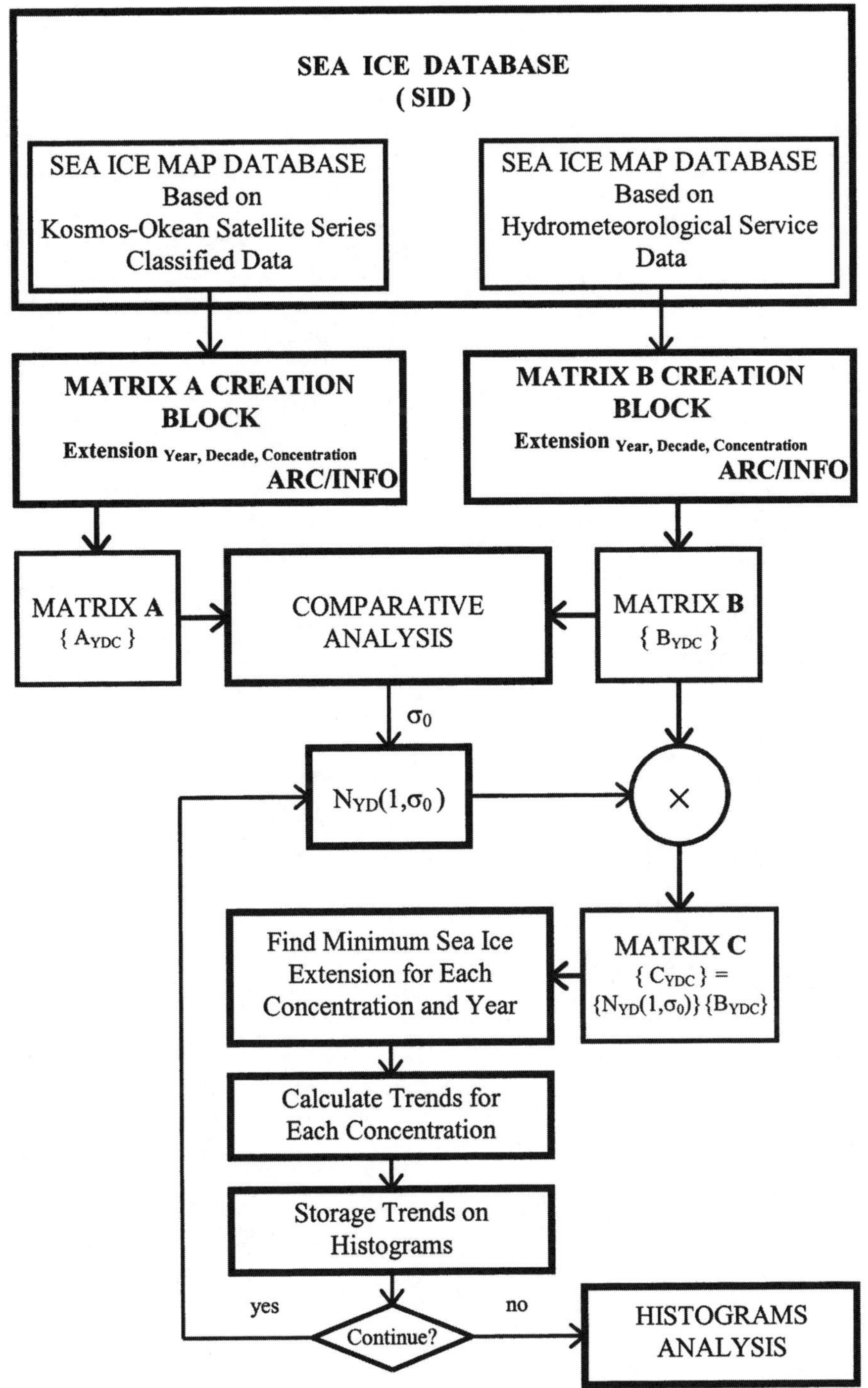

Figure 4.4 The logic scheme of Monte-Carlo analysis.

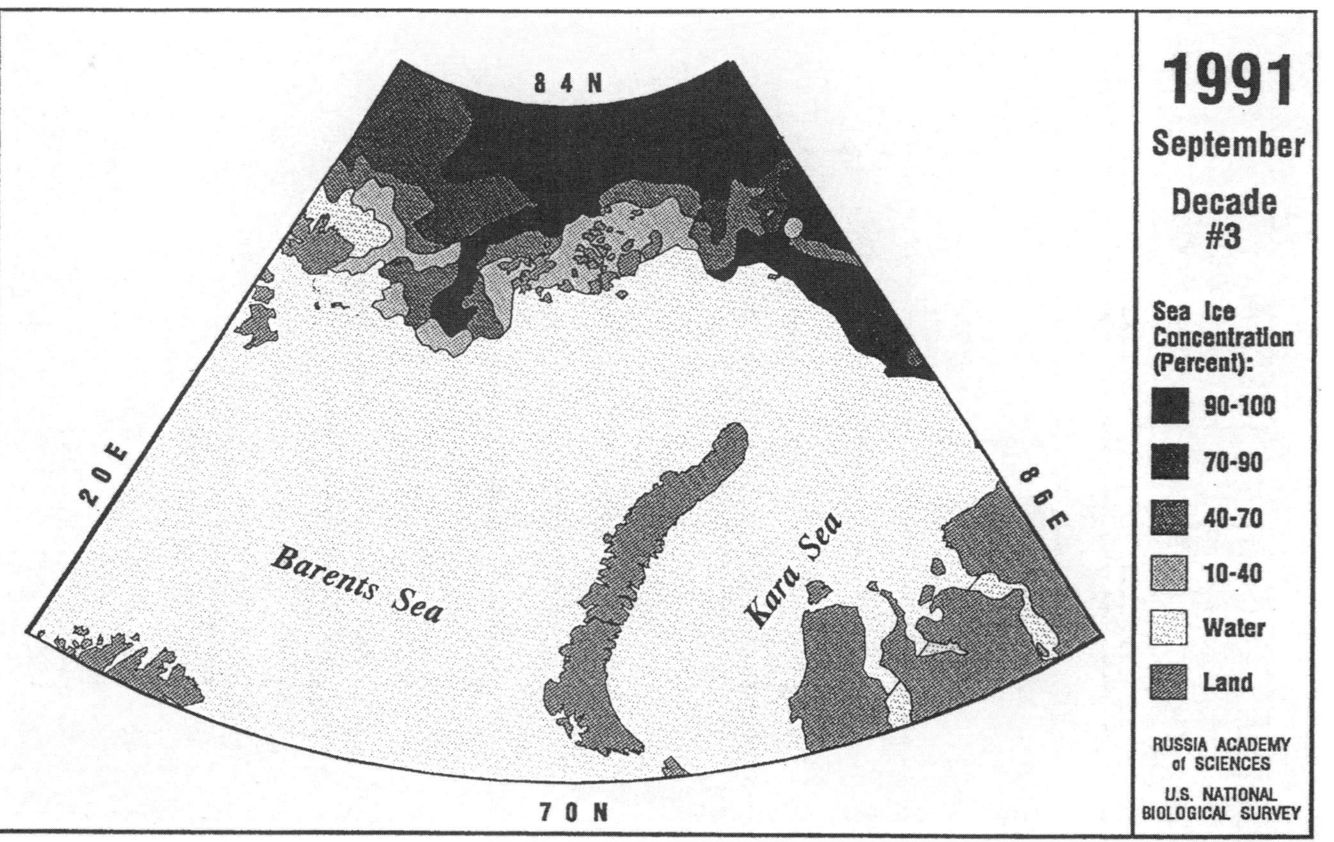

Figure 4.5 Example of quantified sea ice area.

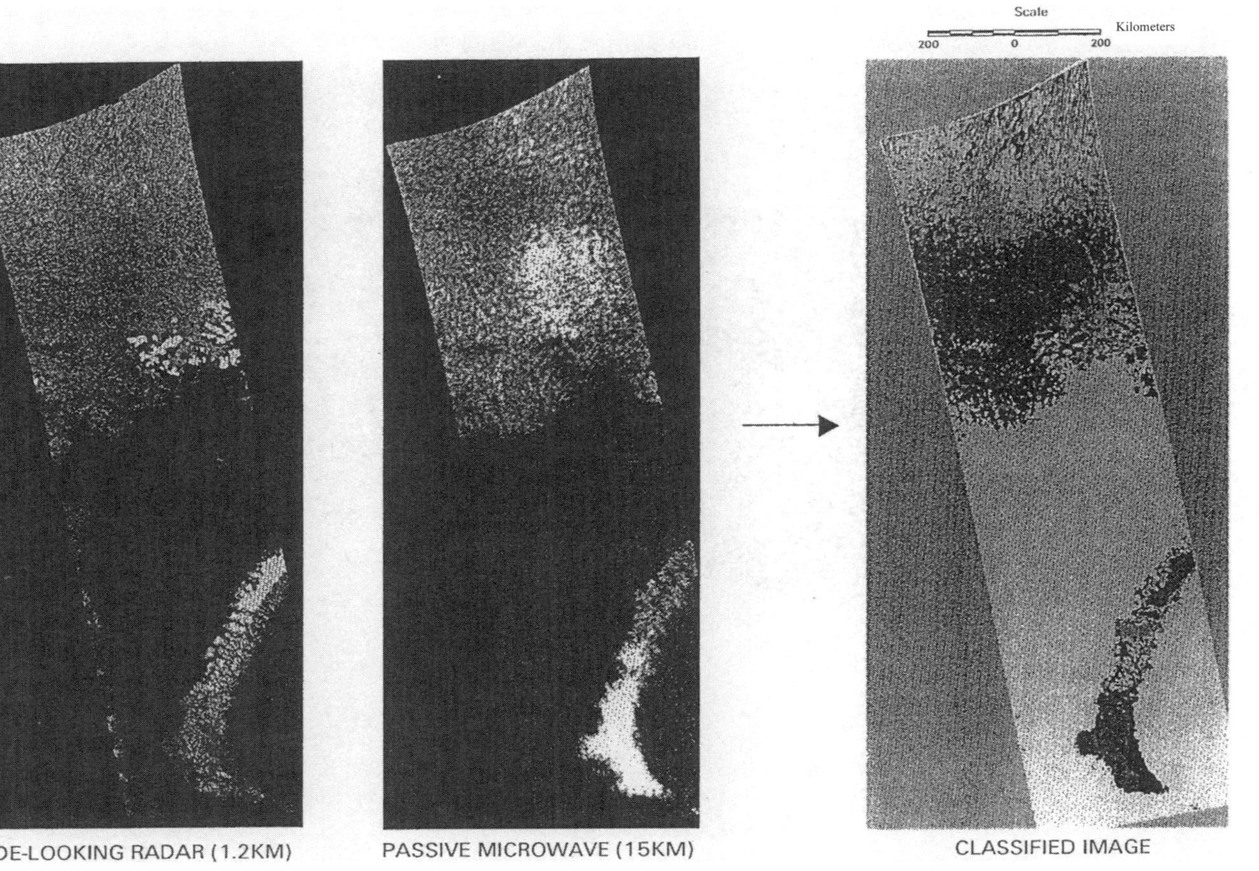

Figure 4.6 Calibrated and georeferenced images of RM-08 (left) and RAR (right).

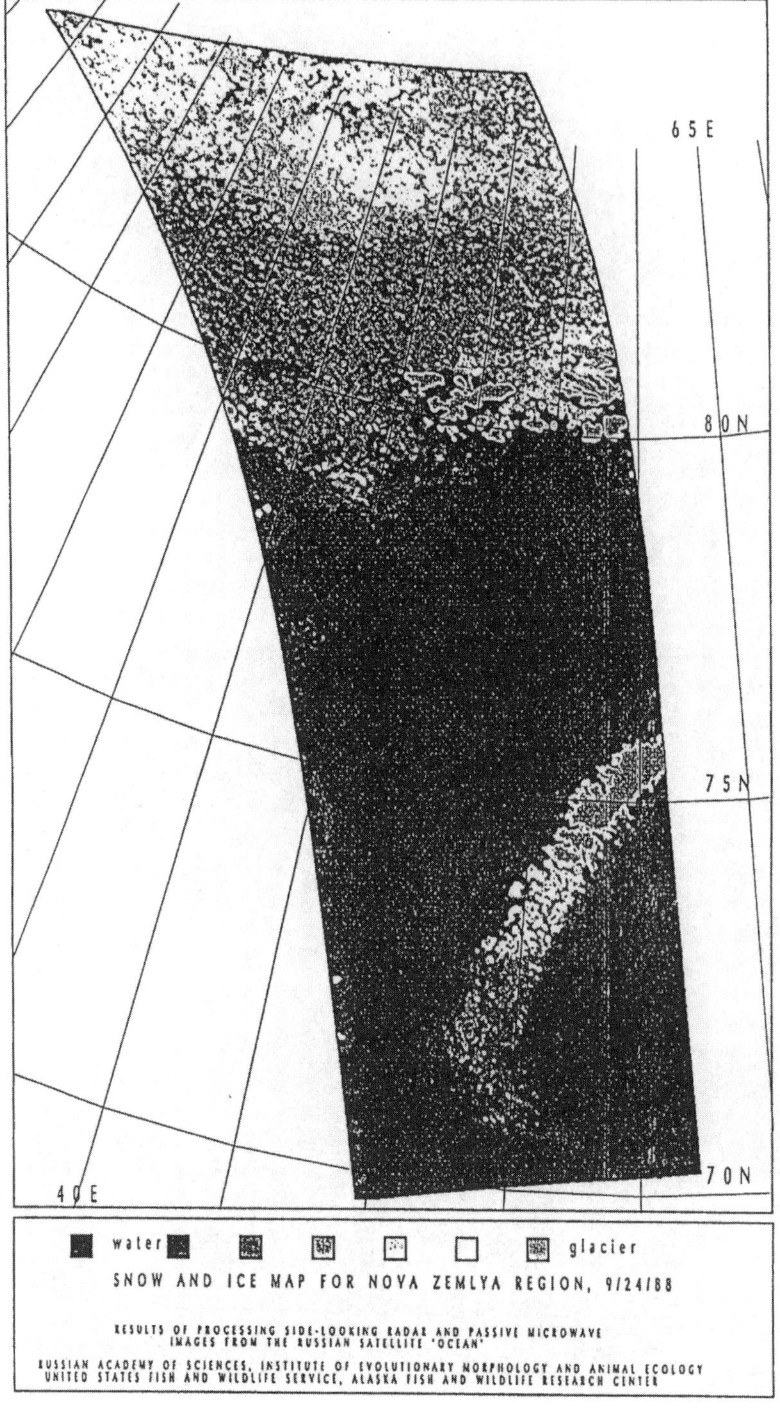

Figure 4.7 Example of OKEAN satellite data classification.

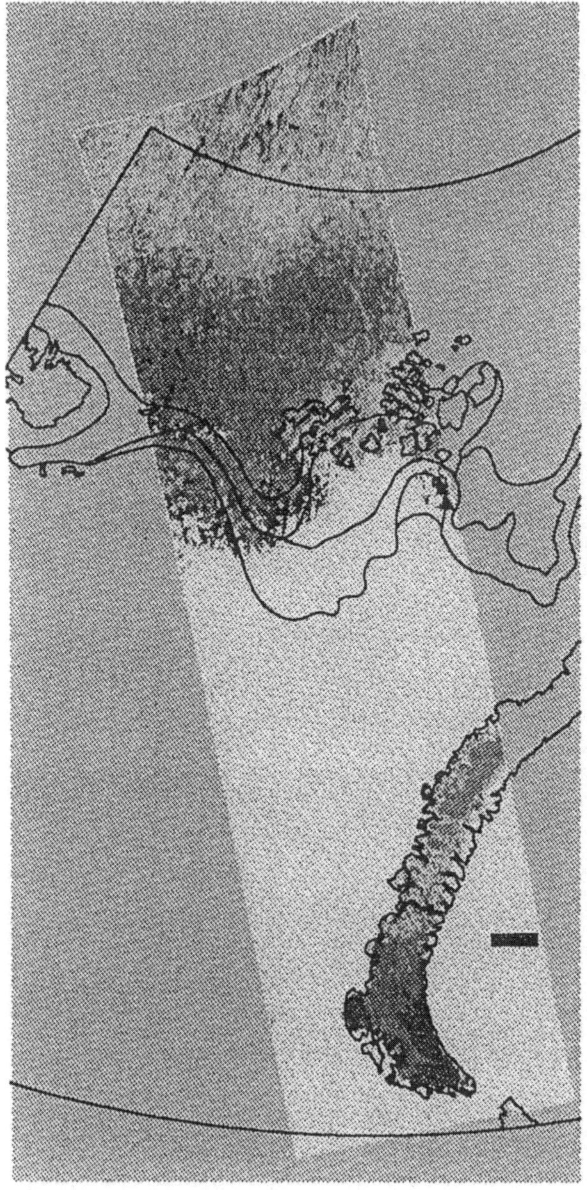

Figure 4.8 Hydrometeorological ice map delineations (white polygons) overlaid on an OKEAN satellite-image ice classification scheme for late September, 1991.

The extension matrices **A** and **B** are formed for statistical modeling, where

$\mathbf{A} = \{A_{YDC}\}$, $\mathbf{B} = \{B_{YDC}\}$, Y — year, D — decade, C — sea ice concentration.

Relative difference ΔE_{YDC}, mean relative difference ΔE_{YDC} and standard deviation σ_0 are calculated in the comparative analysis block:

$$\Delta E_{YDC} = \frac{A_{YDC} - B_{YDC}}{B_{YDC}};$$

$$\Delta \bar{E} = \frac{1}{N} \sum_{Y, D, C} \Delta E_{YDC};$$

$$\sigma_0 = \sqrt{\frac{1}{N-1} \sum_{Y, D, C} (\Delta E_{YDC} - \Delta \bar{E})^2},$$

where: $N = N_M N_C$; N_M — quantity of sea ice maps being compared, N_C — quantity of parameters being compared on map.

The statistical modeling process includes:

- Generating the normally distributed random value ($N_{YD}(1, \sigma_0)$) for every decade D in a year Y of the study period.
- Creating the matrix **C**: $\{C_{YDC}\} = \{N_{YD}(1, \sigma_0)\}\{B_{YDC}\}$.
- Finding the absolute minimum (AM), the average monthly minimum (AMM) and the minimum at the end of August (MEA) of extension for every year Y and ice concentration C in the matrix $\{C_{YDC}\}$.
- Calculating the extension alternation trends for AM, AMM and MEA for every concentration C during a 20-year period by means of the least-squares approximation based on the values received.
- Constructing the distribution histograms using obtained trends values.
- The described process is repeated 10 000 times.

Mean trends values and their limit values for different probabilities are presented in the histograms analysis block.

4.5 RESULTS

Image Classification. Optimal ice classification procedures for the images acquired by OKEAN-01 N3 and OKEAN-01 N6 used RM-08 and PC1; RAR and MSU-M were used for OKEAN-01 N5 and KOSMOS-1500 images; and RAR was used for the KOSMOS-1766 images. The satellite-derived ice concentration classes had, on average, a 75% agreement with the ice-map training polygons. Root-mean-square differences between ice maps and the satellite-derived classes were 15.5%, 19.3%, 18.8% and 11.5% for ice extensions E1, E2, E3, and E4, respectively.

Ice Extent Trends. Minimum monthly-mean sea ice extent in the Barents Sea showed an average increasing trend during the 1974–1993 (Figure 4.9a). Results of Monte-Carlo analysis established upper and lower confidence limits ($P = 0.8$) that corresponded to the increasing trend (Table 4.3). Over the 28-year period, 1966–1993, this region showed an average 11.8% increasing trend in minimum monthly-mean sea ice extent (Figure 4.9b). Minimum monthly-mean sea ice extent in the western Kara Sea showed an average

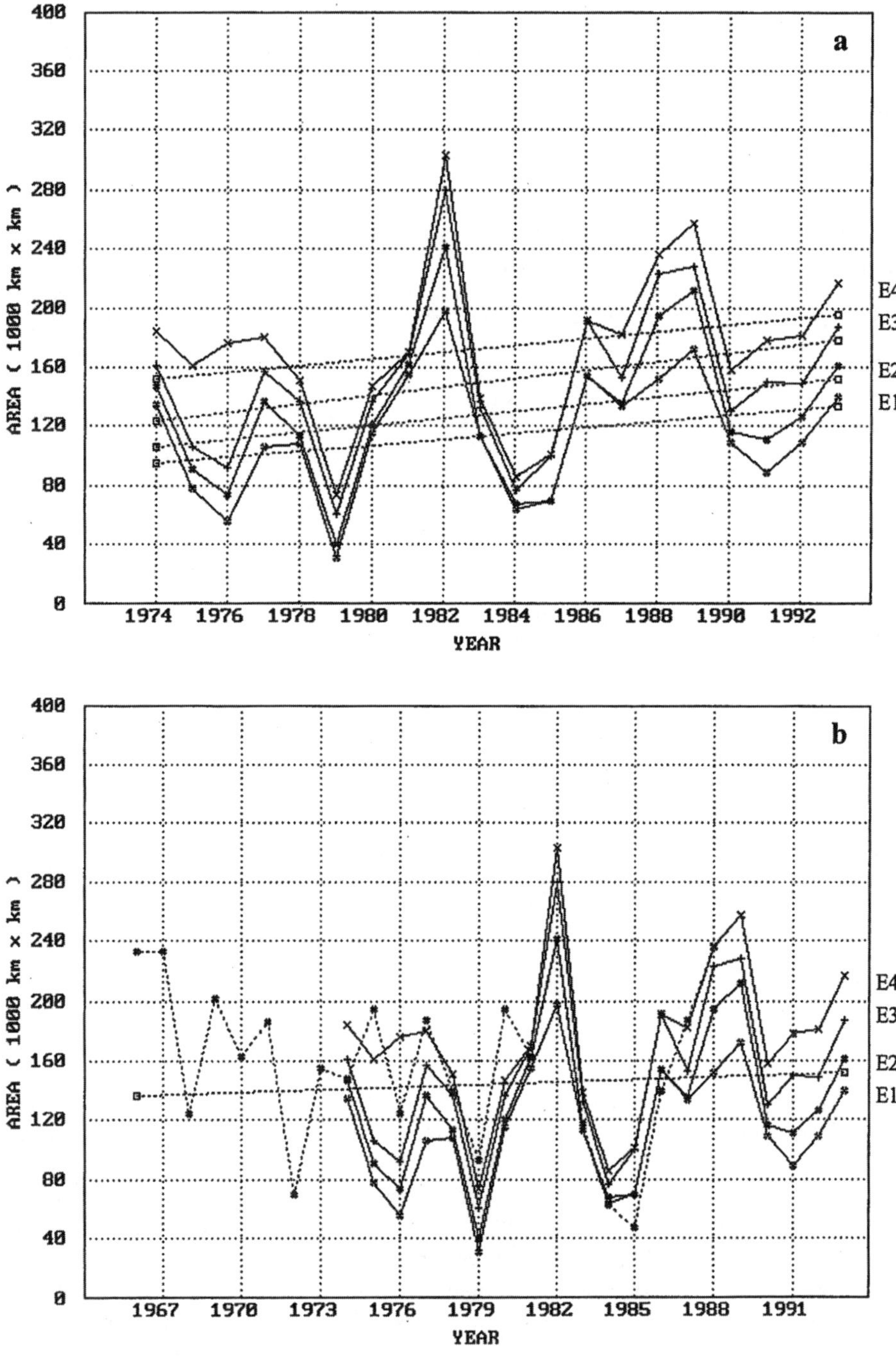

Figure 4.9 Minimum monthly-mean sea ice extent in the Barents Sea.

decreasing trend during the 1974–1993 (Figure 4.10a). Results of Monte-Carlo analysis established upper and lower confidence limits ($P = 0.8$) that corresponded to the decreasing

Table 4.3 20-Year Average Change Trends in the Extent of the Minimum Monthly Mean, the Absolute Annual Minimum Sea Ice Cover and Ice Cover in the End of August in the Barents, Western Kara, and Combined Barents–Kara Seas During the 1974–1993 Period

Sea Ice Extent Definition (% Ice Concentration)							
>90% (E1)		>70% (E2)		>40% (E3)		>10% (E4)	
Trend %	$P = 0.8$ %	Trend %	$P = 0.8$ %	Trend %	$P = 0.8$ %	Trend %	$P = 0.8$ %
MINIMUM-MOUTHLY-MEAN:							
BARENTS SEA							
40.0	30–60	43.4	29–61	45.5	31–63	28.3	18–48
KARA SEA							
–14.7	(–24)–(–4)	–19	(–27)–(–9)	–26.9	(–32)–(–15)	–24.7	(–33)–15
BARENTS and KARA SEAS							
6.9	(–30)–19	6.5	(–4)–18	1.9	(–6)–16	3.0	(–7)–15
ANNUAL MINIMUM:							
BARENTS SEA							
36	19–71	38.6	15–67	37.1	18–69	28.2	7–55
KARA SEA							
–4.1	(–25)–12	–23.7	(–34)–(–4)	–31.1	(–39)–(–9)	–26.2	(–38)–(–8)
BARENTS and KARA SEAS							
11.6	(–8)–29	5.6	(–11)–26	2.5	(–11)–24	5.2	(–12)–24
END OF AUGUST							
BARENTS SEA							
58.5	31–92	52.6	25–85	55.3	27–88	28.1	4–56
KARA SEA							
–29.2	(–43)–(10)	–30.1	(–43)–(–13)	–19	(–34)–(–4)	–18.7	(–33)–(–3)
BARENTS and KARA SEAS							
1.5	(–16)–24	–0.6	(–18)–21	6.7	(–12)–30	4.8	(–14)–28

Trends and estimated confidence intervals ($P = 0.8$) are presented for four definitions of ice-extension thresholds (E1–E4).

trend (Table 4.3). Over the 28-year 1966–1993 period this region showed an average 47.4% decreasing trend in minimum monthly-mean sea ice extent (Figure 4.10b). The recent 20-year changes in minimum monthly-mean sea ice extent appeared consistent with annual variations and periodic cycles characteristic of the extended 28-year database (Figures 4.9 and 4.10). In general, within year comparisons between the Barents and western Kara Seas showed little consistency with respect to the magnitude of ice extension (e.g., relatively high and low ice years in the Barents Sea were not closely associated with similar conditions in the western Kara Sea. Minimum monthly-mean sea ice extent in the Barents and Kara Seas region showed an average very little increasing trend during the 1974–1993 (Figure 4.11).

During 1974–1993, the absolute annual minimum sea ice extension in the Barents and western Kara Seas, and Barents and Kara Seas region also showed increasing and decreasing trends (Figures 4.12, 4.13). The pattern of annual variations was very similar to that observed for the minimum monthly mean, however, the rates of increase and decrease

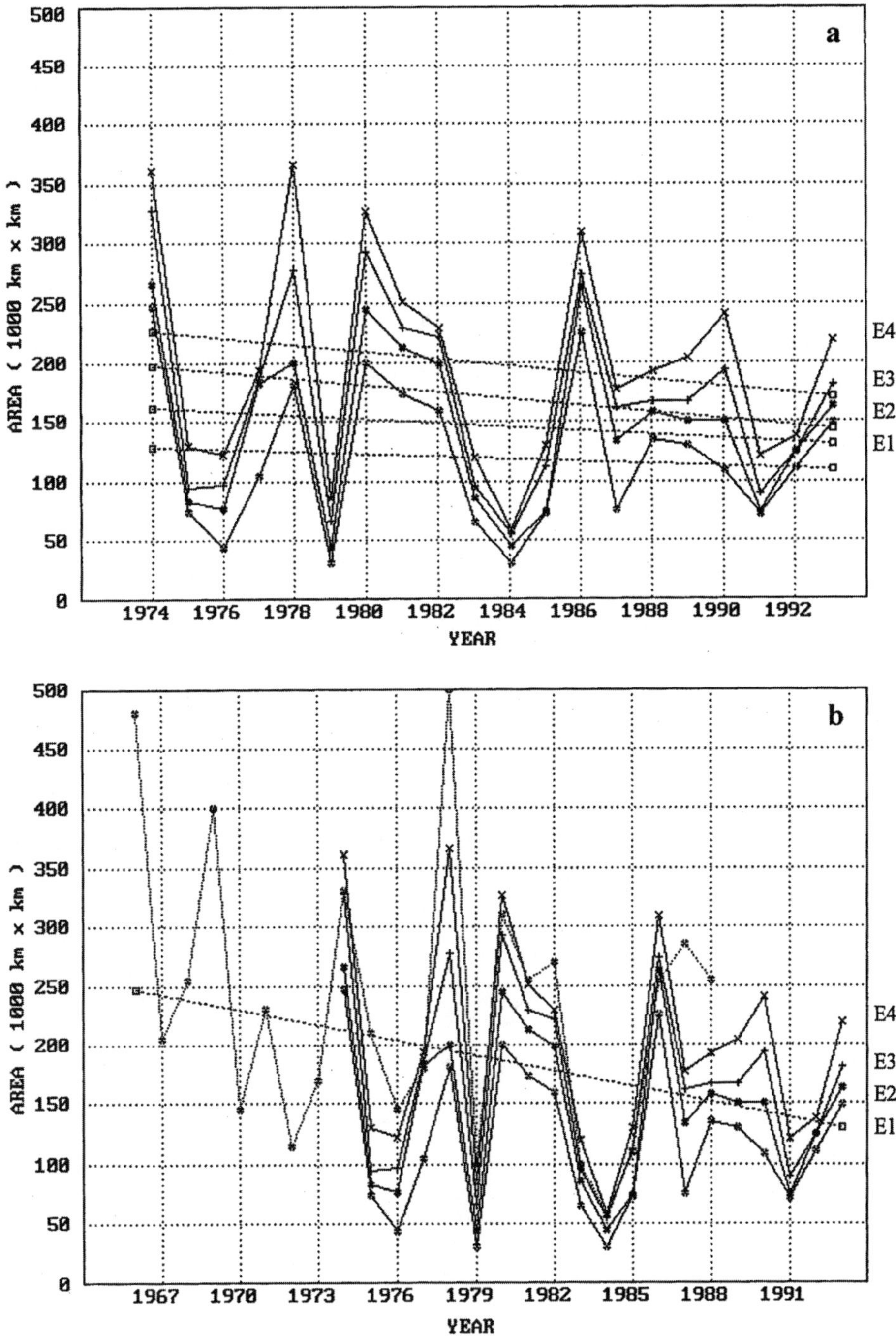

Figure 4.10 Minimum monthly mean sea ice extent in the western Kara Sea.

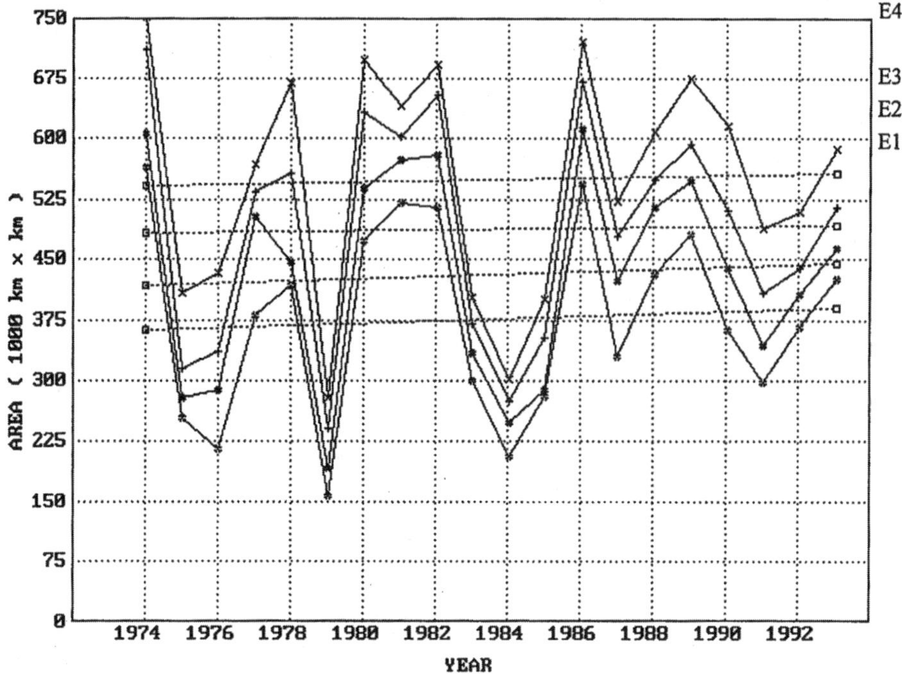

Figure 4.11 Minimum monthly mean sea ice extent in the Barents and Kara Sea region.

for the derived 20-year trends (E1–E4) were higher for the minimum monthly mean ice extent (Table 4.3). The greatest 20-year increase was observed in the Barents Sea for ice extension E3 (>70% concentration).

The slowly increasing trend in sea ice extension (E1, E2, E3, E4), 1974–1993, persisted when data were combined for the Barents–Kara Seas and neighboring areas of the Arctic Ocean (Table 4.3, Figure 4.14). For the combined study area, 20-year trends had less variability among the four ice extensions (E1–E4) compared with the trends detected for the Barents and western Kara Seas individually. Also for the combined study area, the rate of increasing ice extent was more pronounced for the absolute annual minimum ice extent compared with the minimum monthly mean. The sea ice extent at the end of August, 1974–1993, also showed an increasing trend in the Barents Sea (Figure 4.15a), decreasing trend in the western Kara Sea, and increasing trends for Barents and Kara Seas. However, when combined with data in Vinje (1991), the 28-year trend between 1966 and 1993 showed an average 4.7% reduction in the Barents Sea (Figure 4.15b).

The period when minimum sea ice extension generally occurs (1974–1993): late August to early September in the Barents Sea, early to mid September — in the Kara Sea, late August to early September — in the Barents and Kara Seas; high interannual spatial variation: July — in the Barents Sea, July and August in the Barents–Kara Seas.

4.5.1 Some Results of OKEAN and ALMAZ SAR Data Analysis

Although a combined system (RAR, RM-08, MSU-M and SAR) takes advantage of different sensitivities of different sensors to different surface types, high resolution of

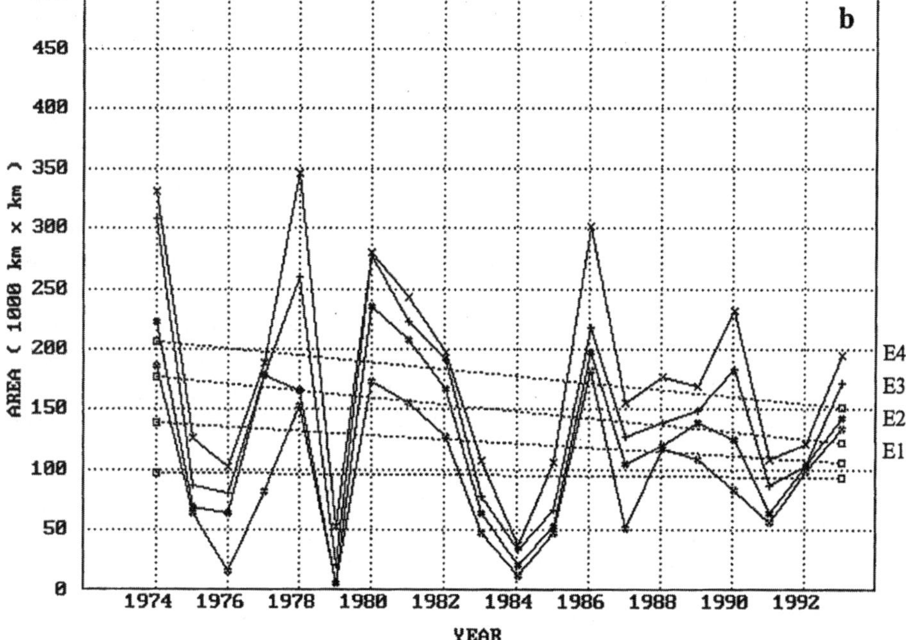

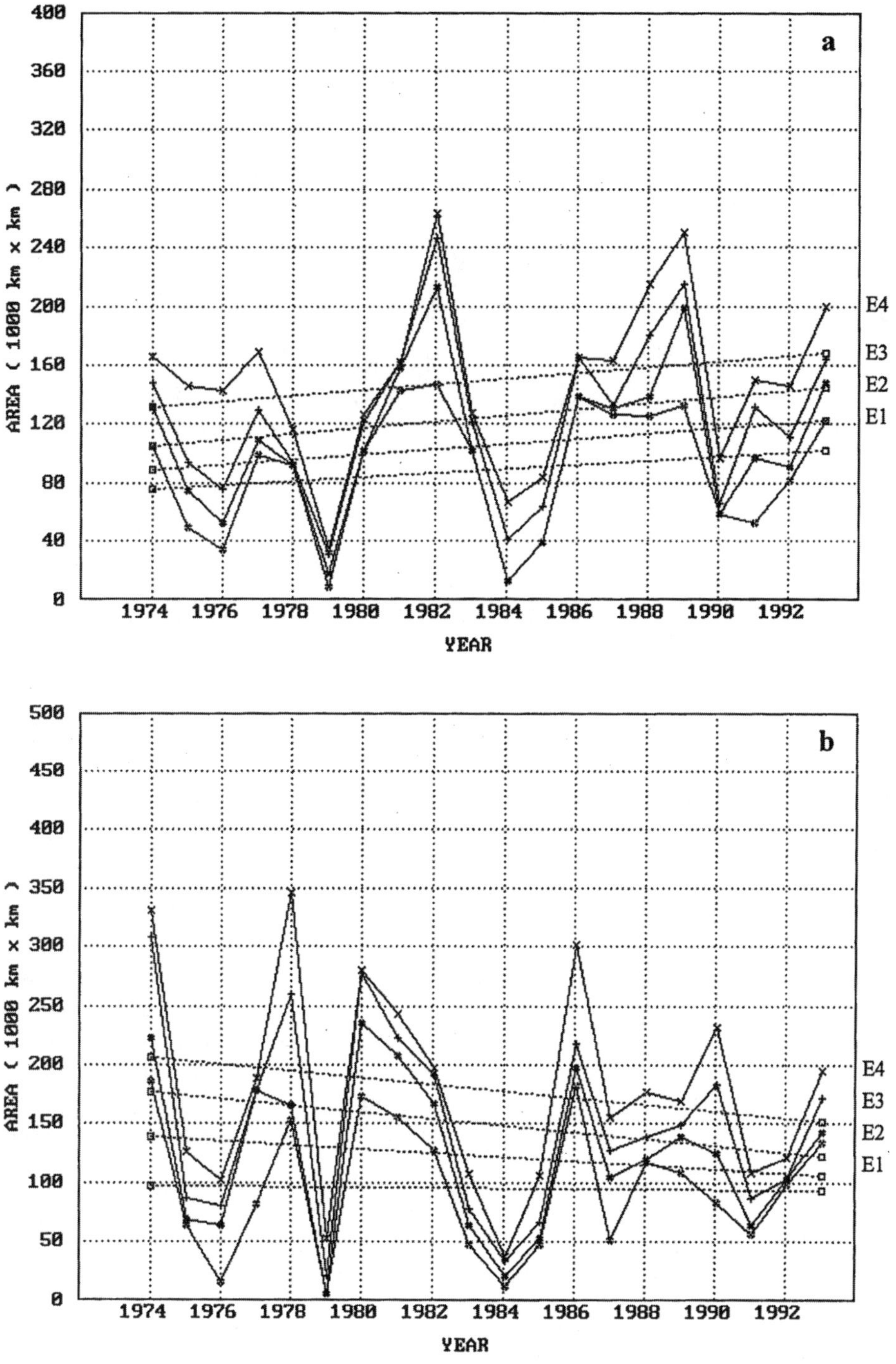

Figure 4.12 Absolute annual minimum sea ice extent in the Barents Sea (a) and western Kara Sea (b).

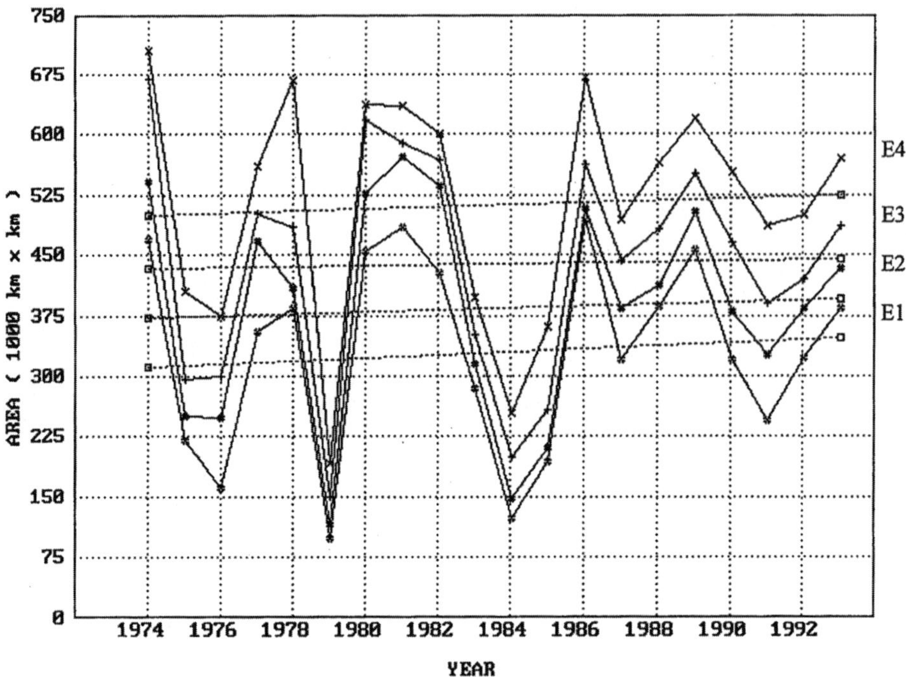

Figure 4.13 Absolute annual minimum sea ice extent in the Barents and Kara Seas region.

ALMAZ SAR data, good temporal . resolution and large scale coverage of the OKEAN-01 data, RAR, RM-08 and SAR have identical problems associated with the inability to define effectively the ice cover during the presence of meltponding, wetness in the snow, flooding and freeze–thaw cycles associated with changes in temperatures. The preliminary comparative analysis of ice conditions in the Kara Sea using OKEAN-01 and ALMAZ SAR data indicates the advantage of good resolution during this period. For instance, the edges of the floes were clearly recognizable and RM-08 brightness temperature data were interpreted as open water because of coarse resolution. The sea ice concentration derived from SAR data was compared with sea ice concentration derived from OKEAN-01 data. The results showed good consistency in a large volume of data compared. High resolution of ALMAZ SAR data is valuable for a good discrimination of sea ice floes and grey ice, FY ice, MY ice. It was very important for more effective OKEAN-01 data interpretation.

4.6 CONCLUSION

This study addressed some important aspects about the use of KOSMOS–OKEAN and ALMAZ SAR microwave data for sea ice monitoring. OKEAN's capability of concurrently acquiring side-looking and passive microwave data, across extensive areas, demonstrated the platform's utility for monitoring regional ice distributions, especially during cloudy periods.

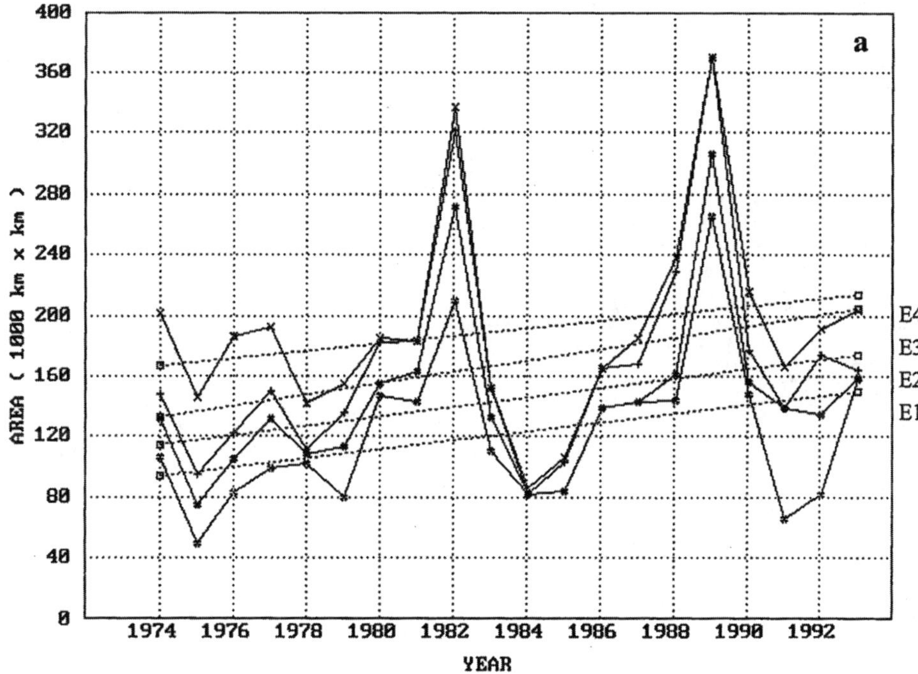

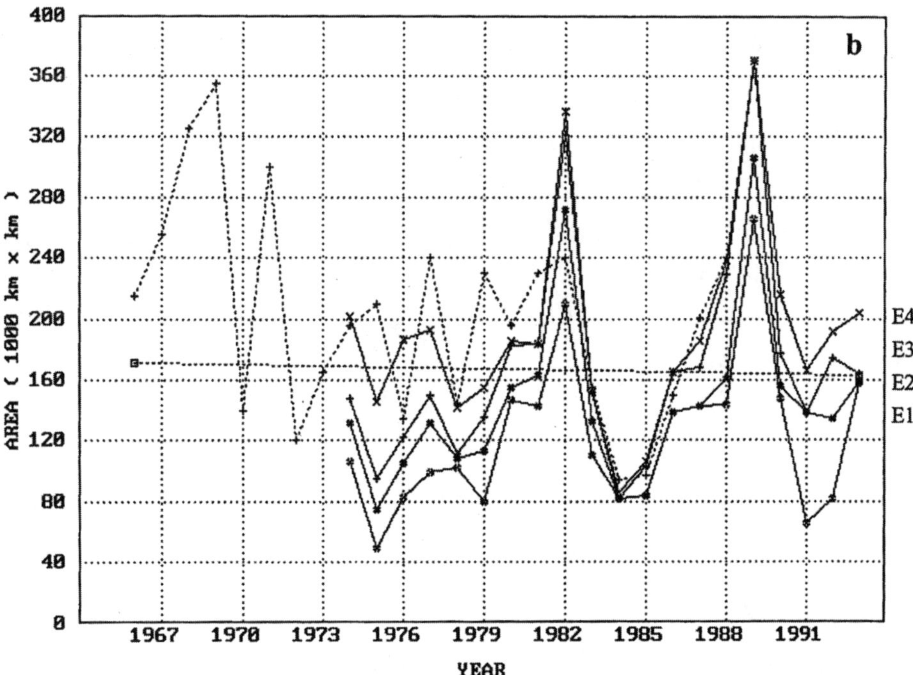

Figure 4.14 Sea ice extent at the end of August in the Barents Sea.

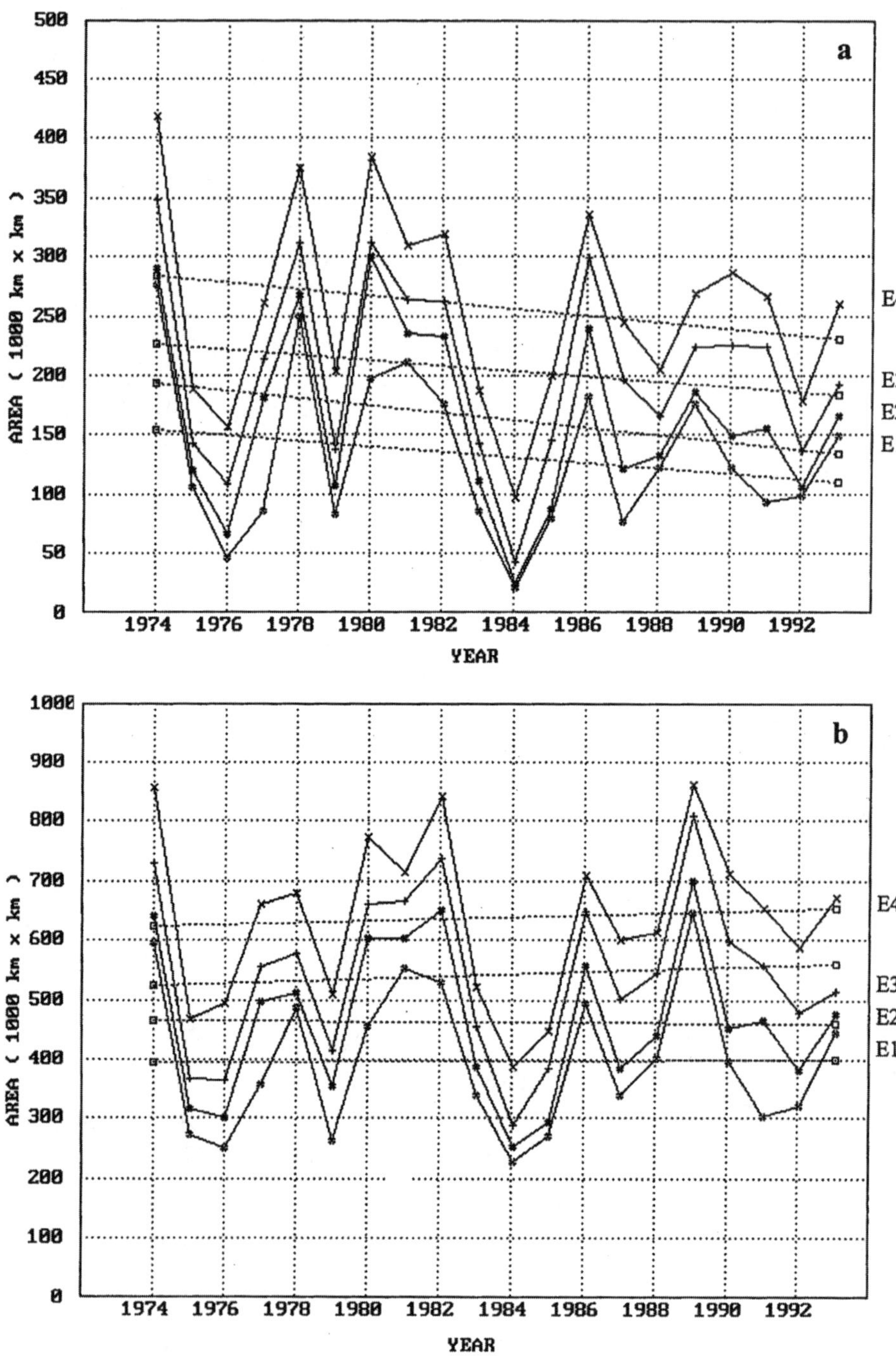

Figure 4.15 Sea ice extent at the end of August in the western Kara Sea (a) and Barents–Kara Seas Region (b).

During the 20 years between 1984 and 1993, all analyses of the Russian Hydrometeorological Center maps indicated increasing and decreasing trends in the extent of minimum sea ice cover for the Barents and Kara Seas. An important constraint to such interpretations lies in the presumed accuracy of the ice maps. Although the magnitude of error inherent to ice maps could not be directly assessed, using satellite remote sensing to generate an alternative measure of ice conditions provided data for estimating one plausible degree of map error. After accounting for disparities between the ice maps and the satellite ice classifications, the trends of minimum ice cover extent in the Barents and Kara Seas remained evident ($P = 0.8$).

The decreasing trend of minimum sea ice extent during the extended 28-year period 1966–1993 demonstrates the importance of long-term databases for interpreting short-term results (e.g., 10-year periods). The increasing 10-year trend in minimum ice extent 1984–1993 appears consistent with respect to the mid-term variability and possible cyclic conditions in the Barents and Kara Seas during the past 28 years. The negative trends in the Barents and Kara Seas for the period of 1978–1987 that were revealed based on SMMR data (Gloersen et al., 1993) were consistent with the results of our studies. Analysis of recalibrated Nimbus-7 scanning multichannel microwave radiometer (SMMR) images and historical U.S. Navy/NOAA ice maps revealed substantial quantitative differences and demonstrated the need for consistently calibrated remote sensing data (Zwally et al., 1991). Furthermore, consistent criteria for defining the limits of ice extent are required for long-term trend analysis (Parkinson, 1991a; Parkinson, 1991b). This study detected notable differences between the four definitions of ice extension (E1–E4) with respect to the magnitude of derived trends (Table 4.3).

The capability of the KOSMOS–OKEAN satellites to concurrently acquire side-looking radar and passive microwave data across extensive areas demonstrated the platform's utility for monitoring regional ice distributions, especially during cloudy periods. Preliminary comparisons between OKEAN-01 and high-resolution ALMAZ SAR images in the Kara Sea showed good consistency for detecting ice concentration classes when large volumes of data were compared (Belchansky et al., 1993). ALMAZ SAR data are valuable for discriminating sea ice flows, grey ice, FY ice and MY ice, and SAR will be important for future refinement of OKEAN-01 data interpretation.

The Electrically Scanning Microwave Radiometer (ESMR) onboard Nimbus-5, Scanning Multichannel Microwave radiometers (SMMR) onboard Nimbus-7, and DMSP (SSM/I) satellites have been the primary sensors used to derive ice concentration, to distinguish FY ice from MY ice, and to determine the interannual spatial variability of north polar sea ice with approximately 50 km resolution (Cavalieri et al., 1984; Parkinson, 1991b; Gloersen et al., 1993). An integrated analysis of these coarse-resolution sensors with higher-resolution data, such as RAR and SAR, could enhance current methodologies for characterizing the Arctic summer sea ice cover and trends.

The most sensitive detection of minimum sea ice extent as an early indicator of global change and as a significant parameter correlated with biological productivity and Arctic marine ecosystems must include routine satellite monitoring with adequate spatial and temporal resolutions for compatibility with climate parameters and models. An integrated multisensor remote sensing approach could take advantage of differing sensitivities to different surface types, enhanced temporal resolution and multiscale spatial resolution that could be used to assess the precision of broad-scale or long-term interpretations.

REFERENCES

Aronoff, S. A., 1984. An approach to optimized labeling of image classes. *Photogrammetric Engineering and Remote Sensing*, 50, pp. 719–721.

Belchansky, G. I. and Pichugin, A. P., 1991. Radar Sensing of Polar Regions. Proceedings of the International Conference on the Role of the Polar Regions in Global Change, Edited by G. Weller, C. L. Wilson, and B. A. Severin, Geophysical Institute, University of Alaska, Fairbanks, and Center for Global Change and Arctic System Research, University of Alaska, Fairbanks, Vol. 1, pp. 47–57.

Belchansky, G. I., Douglas, D. C., Ovchinnikov, G. K., Pank, L. F. and Petrosyan, V. G., 1992. Processing of space monitoring data for studying large mammals in arctic environment. *Earth Research from Space, Russian Academy of Sciences*, No. 2, pp. 75–81.

Belchansky, G. I., Douglas, D. C. and Ovchinnikov, G. K., 1994. Processing of Space-monitoring Data to Document Parameters of the Habitat of Arctic Mammals. *Soviet Journal of Remote Sensing*, Vol. 11, No. 4, pp. 623–636.

Belchansky, G. I., Douglas, D. C., Mordvintsev, I. N. and Ovchinnikov, G. K., 1995. Assessing trends in Arctic sea ice distribution in the Barents and Kara seas using Kosmos–Okean satellite series. *Polar Record* 31(177), pp. 129–134.

Belchansky, G. I., Mordvintsev, I. N. and Douglas, D. C., 1996. Assessing variability and trends in Arctic sea ice distribution using satellite data. Proceeding of IGARSS'96, IEEE, Nebraska, USA, pp. 642–644.

Cavalieri, D. J., Gloersen, P. and Campbell, W. J., 1984. Determination of sea ice parameters with Nimbus-7 SMMR. *J. Geophys. Res.*, No. 89, pp. 5355–5369.

Gloersen, P. and Campbell, W. J., 1991. Variations of extent, area, and open water of the polar sea ice covers: 1978–1987. Proceedings of the International Conference on the Role of the Polar Regions in Global Change, Edited by G. Weller, C. L. Wilson, and B. A. Severin, Geophysical Institute, University of Alaska Fairbanks, and Center for Global Change and Arctic System Research, University of Alaska, Fairbanks, Vol. 1, pp. 28–34.

Grody, N.C. and Basist, A. N., 1997. Interpretation of SSM/I measurements over Greenland, IEEE transaction. *Geoscience and remote sensing*, Vol. 35, No. 2, March, pp.360–365.

Parkinson, C. L., 1991a. Strengths and weaknesses of sea ice as a potential early indicator of climate change. Proceedings of the International Conference on the Role of the Polar Regions in Global Change, Edited by G. Weller, C. L. Wilson, and B. A. Severin, Geophysical Institute, University of Alaska Fairbanks, and Center for Global Change and Arctic System Research, University of Alaska, Fairbanks, Vol. 1, pp. 17–21.

Parkinson, C. L., 1991b. Interannual variability of monthly sea ice distributions in the north polar region. Proceedings of the International Conference on the Role of the Polar Regions in Global Change, Edited by G. Weller, C. L. Wilson, and B. A. Severin, Geophysical Institute, University of Alaska Fairbanks, and Center for Global Change and Arctic System Research, University of Alaska, Fairbanks, Vol. 1, pp. 71–78.

Smith, T. G., Hammill, M. O. and Taugbol, G. A., 1991. Review of the developmental, behavioral, and physiological adaptations of the ringed seal, *Phoca hispida*, to life in the arctic winter. *Arctic*, Vol. 44, pp. 124–131.

Stirling, I., 1980. The biological importance of polynyas in the Canadian Arctic. *Arctic*, Vol. 33, pp. 303–315.

Stirling, I. and Derocher, A. E., 1993. Possible impacts of climatic warming on polar bears. *Arctic*, Vol. 46, No. 3, pp. 240–245.

Stirling, I., 1996. The importance of polynyas, ice edges, and leads to marine mammals and birds. *Journal of Marine Systems*, 298, pp. 1–13.

Vinje, T., 1991. Sea ice variability in the Nordic Seas. Proceedings of the International Conference on the Role of the Polar Regions in Global Change, Edited by G. Weller, C. L. Wilson, and B. A. Severin, Geophysical Institute, University of Alaska Fairbanks, and Center for Global Change and Arctic System Research, University of Alaska, Fairbanks, Vol. 1. pp. 23–27.

Zwally, H. J., Comiso, J. C. and Walsh, J. E., 1991. Variability of Antarctic sea ice. Proceedings of the International Conference on the Role of the Polar Regions in Global Change, Edited by G.

Weller, C. L. Wilson, and B. A. Severin, Geophysical Institute, University of Alaska Fairbanks, and Center for Global Change and Arctic System Research, University of Alaska, Fairbanks, Vol. 1, pp. 22.

5

Comparative Analysis of OKEAN-01, SSM/I and AVHRR Satellite Data for Monitoring of Arctic Sea Ice Habitat

5.1 INTRODUCTION

Multisensor satellite monitoring of Arctic sea ice habitat is critical for understanding marine mammal ecology. However, periods of Arctic darkness and cloud cover preclude systematic data acquisition from multispectral optical systems. However, microwave sensors are not limited by illumination, clouds or fog. The measured radio brightness temperature (T_b) and the radar backscatter coefficient (σ^0) depend only on the geophysical characteristics of sea ice (Comiso, 1983; Cavalieri et al., 1984; Kwok et al., 1992; Kwok and Cunningham, 1994; Swift et al., 1985; Fung, 1994).

Multispectral passive microwave sensors such as Special Sensor Microwave/Imager (SSM/I) are reliably used to compute sea ice concentration and to distinguish FY ice from MY ice (Cavalieri, 1994; Kwok et al., 1996). But because of their relatively coarse resolution (near 25 km), ecological application has been limited to broad-scale investigations. Global, daily (1978–present), 25 km-resolution, sea ice concentration and ice-type data sets (National Snow and Ice Data Center, Boulder, Colorado) are a primary source of data for global sea ice and climate studies (Gloersen et al., 1993).

Compared with passive microwave instruments, SAR satellite systems (Russian ALMAZ, European ERS, Japanese JERS and Canadian RADARSAT) collect high-resolution (20 m) data over much smaller geographic areas. The OKEAN-01 satellites collect simultaneously intermediate-resolution RAR, optical and passive microwave imagery. Acquiring both passive and active data provides more distinct, integrated signatures for sea ice.

This chapter describes the comparative analysis of multisensor satellite monitoring of Arctic sea ice habitat from the OKEAN-01 satellite series, the SSM/I and the AVHRR satellite observations. Image processing and classification algorithms were used to equalize, calibrate, georeference and classify the raw microwave OKEAN-01 data, and to correct the microwave radiometer calibration wedge based on SSM/I brightness temperature. Sea ice type and concentration algorithms utilize a radar and passive microwave information, *a priori* knowledge about the scattering and emission parameters of the basic sea ice types and a linear mixture model for measured values of the brightness temperature and the radar backscatter coefficient. To conduct a comparative assessment, ice concentrations were estimated using SSM/I ice concentrations computed with NASA Team algorithms and visible and thermal-infrared AVHRR data.

5.2 STUDY AREA

The study area (Figure 5.1) covers the Barents, Kara and Laptev Seas and adjacent parts of the Arctic Ocean (70–85 degrees north latitude; 10–120 degrees east longitude). As was mentioned in Chapter 3, the climate is continental, with frequent strong winds, fogs and storms and very low winter temperatures. During the winter, the prevailing wind directions determine the conditions of snow accumulation and thaw. From March or April till August a period of steady windless weather is observed. The main peculiarities of the study area are presented in Chapter 3. Extended winter darkness (from 23 days at 68 degrees north latitude to 176 days at the North Pole) and high cloudiness during the year (Barents Sea: 21–26 days per month for May–October, and 14–18 days for November–April; Kara Sea: 19–23 days per month for May–October, and 8–10 for November–April; Laptev Sea: 21–26 days per month for May–October and 6–12 for November–April) limit using the optical remote sensing imagery.

5.3 DATA DESCRIPTION

The OKEAN-01 satellite passive microwave raw data include the set of brightness calibration wedges and imagery. Every calibration wedge has eight temperature gradations in degrees Kelvin (1 — 110 K, 2 — 142 K, 3 — 174 K, 4 — 194 K, 5 — 226 K, 6 — 252 K, 7 — 280 K, 8 — 310 K). Every wedge gradation consists of 48 lines corresponding to the definite lines on the image and provides calculating brightness temperature and correcting amplification coefficient. A radiometric image is constructed by means of an antenna system that includes a cutout parabolic reflector and linear scanning irradiator. Data from a single scan are converted to 728 pixels and transmitted at 466.5 MHz frequency. An error of absolute calibration is about 4 K (Chapter 2).

The OKEAN-01 satellite radar raw data includes the set of brightness calibration wedges and imagery. The wedge structure is similar to the microwave instrument wedge. The wedge is generated by a radar pulse and an attenuator, is used for backscatter coefficient (σ^0) calculation and for image correction. A string data (596 pixels) is transmitted at 466.5 MHz frequency. RM-0.8 and RAR technical specifications and raw imagery characteristics are presented in Chapter 2. The SSM/I detects microwave radiation on four frequencies: 19.35 GHz, 22.235 GHz, 37.0 GHz, and 85.5 GHz (19, 37, and 85 GHz channels with dual polarization, 22 GHz channel with only vertical polarization). One swath width is approximately 1400 km, pixel size is 25×25 km for 19 GHz, 22 GHz, 37 GHz and 12.5×12.5 km — for 85 GHz. An error of absolute calibration is less than 3 K for all channels. The AVHRR/2 (NOAA-12 and NOAA-14 satellites) collects five-channel data:

1. Visible (0.58–0.68 μm)
2. Near IR (0.725–1.10 μm)

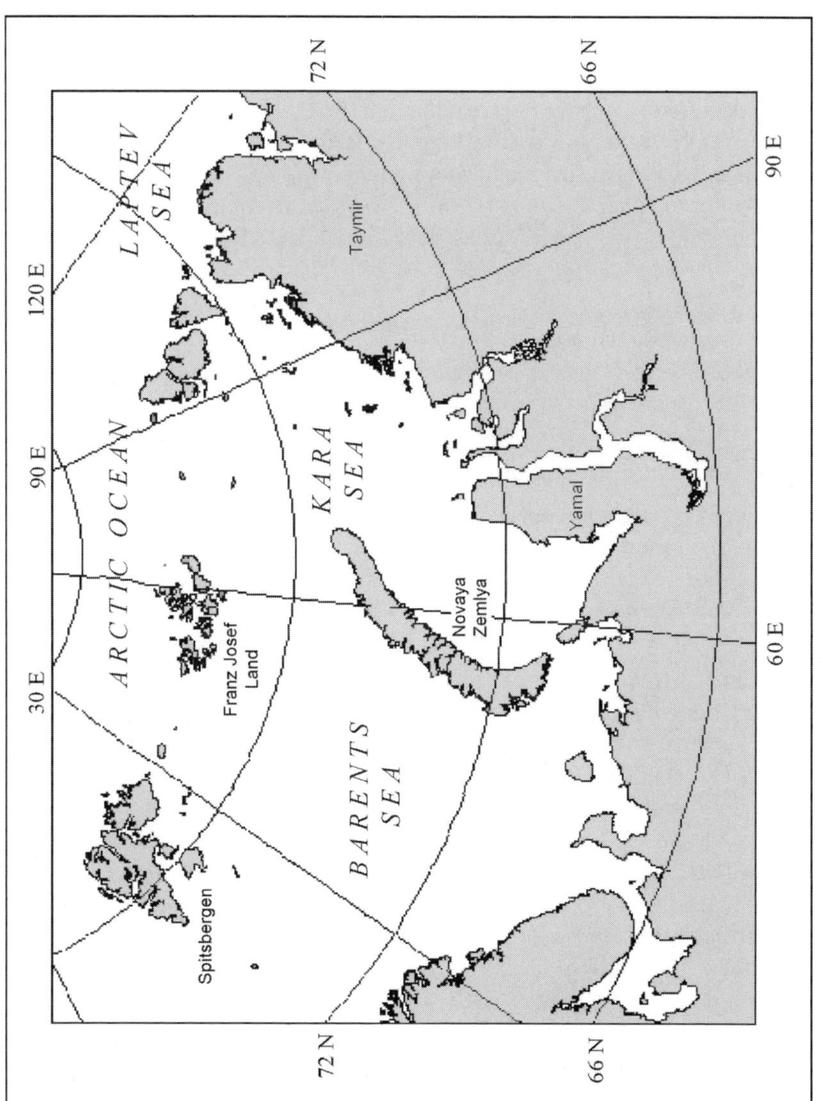

Figure 5.1 Location of study area.

3. Thermal (3.55–3.93 μm)
4. Thermal (10.30–11.30 μm)
5. Thermal (11.50–12.50 μm)

with spatial resolution of 1.1 km, swath width of 2900 km.

Nearly coincident OKEAN-01 N7, N8 satellites, SSM/I and AVHRR data were used to develop three (OKEAN-01, SSM/I, and AVHRR) databases.

The OKEAN-01 database contains digital radar (RAR) and passive microwave (RM-08) imagery acquired at weekly intervals through the Scientific Research Center for Natural Resources (Dolgoprudny, Moscow Region), from October 1995 through early January 1998. Although the downlink included a simultaneous 960 km-swath width of 1.0 km resolution near-infrared imagery (0.8–1.1 μm), these data were typically contaminated by clouds or darkness and were not used for ice analysis. Each OKEAN-01 satellite pass is presented by three types of files:

1. Raw RM-08 and RAR imagery files (intensity value ranges 0–255, step 1, format byte/pixel, pixel size 1×1 km, and documentation files (orbital parameters, time and conditions of survey, image size)
2. Geometrically and radiometrically corrected and calibrated imagery (pixel size 1×1 km, format 2 bytes/pixel, polar stereographic projection
3. Sea ice type and concentration images (band 1: water, band 2: FY ice, band 3: MY ice, band 4: deformed ice, band 5: other ice, pixel size: 3×3 km, format: byte/pixel, sea ice concentration value ranges from 0 to 100 with 0.1 increment, polar stereographic projection).

The SSM/I database contains three subsystems. The first subsystem includes the SSM/I Daily Polar Gridded Brightness Temperatures from the National Snow and Ice Data Center (NSIDC), University of Colorado, for the period of May 1995–June 1997. The data are presented in the Hierarchical Data Format (HDF) (values range from 50.0 K to 350.0 K with 0.1 K increment, format: 2 bytes/pixel, polar stereographic projection). The second subsystem contains the SSM/I Gridded Brightness Temperature from Global Hydrology Resource Center (GHRC) associated with Marshall Space Flight Center (MSFS) for period of October 1995–September 1996. Each scene is presented in HDF files (T_b values scale is 100, with 0.01 K increment, format: 2 bytes/pixel), high resolution (85.5 GHz) and low resolution geolocation files, browse files. The third subsystem contains the SSM/I Daily Polar Gridded Sea Ice Concentrations from NSIDC for the period of October 1, 1995–September 30, 1997. Daily total sea ice concentration and MY ice fractions were provided using the NASA Team algorithm (pixel size — 25×25 km, data range from 0 to 100, with increment of 1, polar stereographic projection).

The AVHRR database contains AVHRR/2 (NOAA-14 satellite) images for the time period of March 1996–April 1997. These data were obtained from the U.S. Geological Survey. The database includes as raw AVHRR data geometrically corrected georeferenced images in polar stereographic projection, pixel size: 1×1 km, channels 3, 4, 5 are radiometrically calibrated, data values are brightness temperature in degrees Celsius multiplied by 10 and data increment is 0.1°.

5.4 SATELLITE DATA PROCESSING

5.4.1 General Description of OKEAN-01 Data Processing

A special software was used to calibrate, equalize, and georeference raw passive microwave and radar imagery (Chapter 2; Belchansky et al., 1993). Brightness temperatures in degrees Kelvin (T_b) were derived for every microwave radiometer pixel based on linear relationships with calibration data that were embedded in the raw imagery during data collection. The radar backscatter coefficient, sigma-0 (σ^0), was derived for each radar image pixel as the ratio of the backscatter return power divided by the original output power. Each image was radiometrically and geometrically corrected and equalized to compensate for sensor look-angle. A set of geographic tiepoints was derived for each image using the satellite's orbital ephemeris data, a single user-defined ground-control point and iterations of user adjustments to the spacecraft attitude (pitch, yaw, roll). The T_b and σ^0 images were converted to ARC/GRID® (Environmental Systems Research Institute, Redlands, California), transformed to polar stereographic projection with 1-km pixel resolution and then averaged to 3-km resolution using a low-pass filter. Let us consider some details of data processing methodology. The example of georeferenced and calibrated satellite data: (A, B) OKEAN-01 N7 RAR and RM-0.8 passive microwave imagery; (C) SSM/I daily gridded brightness temperature; (D, E) AVHRR/2 visible (0.58–0.68) and thermal (10.30–11.30) imagery are presented in Figure 5.2.

5.4.2 Brightness Temperature and Radar Backscatter Coefficient Algorithms for OKEAN-01 Passive Microwave and Radar Data

Brightness temperature (T_b) calculation was performed after georegestration and included estimation and correction of T_b across the image lines based on the calibration wedge, and correction of T_b along the image lines based on the reference objects on the image. Each byte of microwave data file describes one image or wedge pixel. The documentation file describes image and wedge parameters and is used for splitting data to the wedge and image and for definition of separate wedge gradation placement.

The wedge correction was performed by averaging intensity values on the gradations and by eliminating rough errors in onboard data processing systems. The table of correspondence between calibration wedge gradations and plates was developed based on partially linear representation. Dependence between intensity $I(r, L)$ from number r for every gradation L was represented as

$$I(r, L) = I_{k1} + (I_{k2} - I_{k1}) \times \frac{r - r_{k1}}{r_{k2} - r_{k1}}, \tag{5.1}$$

where $k1, k2$ = numbers of calibration wedge plates, corresponding to the gradation number L; I_{ki} = intensity of plate number ki ($i = \{1, 2\}$), r_{ki} = number of plate ki middle line ($i = \{1, 2\}$). Numbers $k1$ and $k2$ were chosen so that those plates were situated as close as possible to line r and desirably before and after it.

Brightness temperature $T_b(I)$ of pixel with intensity I in line r was calculated as

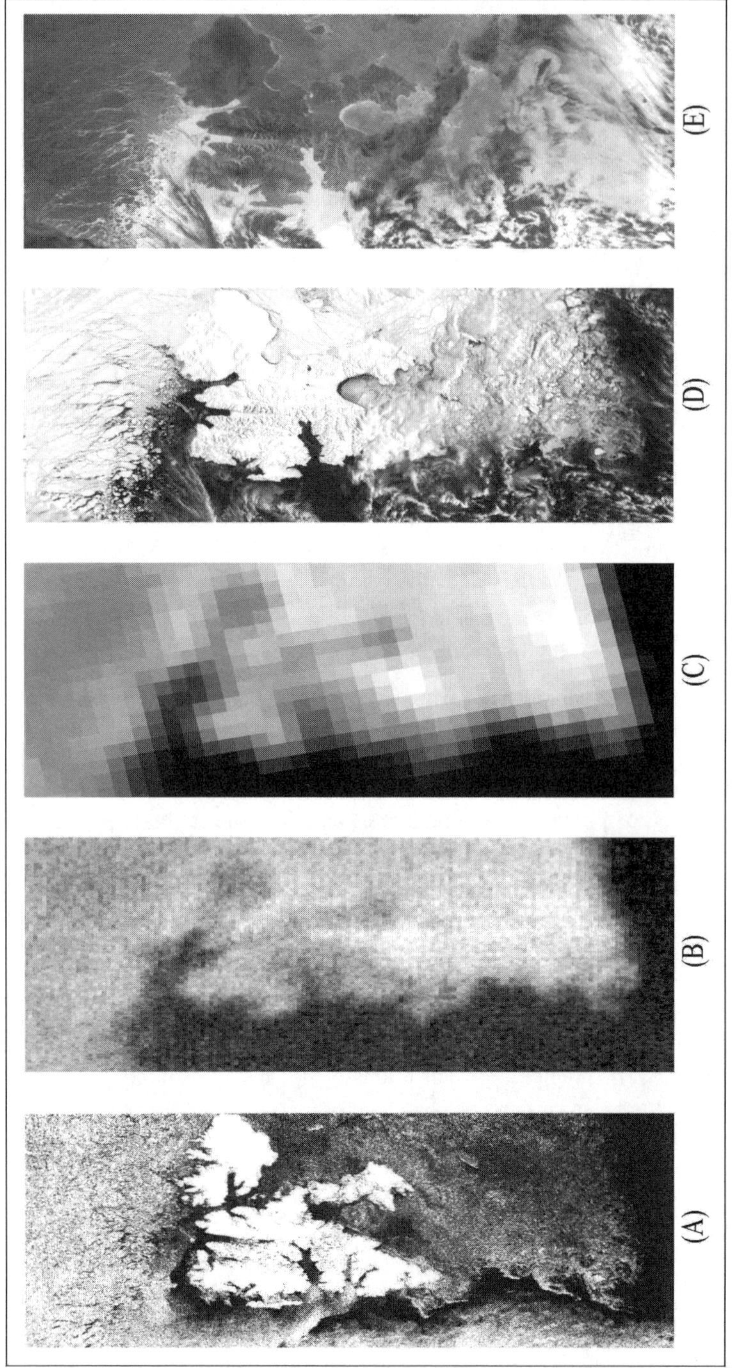

Figure 5.2 An example of georeferenced and calibrated satellite data of the Spitsbergen region in the Barents Sea: (A, B) – OKEAN-01 N7 RAR and RM-0.8 passive microwave imagery (May 07, 1996, 01:12 UT); (C) – DMSP (F-13) SSM/I daily gridded brightness temperature (May 07, 1996); (D, E) – NOAA-14 AVHRR/2 visible (0.58–0.68) and thermal (10.30–11.30) imagery (May 07, 1996, 09:00 UT).

$$T_b(I) = T^*(L) + (T^*(L+1) - T^*(L)) \times \frac{I - I(r, L)}{I(r, L+1) - I(r, L)}, \tag{5.2}$$

where $T^*(L)$ = brightness temperature corresponding to the gradation number L of calibration wedge. Values of L were defined by the inequality $I(r, L) \le I < I(r, L+1)$.

The correction of the amplifying coefficient along lines was performed using a reference object on the image. The equalization function was presented as

$$f(c) = \frac{\sum_{r=r1}^{r2} T(c, r)}{\max_c \left(\sum_{r=r1}^{r2} T(c, r) \right)}, \tag{5.3}$$

where c = a number of column (pixel in the line), $r1$, $r2$ = first and last lines of image fragment.

The calculation of radar backscatter coefficient (σ^0) was performed based on georegistered data. The algorithms performed the approximation and correction of the calibration wedge using similar to passive microwave methodology; the image correction along the image lines; the calculation of the radar backscatter coefficient for each RAR pixel as the ratio of the backscatter return power divided by the original output power and the equalization of backscatter coefficient based on image reference objects.

5.4.3 RM-0.8 Calibration Wedge Correction Algorithms for OKEAN-01 and SSM/I Microwave Data

Nearly coincident data from the SSM/I and RM-0.8 instruments were used to correct the RM-0.8 calibration wedge based on the scatterplot of the comparison values between T_b (RM-08, 36.62 GHz, H channel) and T_b (SSM/I, 37 GHz, H channel). The algorithms included: georegistering the RM-0.8 image to the polar stereographic projection (pixel size 25×25 km); developing the scatterplot T_b (RM-08), T_b (SSM/I); approximating the scatterplot by the partially linear function and the least squares method (the nodes T_0, T_1,..., T_7 of this function correspond to the RM-08 wedge temperature gradations) and calculating the values of the new calibration wedge temperature gradations by substituting the old values in the approximating function.

The structure of the approximation function and restrictions were presented as

$$\varphi(x) = \begin{cases} a_0 x + b_0, \ x \in S_0, \ \text{where } S_0 = \{T_0 \le x < T_1\} \\ a_1 x + b_1, \ x \in S_1, \ \text{where } S_1 = \{T_1 \le x < T_2\} \\ \dots\dots\dots\dots\dots\dots\dots\dots\dots\dots \\ a_6 x + b_6, \ x \in S_6, \ \text{where } S_6 = \{T_6 \le x < T_7\} \end{cases}, \tag{5.4}$$

where $x = T_b$ (PM-08), $y = T_b$ (SSM/I).

The procedure of iteration wedge correction was used to improve the correction accuracy. This procedure included repeated processing raw RM-08 data to calculate new brightness temperature image, new approximating function and new calibration wedge gradation values. The iteration process was finished when the square root of calibration wedge gradations was not greater than a given level.

5.4.4 Sea Ice Concentration Algorithms for OKEAN-01 Active and Passive Microwave Data

The concentrations of MY sea ice (*MY*), Y/FY ice (*Y/FY*), and open water (*OW*) within each pixel are estimated using a linear mixture model for the measured values of T_b and σ^0 (Chapter 3, Belchansky et al., 1997). By omitting landmasses from the analysis, the following equation was assumed for each pixel:

$$C(OW) + C(Y/FY) + C(MY) = 1, \qquad (5.5)$$

where $C(OW)$, $C(Y/FY)$ and $C(MY)$ are the respective area fractions of open water, young or FY ice, and MY ice. The composite upwelling brightness temperature (T_b) and composite backscatter (σ^0) for each pixel were represented by the following equations:

$$T_b = [1 - C(Y/FY) - C(MY)] \times T_b(OW) + C(Y/FY) \times T_b(Y/FY) + C(MY) \times T_b(MY), \qquad (5.6)$$

$$\sigma^0 = [1 - C(Y/FY) - C(MY)] \times \sigma(OW) + C(Y/FY) \times \sigma(Y/FY) + C(MY) \times \sigma(MY), \qquad (5.7)$$

where $T_b(OW)$, $T_b(Y/FY)$, $T_b(MY)$, $\sigma(OW)$, $\sigma(Y/FY)$, and $\sigma(MY)$ are the respective brightness temperature and backscatter coefficients of open water, young or FY ice, and MY ice. The ice-type coefficients were estimated for each OKEAN-01 image at the centroids of three data clusters evident on a two-dimensional plot of the σ^0 and T_b pixel values (Figure 5.3). The three clusters represented the three primary ice types (*OW*, *Y/FY*, *MY*), and their relative positions implied their identification based on published emissivity and backscatter characteristics of the respective pure ice types (Onstott et al., 1982; Eppler et al., 1986). The two-dimensional representation of the σ^0 and T_b image data (Figure 5.3) illustrates how simultaneous analysis of both active radar and passive microwave data improves quantitative separation between ice types. Area fractions for each of the three types [$C(OW)$, $C(Y/FY)$, $C(MY)$], and total ice concentration [$C(Y/FY) + C(MY)$], were calculated for every image pixel by simultaneous solution of Equations (5.5)–(5.7). When open water was not present within the extent of an OKEAN-01 image, $C(Y/FY)$ and $C(MY)$ were estimated using the least squares method. If an area fraction solution was less than zero, we assigned zero concentration to the respective type and proportionally recalculated the area fractions for the remaining types. This analysis was applied to each image pixel to create maps of sea ice types and concentrations (Figure 5.8). The two most significant problems, arising when environmental conditions caused confounding between the ice type signatures, were common to all radar and microwave sea ice classification methods. One problem occurred when open water surfaces were roughened by the wind, which often returned sufficient radar backscatter to become confounded with sea ice. The second problem occurred on the pack ice during the summer when the presence of melting snow and surface ponding caused the

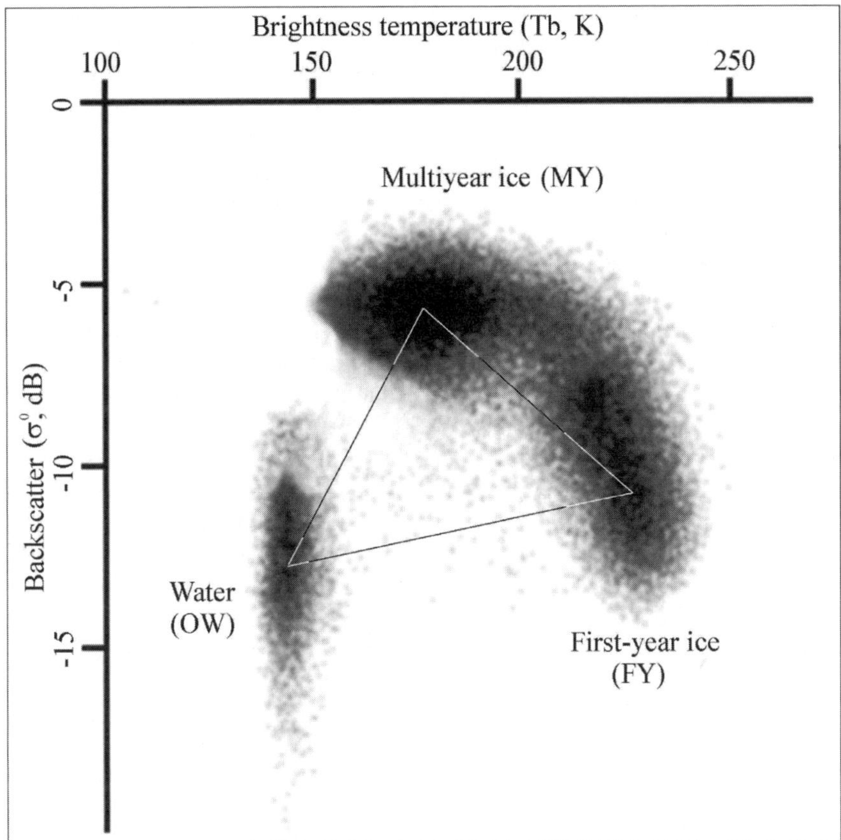

Figure 5.3 Scatterplot of the σ^0 and T_b pixel values of OKEAN-01 N7 image (May 07, 1996, 01:12 UT).

backscatter and brightness temperature signatures to have water-like characteristics (Cavalieri et al., 1990; Comiso, 1990; Kwok et al., 1992; Emery et al., 1994). Melt conditions typically prevented the discrimination of FY and MY ice types, and sometimes led to underestimates of total ice concentration. When necessary, we corrected misclassified ice caused by wind-roughened water by identifying suspicious pixels using T_b and σ^0 thresholds, and then substituting an estimate of *C*(*OW*) derived solely from the RM-08 microwave data. During summer melt conditions, we estimated total ice concentration without discrimination between MY and FY ice. A third condition that sometimes confused ice type signatures occurred in areas of highly deformed FY ice or areas of very young ice, both of which returned sufficient backscatter to suggest a predominance of MY ice. For these areas, we used a manually interpreted mask to reassign the MY area fractions to the FY ice class.

5.4.5 Sea Ice Concentration Algorithms for AVHRR/2 Data

Both surface albedo (channel 1 or 2) and thermal brightness temperature (channel 4 or 5) can be used for estimating ice concentration when solar illumination is sufficient, and only thermal brightness temperature (channel 4 or 5) can be used in the period of Arctic darkness. The surface albedo (channel 1) is used first of all during midsummer because temperatures of the melting snow and sea ice are indistinguishable from temperatures of open water. During summer melt conditions, the surface albedo drops because of higher water content in the surface snow cover (Emery et al., 1994).

The surface albedo (channel 1) and the two-channel data (channel 1s and 4) were used in this study for estimating ice concentrations. The two-channel AVHRR scheme might be useful for estimating total ice concentration without discrimination among MY, young and FY sea ice in conditions of summer melt.

The AVHRR/2 data processing included ingesting, calibrating, remapping to a polar stereographic projection, land masking, cloud cover identification and masking and estimating the ice concentration (Figure 5.2). For cloud masking the surface temperatures ($T3$, $T4$) in Kelvin from channels 3, 4 were used to develop a scatterplot of channels 3–4 versus channel 4 (Massom and Comiso, 1994). The threshold values of cloud-covered regions were determined empirically by adjusting the difference interactively until cloud-covered areas in the image as displayed on a computer screen were completely masked.

A simple linear relationship was used to compute ice concentrations as a function of surface albedo (channel 1) (Emery et al., 1994). This relationship utilized threshold values for albedo to represent 100% ice concentration and 0% ice concentration. Based on histogram analysis and visual interpretation of the imagery, an albedo of 72.5% was chosen as the threshold for 100% concentration, while open water was taken as albedo below 6%.

To estimate the ice concentrations from two-channel data, the concentration of ice type 1 ($ICE1$), ice type 2 ($ICE2$), and open water (OW) within each pixel were determined consuming the linear mixture models for the measured surface albedo A (channel 1) and thermal brightness temperature T (channel 2). The following equation was assumed for each pixel:

$$C(OW) + C(ICE1) + C(ICE2) = 1, \tag{5.8}$$

where $C(OW)$, $C(ICE1)$ and $C(ICE2)$ are the respective area fractions of open water, sea ice (1), sea ice (2). The composite upwelling thermal brightness temperature (T) and composite surface albedo (A) for each pixel were represented by the following equations:

$$T = C(ICE1) \times T(ICE1) + C(ICE2) \times T(ICE2) + [1 - C(ICE1) - C(ICE2)] \times T(OW), \tag{5.9}$$

$$A = C(ICE1) \times A(ICE1) + C(ICE2) \times A(ICE2) + [1 - C(ICE1) - C(ICE2)] \times A(OW), \tag{5.10}$$

where $T(OW)$, $T(ICE1)$, $T(ICE2)$, $A(OW)$, $A(ICE1)$, and $A(ICE2)$ are the respective temperature and albedo of OW, $ICE1$, $ICE2$. The ice-type coefficients were estimated for each AVHRR image at the centroids of three data clusters evident on a two-dimensional plot of the thermal brightness temperature (T) and composite surface albedo (A) pixel values (Figure 5.4). The three clusters represented the three primary ice-types (OW, $ICE1$, $ICE2$). The two-dimensional representation of T and A data (Figure 5.4) illustrates how simultaneous analysis of two AVHRR channels improves quantitative separation between ice and water. Area fraction of open water OW [$C(OW)$], and total ice concentration [$C(ICE1) + C(ICE2)$], were calculated for every image pixel by simultaneous solution of Equations (5.8)–(5.10).

An efficiency of linear mixture models was estimated when the number of nodes on a scatterplot was more than three. For comparative analysis the total ice concentrations were estimated using the linear mixture models, both triangle "T1"(nodes 1-2-3) and triangle

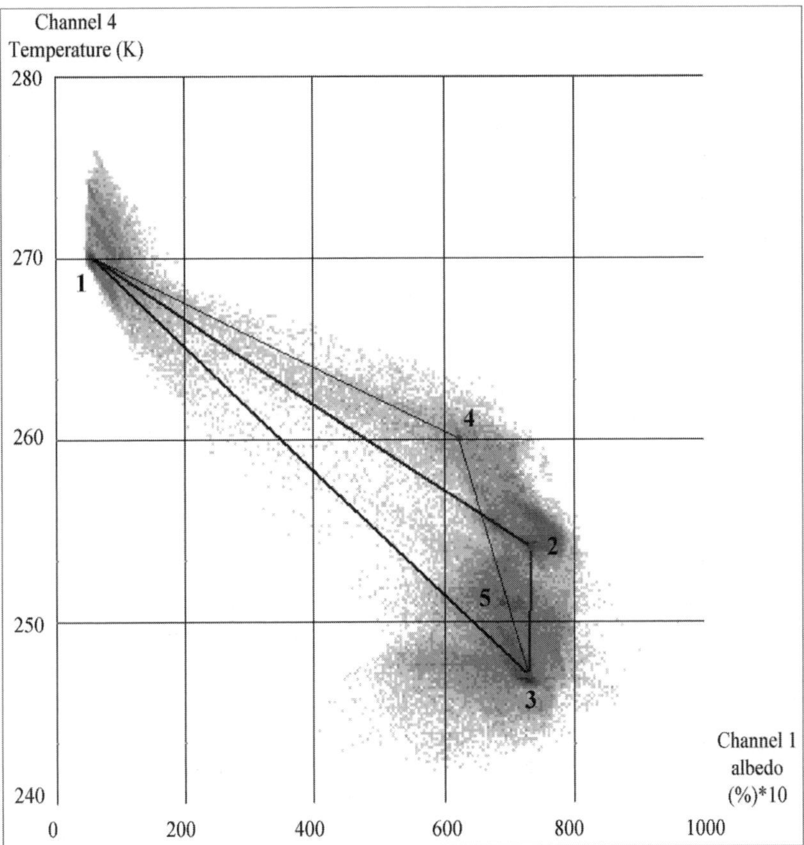

Figure 5.4 Scatterplot of the thermal brightness temperature (T, channel 4) and surface albedo (A, channel 1) pixel values of the AVHRR data (AVHRR/2 NOAA-14, pass # A9710109 – April 11, 1997, 09:00 UT).

"T2" (nodes 1-4-3), (Figure 5.4), and water pixel distribution based on the sea ice classification algorithm assuming normal distribution of sea ice types on the scatterplot.

The distribution of water pixels was estimated utilizing two-dimensional scatterplot, and nodes (Figure 5.4). Every pixel $V(x, y, p)$ on the scatterplot image was characterized by its coordinates (x, y) and pixel value p. The two-dimension normal distribution of sea ice type i was characterized by the weight $P[i]$ (sum of pixel values of type i), mean vector-line $M[i]$ $\{M(Ai), M(Ti)\}$ and dispersion matrix $S[i]$ (2×2). The $P[i]$, $M[i]$, $S[i]$ values were determined only by pixels related to the sea ice type i. For initial filling of the class i the node and non-zero pixels from eight pixels (surrounding the node) were related to the corresponding sea ice type i. For each unclassified pixel and every sea ice type i the values $F[i]$ were calculated (Patric, 1980)

$$F[i] = P[i] \times p \times \left[\frac{1}{\sqrt{\det(S[i])}}\right] \times e^{0.5(N - M[i])(S[i])^{-1}(N - M[i])^T},$$

where N is vector-line containing the pixel V coordinates, p is the weight of pixel V, $F[i]$ is proportional to the probability distribution density) and the rectangular matrix of $F[i]$ was constructed. The maximum element in the matrix determined the pixel to be classified as i sea ice type. This pixel was added to the sea ice type i. The sea ice type i weight, mean vector-line and dispersion matrix were recalculated and the described procedure was repeated until all pixels are classified.

5.5 RESULTS

5.5.1 Comparative Analysis of the 36.62 H RM-0.8 and 37 H SSM/I Channel Brightness Temperatures, and Correction of the RM-08 Calibration Wedge

The comparisons of the similarities and differences in the brightness temperatures computed from 36.62 H RM-0.8 (pass), 37 H SSM/I daily gridded brightness temperatures and 37 H SSM/I channel brightness temperatures (pass) suggested that differences among SSM/I data were not essential (2–4 K for sea ice) for the comparisons among the OKEAN-01, SSM/I and AVHRR derived sea ice types and concentrations. So, for the analysis and the calibration wedge correction we used the 36.62 H RM-0.8 and 37 H SSM/I daily gridded brightness temperatures. The results illustrated the output from the two sensors to be highly correlated for every month of the year (0.93–0.94), and the tendency for the 36.62 H RM-0.8 to produce a higher brightness temperature than those estimated with the 37 H SSM/I channel data. A scatterplot (Figure 5.5A) relating the 36.62 H RM-08 and 37 H SSM/I channel daily gridded brightness temperatures for April 11, 12, 1997 illustrates the example of the differences in brightness temperatures (Barents Sea: OKEAN-01 pass #13444, 13458; mean = –1.2, RMS = 9.8, r = 0.9451, number of observations = 2666). The RM-08 data are more efficient than the 37 H SSM/I channel data in terms of spatial resolution. To come to a conclusion about which brightness temperature might be more accurate requires comparison with ground observations. However the documentation on SSM/I and RM-0.8 sensors shows that SSM/I accuracy is better than RM-0.8. So the received results may be considered as results of RM-0.8 validation. The SSM/I brightness temperatures were used for correction of the RM-0.8 calibration wedge. Some aspects of the methodology and approximating partially linear function are presented in Figure 5.5A, B.

A scatterplot (Figure 5.5B) relating the RM-0.8 brightness temperature (computed with the OKEAN-01 wedge corrected based on 37 H SSM/I channel brightness temperature) and 37 H SSM/I brightness temperatures illustrates the efficiency of RM-0.8 wedge correction algorithms (mean = –1.0; RMS = 6.2; r = 0.9520). Figure 5.7 shows the changes in the differences in brightness temperatures (T_b) between RM-0.8 36.62 H computed from OKEAN-01 data (RM-0.8 old — without correction and new — with correction of the calibration wedge) and 37 H SSM/I channel daily gridded brightness temperatures for transect #1, (Barents Sea: OKEAN-01 N7, pass #13444, April, 11, 1997, 04:30 UT), and the tendency for the 36.62 H RM-0.8 instrument to produce higher brightness temperatures than the 37.0 H SSM/I channel.

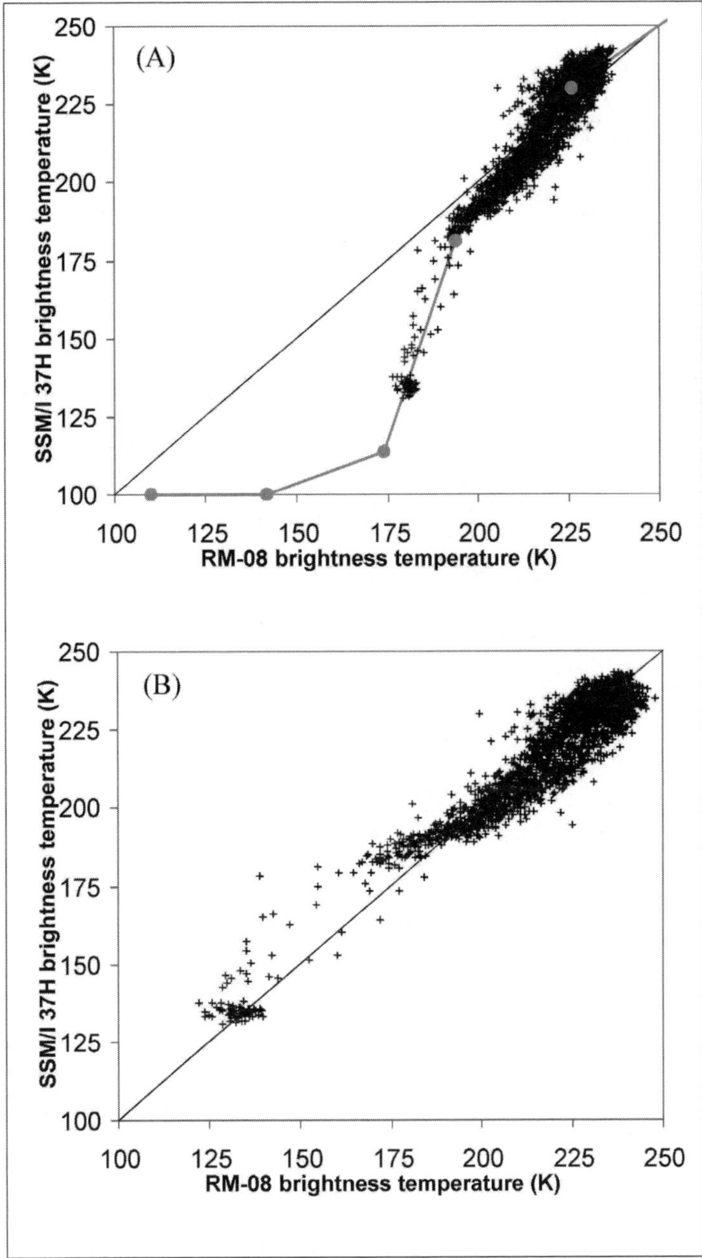

Figure 5.5 A scatterplot of the comparison values between the 36.62 H RM-08 and 37 H SSM/I channel brightness temperatures: A) RM-0.8 brightness temperatures computed using the OKEAN-01 calibration wedge; B) RM-0.8 brightness temperature computed using the OKEAN-01 calibration wedge corrected based on 37 H SSM/I channel brightness temperature.

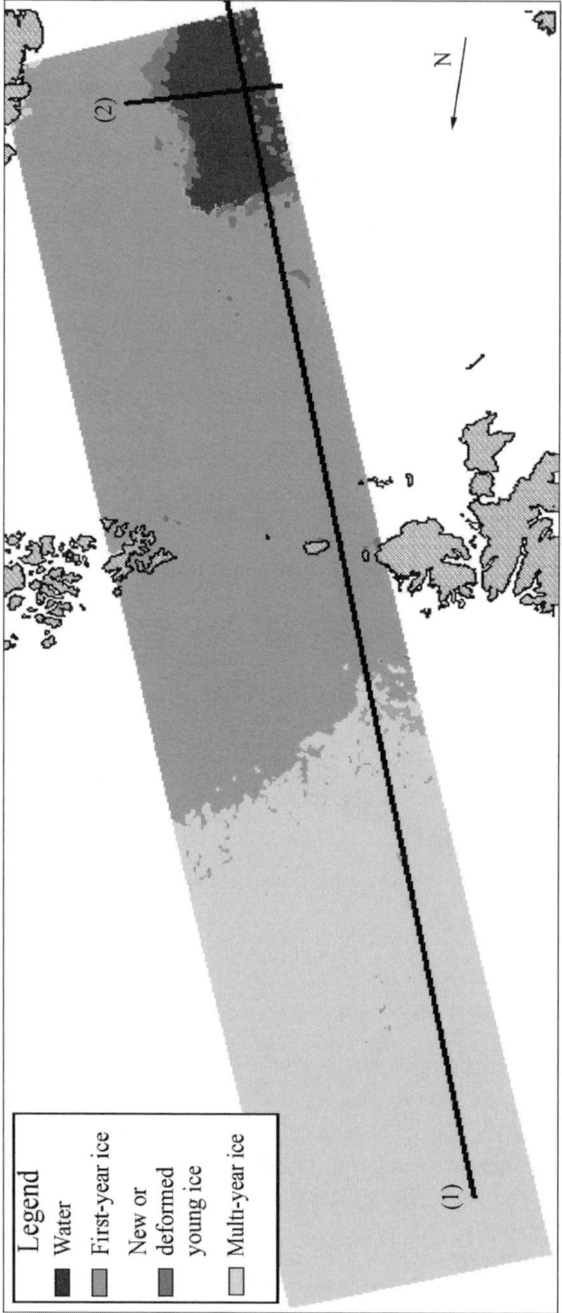

Figure 5.6 Example of OKEAN-01 sea ice type classification with transects.

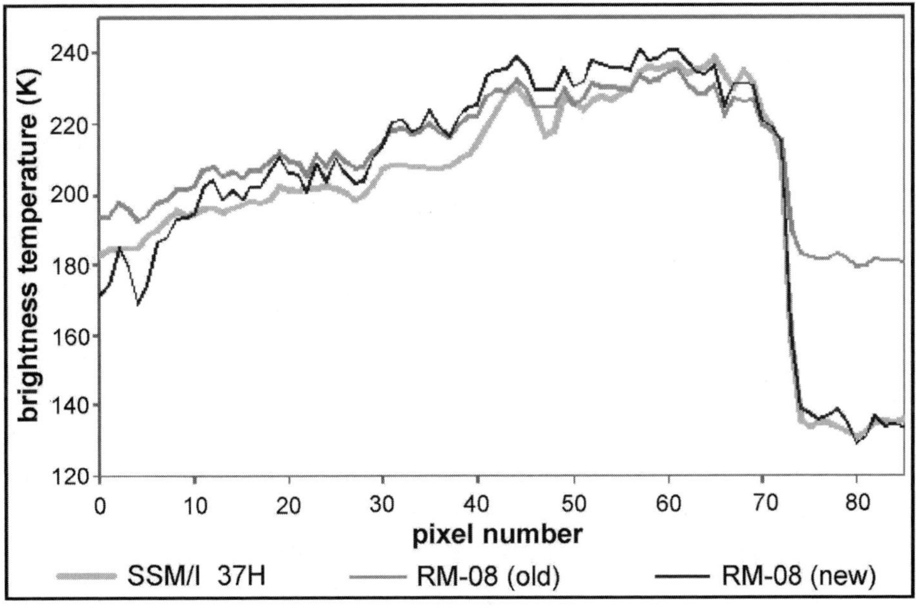

Figure 5.7 The changes of the differences in brightness temperatures (T_b) between RM-0.8 36.62 H (RM-0.8 old – without correction and new – with correction of calibration wedge) and 37 H SSM/I channel daily gridded brightness temperatures for transect #1.

5.5.2 Sea Ice Mapping Using OKEAN-01 Active and Passive Microwave Data

Figure 5.8 shows an example of a composite sea ice map in a polar projection derived from 7 OKEAN-01 satellite imagery dated (left to right) December 21, 1995–December 27, 1995. The mosaic maps were analyzed jointly with weekly ice maps from the U.S. Navy/NOAA Joint Ice Center. The results of a comparison of this and other sea ice maps derived from OKEAN-01 data with the weekly sea ice map of the U.S. Navy/NOAA Joint Ice Center illustrated considerable similarity.

5.5.3 Comparisons Between OKEAN-01, SSM/I and AVHRR Derived Sea Ice Concentration

The OKEAN-01 concentrations of MY sea ice (*MY*), Y/FY sea ice (*Y/FY*), and total sea ice within each pixel were estimated using linear mixture models and OKEAN-01 calibration wedge. A scatterplot (Figure 5.9) relating the OKEAN-01 and SSM/I sea ice concentrations (Barents Sea; OKEAN-01 N7: pass #13444; SSM/I: daily sea ice concentrations, NASA Team algorithm April 11, 1997) illustrates the tendency for the OKEAN-01 algorithm and data to produce: lower concentrations of young or FY sea ice (*Y/FY*) for *C*(*Y/FY*) less than or equal to 60% and to produce higher concentrations of MY sea ice (*MY*) for *C*(*MY*) greater than or equal to 40% than those produced from SSM/I data. The character of the similarity is illustrated in Table 5.1.

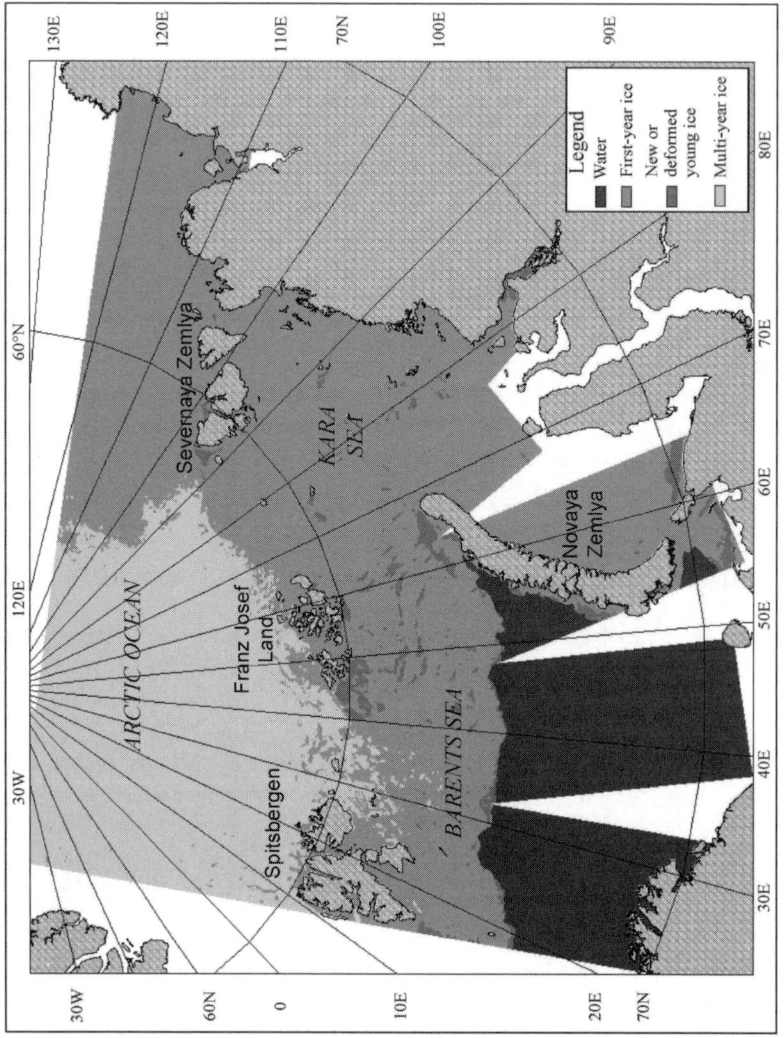

Figure 5.8 An example of the sea ice maps derived from 7 OKEAN-01 satellite scenes for the time period of December 21, 1995 – December 27, 1995.

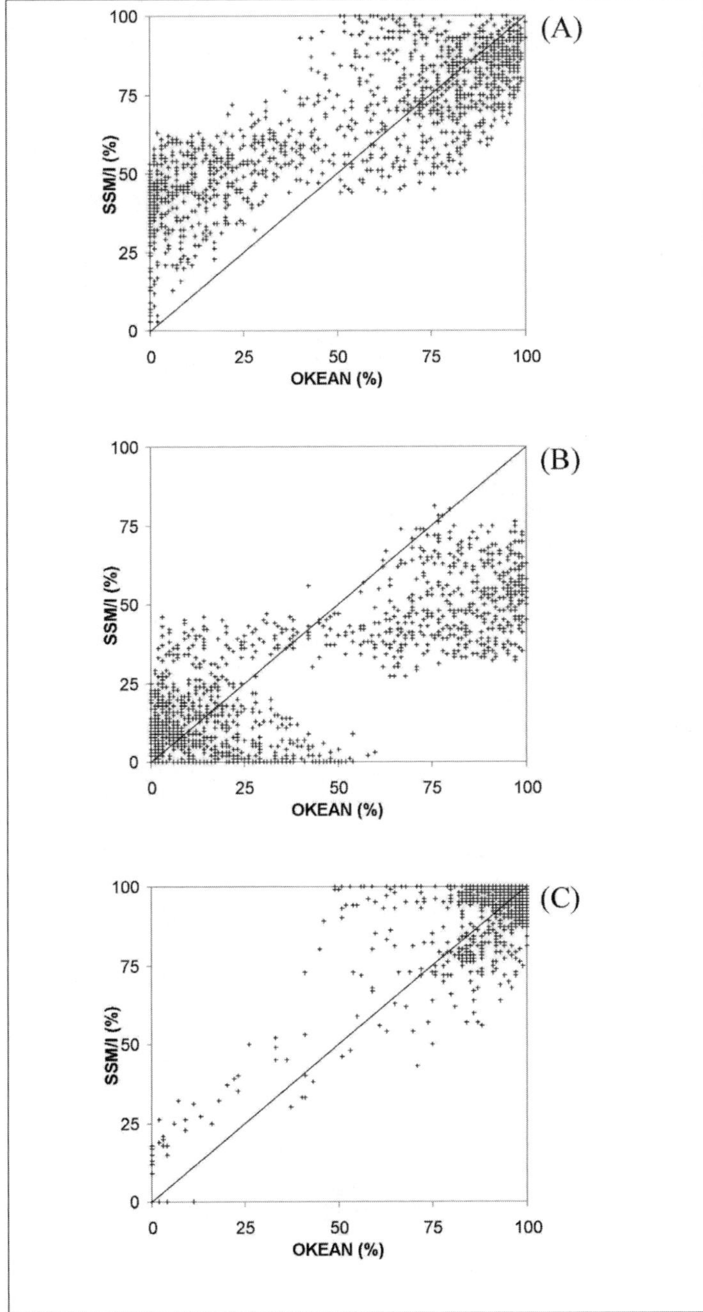

Figure 5.9 A scatterplot of the comparison values between the SSM/I and OKEAN-01 sea ice concentrations (A – *Y/FY*; B – *MY*; C – *Total*).

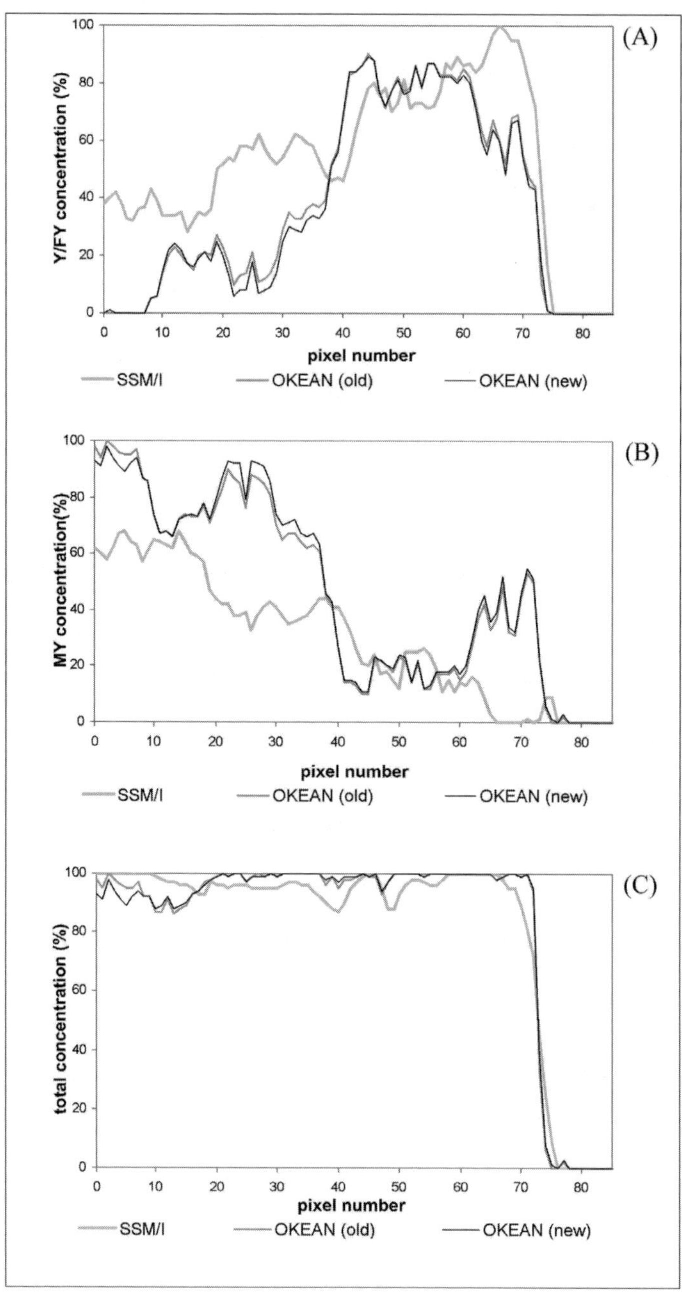

Figure 5.10 The comparison values between the SSM/I and OKEAN-01 sea ice concentrations along the transect #1.

Table 5.1 Comparisons between OKEAN-01 and SSM/I Sea Ice Type Concentrations

Pass #, Date	OKEAN-01–SSM/I								
	OKEAN-01 N7: pass #13444, April 11, 1997, 04:30 UT			OKEAN-01 N7: pass #15261, August 12, 1997, 12:38 UT			OKEAN-01 N7: pass #15832, September 20, 1997, 06:49 UT		
	Y/FY	MY	Total	Y/FY	MY	Total	Y/FY	MY	Total
Pixel (km)	25	25	25	25	25	25	25	25	25
Mean (%)	13.0	–12.6	0.3	–1.3	2.4	1.0	5.1	–7.3	–2.2
RMS (%)	25.1	25.7	9.6	12.9	8.1	10.4	18.5	18.3	14.2
S.D. (%)	21.5	22.4	9.6	12.9	7.7	10.3	18.1	17.9	14.0
R	0.82	0.80	0.91	0.89	0.86	0.92	0.85	0.91	0.94
# Obs.	1414	1414	1414	919	919	919	1013	1013	1013

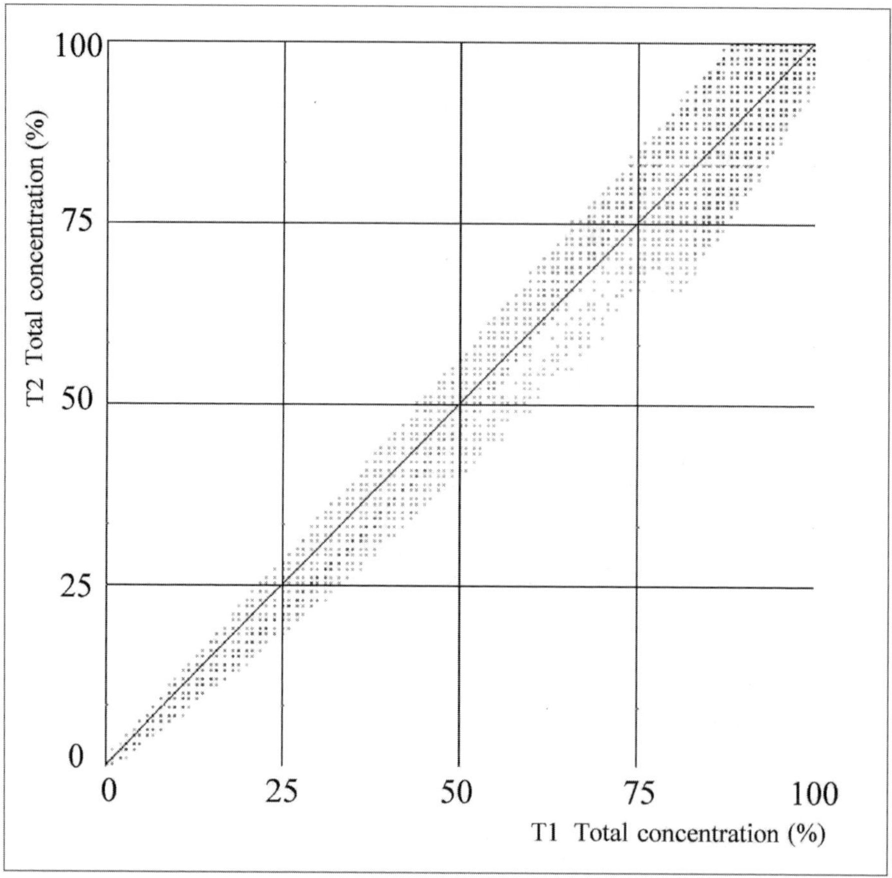

Figure 5.11 Scatterplot of the comparison values between the total sea ice concentration of scheme T1 and scheme T2.

Figure 5.10 shows the changes of the differences in concentrations between OKEAN-01 and SSM/I daily gridded concentrations for transect #1 (Barents Sea: OKEAN-01 N7, pass #13444, April 11, 1997, 04:30 UT).

Unfortunately, periods of prolonged darkness and persistent cloud cover precluded systematic AVHRR data acquisition for the study area and limited the analysis of the coincident OKEAN-01 and AVHRR. An efficiency of linear mixture models was estimated when five nodes were defined on the scatterplot (1. 55270; 2. 734254; 3. 729247; 4. 622260; 5. 689251) (Figure 5.4). For triangle "T1" the computed total ice concentration was 82.2%, for triangle "T2"—82.9%. The differences between ice concentrations calculated utilizing the triangle T1 and the triangle T2 were characterized by the mean value of 0.7%, standard deviation of 4.5%, RMS of 4.6% and correlation of 0.9874. These results show only slight differences between two schemes of ice concentration estimations based on linear mixture models (Figure 5.11).

The distribution of water pixels (accuracy of computing of total ice concentration) was estimated for a set of classification schemes and nodes. Table 5.2 represents the results of sea ice classification using a two-dimensional scatterplot (pixel size 12×12 km; number of

Table 5.2 The Results of AVHRR Data Classification Using the Two-channel Data and Scatterplot

Class	Normal, 3 classes	Normal, 5 classes	Linear, 3 classes
1	1533	1542	1409
2	2963	3026	3820
3	4746	2929	4013
4		34	
5		1711	

Table 5.3 Comparisons between OKEAN-01, AVHRR and SSM/I Derived Sea Ice Concentration

	AVHRR (1)–OKEAN-01		AVHRR (1,4)–OKEAN-01		SSM/I–AVHRR (1)	OKEAN-01–SSM/I
Pixel Size (km)	25	3	25	3	25	25
Mean (%)	–2.8	–3.9	–1.8	–2.7	2.0	0.3
RMS (%)	12.0	12.3	11.3	11.2	11.8	9.6
S.D. (%)	11.6	11.7	11.1	10.9	11.7	9.6
R	0.84	0.85	0.84	0.86	0.93	0.91
# Obs.	445	32019	445	32019	1953	1414

pixels 9242) and linear mixture models. This table shows changes in the distribution of water pixels for total ice concentration estimation in every scheme of classification using all possible nodes in Figure 5.4. In Table 5.3 "Linear, 3 classes" means that the dominant types were estimated using the linear mixture model algorithm and "T1" scheme (nodes 1-2-3), "Normal, 3 classes" means that the sea ice data were classified assuming normal sea ice type distributions and "T1" scheme (nodes 1-2-3), "Normal, 5 classes" means that the sea ice data were classified assuming normal sea ice type distributions and nodes 1-2-3-4-5.

Scatterplots (Figure 5.12) relating (a) the OKEAN-01 N7 (pass #13444, April 11, 1997, 04:30 UT) and AVHRR 1 (channel 1, pass # a9710109 – April 11, 1997, 9:00 UT) ice concentrations, (b) the OKEAN-01 N7 and AVHRR 1, 4 (channel 1, 4) ice concentration, (c) the SSM/I (daily sea ice concentrations, NASA Team algorithm, April 11, 1997) and AVHRR 1 ice concentration and Table 5.3 illustrate the character of a similarity of the concentrations computed using different systems.

The results of comparisons between OKEAN-01, AVHRR and SSM/I derived sea ice concentrations, brightness temperatures, albedo, channel 4 brightness temperatures along transect #2 on near coincident OKEAN-01, SSM/I and AVHRR data are in Figure 5.13.

5.6 CONCLUSION

The comparative analysis of the nearly coincident OKEAN-01 N7, N8 satellites, SSM/I and AVHRR data for Arctic sea ice habitat studies shows that OKEAN-01-, SSM/I-, and AVHRR-derived sea ice habitat parameters are highly correlated, with uniform biases, and yielded a realistic estimate of total ice concentration. The intercomparison of sea ice types and concentrations derived using OKEAN-01, SSM/I and AVHRR sensors is very critical for interpretation of the physical processes affecting the algorithms performances.

The OKEAN-01 satellite microwave data provide unique information for developing new methods of sea ice habitat studies. These data have similar spatial resolution to the AVHRR, but the persistence of clouds and darkness preclude applications of AVHRR for

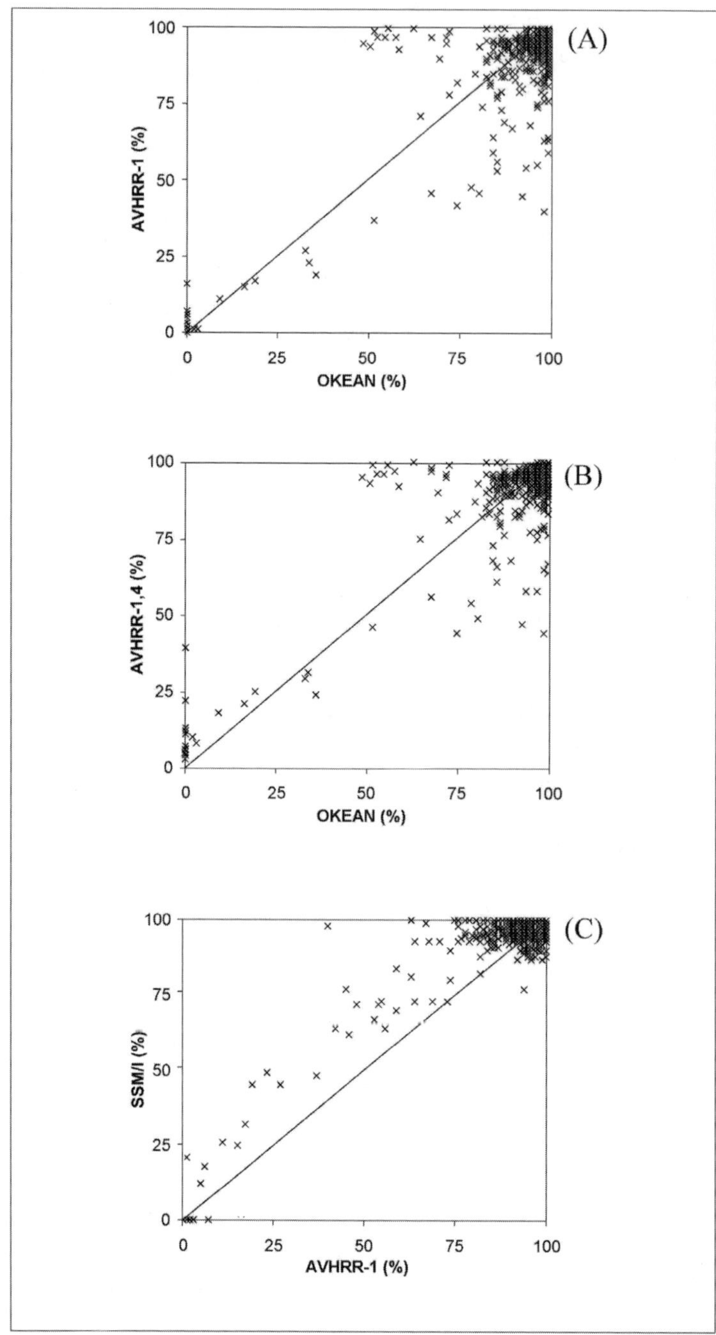

Figure 5.12 A scatterplot of the comparison values between the AVHRR and OKEAN-01, SSM/I and AVHRR sea ice concentrations.

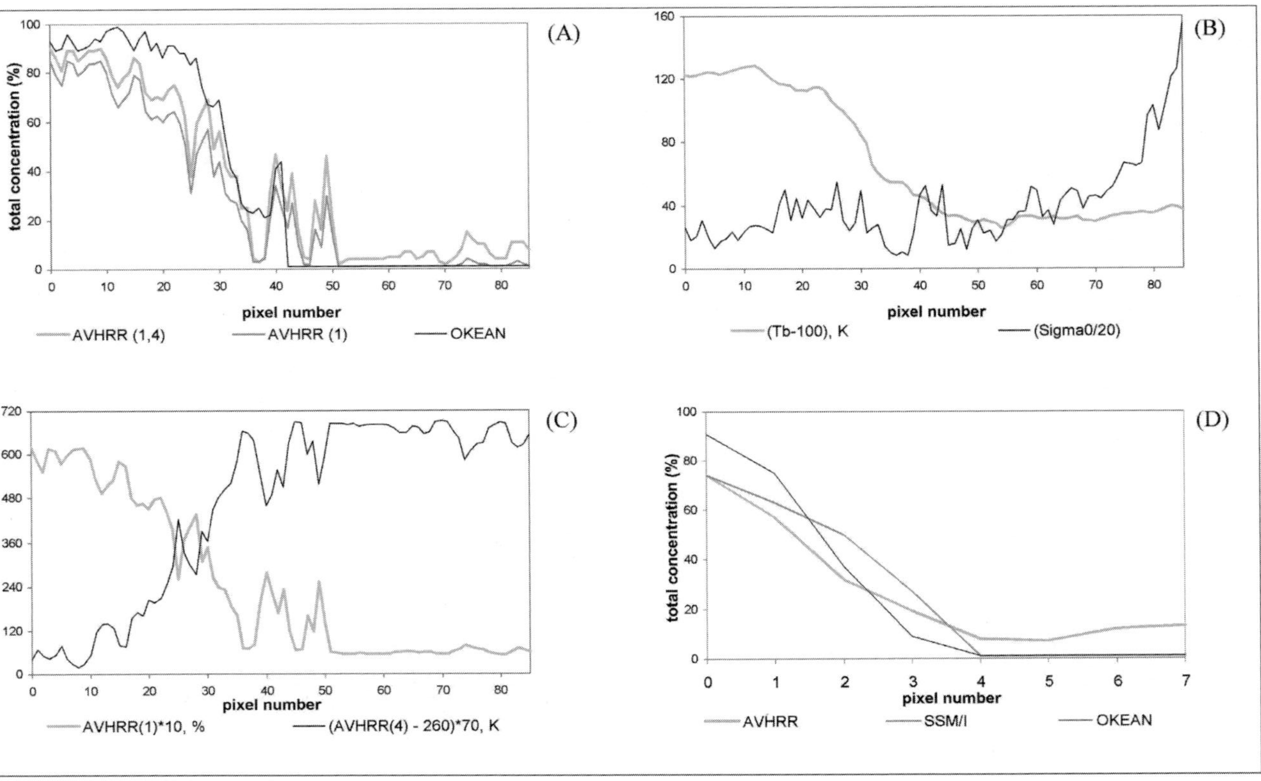

Figure 5.13 OKEAN-01, AVHRR and SSM/I derived sea ice parameters: sea ice concentrations, brightness temperatures (T_b), albedo – channel 1, brightness temperatures – channel 4 along the transect #2.

systematic sea ice habitat studies. The clouds and absence of solar radiation do not affect active radar and passive microwave instruments. However, the radar data are prone to problems associated with wind-roughened water surfaces, which can return sufficient backscatter to incorrectly suggest the presence of sea ice.

The passive microwave data, being much less affected by wind-roughened water surfaces, provided information to correctly detect the presence of open water. The SSM/I instrument takes advantage of the good temporal resolution and large-scale global coverage of the data. However, both active and passive microwave sensors suffer from problems associated with inability to accurately estimate the sea ice cover during the presence of meltponding, wetness in the snow cover, flooding, and freeze–thaw cycles.

To this end, a multisensor approach takes advantage of differing sensitivities to different sea ice habitat parameters, enhanced temporal resolution, and multi-scale spatial resolutions that can be used to assess the precision of broad-scale or long-term interpretations.

REFERENCES

Belchansky, G. I., Douglas, D. C. and Ovchinnikov, G. K., 1994. Processing of space monitoring data to document parameters of the habitat of Arctic mammals. *Soviet Journal of Remote Sensing*, 11(4), pp. 623–636.

Belchansky, G. I., 1995. Assessment of sea ice distribution with radar remote sensing data. *J. Proceedings of Russian Academy of Sciences*, 343: pp. 130–135.

Belchansky, G. I., Douglas, D. C., Mordvintsev, I. N. and Ovchinnikov, G. K., 1995. Assessing trends in Arctic sea ice distribution using Kosmos–Okean satellite series. *Polar Record*, 31(177), pp. 129–134.

Belchansky, G. I., Douglas, D. C. and Kozlenko, N. N., 1997. Determination of the types and concentration of sea ice from satellite data from two-channel microwave active and passive observation systems. *Earth Observation and Remote Sensing*, 14(6), pp. 891–905.

Cavalieri, D. J., 1994. A microwave technique for mapping thin sea ice. *J. Geophys. Res.*, 99(C6), pp. 12561–12572.

Cavalieri, D. J., Gloersen, P. and Campbell, W. J., 1984. Determination of sea ice parameters with Nimbus-7 SMMR. *J. Geophys. Res.*, 89, pp. 5355–5369.

Cavalieri, D. J., Burns, B. A. and Onstott, R. G., 1990. Investigation of the effects of summer melt on the calculation of sea ice concentration using active and passive microwave data. *J. Geophys. Res.*, 95(C4), pp. 5339–5369.

Comiso, J. C., 1983. Sea ice effective microwave emissivities from satellite passive microwave and infrared observations. *J. Geophys. Res.*, 88(C12), pp. 7686–7704.

Comiso, J. C., 1990. Arctic multiyear ice classification and summer ice cover using passive microwave satellite data. *J. Geophys. Res.*, 95(C8), pp. 13411–13422.

Comiso, J. C., 1991. Top/bottom multisensor remote sensing of Arctic sea ice. *J. Geophys. Res.*, 96(C2), pp. 2693–2709.

Comiso, J. C., 1994. Surface temperatures in the polar regions from Nimbus-7 temperature humidity infrared radiometer. *J. Geophys. Res.*, 99(C3), pp. 5181–5200.

Emery, W. J., Radebaugh, M., Fowler, C. W., Cavalieri, D. J. and Steffen, K., 1991. A comparison of sea ice parameters computed from advanced very high resolution radiometer, Landsat imagery and from airborne passive microwave radiometer. *J. Geophys. Res.*, 96(C12), pp. 22075–22085.

Emery, W. J., Fowler, C. and Maslanik, J. K., 1994. Arctic sea ice concentrations from special sensor microwave imager and advanced very high resolution radiometer satellite data. *J. Geophys. Res.*, 99(C9), pp. 18329–18342.

Eppler, D. T., Farmer, L. D. and Lohanick, A. W., 1986. Classification of sea ice types with single-band (33.6 GHz) airborne passive microwave imagery. *J. Geophys. Res.*, 91(C9), pp. 10661–10695.

Fung, A. K., 1994. Microwave scattering and emission models and their applications. Artech House, Boston.

Gloersen, P. and Campbell, W. J., 1988. Variations in the Arctic, Antarctic and global sea ice covers during 1978–1987 as observed with Nimbus-7 SMMR. *J. Geophys. Res.*, 22(C9), pp. 10666–10674.

Gloersen, P. and Campbell, W. J., 1991. Variation of extent, area, and open water of the polar sea ice cover:1978–1987. In: G. Weller, C.L. Wilson and B.B. Severin (Editors), Proc. Int. Conf. on the role of polar regions in global change. Geophysical Institute, University of Alaska, Fairbanks, 1, pp. 28–34.

Gloersen, P., Campbell, W. J., Cavalieri, D. J., Comiso, J. C., Parkinson, G. L. and Zwally, H. J., 1993. Satellite passive microwave observations and analysis of Arctic and Antarctic sea ice, 1978–1987. *Annals of Glaciology*, 17, pp. 149–154.

Kwok, R., Rignot, E., Holt, B. and Onstott, R., 1992. Identification of sea ice types in space borne synthetic aperture radar data. *J. Geophys. Res.,* 97(C2), pp. 2391–2402.

Kwok, R. and Cunningham, G. F., 1994. Backscatter characteristics of the winter ice cover in the Beaufort Sea. *J. Geophys. Res.*, 99(C4), pp. 7787–7802.

Kwok, R., Comiso, J. C. and Cunningham, G. F, 1996. Seasonal characteristics of the perennial ice cover of the Beaufort Sea. *J. Geophys. Res.*, 101(C12), pp. 28417–28439.

Massom, R. and Comiso, J. C., 1994. The classification of Arctic sea ice types and the determination of surface temperature using advanced very high resolution radiometer data. *J. Geophys. Res.*, 99(C3), pp. 5201–5218.

Onstott, R. G., Moore, R. K., Gogineni, S. and Delker, C., 1982. Four years of low-altitude sea ice broad-band backscatter measurements. *IEEE J. Oceanic Eng.*, 7(1), pp. 44–50.

Patric, E. A., 1980. Fundamentals of pattern recognition. Department of Electrical Engineering Perdue University, Prentice-Hall, Inc., Englewoood Cliffs, N. J., 1972.

Swift, C. T., Fedor, L. S. and Ramseier, R. O., 1985. An algorithm to measure sea ice concentration with microwave radiometers. *J. Geophys. Res.*, 90(C1), pp. 1087–1099.

6

Boreal Forest Habitat Studies Using OKEAN-01 Satellite Data

6.1 INTRODUCTION

Multispectral optical satellite scanners provide synoptic data of large areas. These data are commonly used for boreal forest ecological studies. However, periods of prolonged darkness and persistent cloud cover in boreal forest regions often preclude data acquisition on desired dates.

The microwave satellite systems essentially increase the potential boreal forest monitoring by affording all-weather data collection. Same aspects associated with classifying boreal forest vegetation using SAR data have been identified (Way et al., 1990). However, more studies are required to evaluate the diversity of methods for processing and analyzing microwave data in the context of specific ecological applications and specific microwave system characteristics.

This chapter demonstrates some results of estimating the efficiencies of boreal forest mapping techniques based on OKEAN-01 satellite data. These studies included elaborating the integrated database of OKEAN-01 satellite and ground polygon data, developing the data processing system, designing a modified ground-derived classification system and assessing the capability of OKEAN-01 satellite data for boreal forest vegetation classification. One goal of this work was to find efficient algorithms that would ensure the classification exactness and to isolate vegetation classes for the purposes of concrete ecological investigations.

Classification problems stem from the fact that physically differentiated classes in classification systems may differ significantly. Therefore, in a more general sense, the task of classification should be considered as the search for a better correspondence between classifications of remote sensing multichannel data and the *in situ* data. Naturally, the class attributes should be generated by a geographic information system (GIS).

The problem of optimal classification in such a case can be formulated in the following way. Let there be M classes obtained after the processing of remote sensing multichannel data on the basis of a certain *a priori* logics (e.g., for classification without instruction using a certain threshold of spectral intensity) and N classes within the frames of a present system for classification of objects of a sounded surface (e.g., vegetation types) with *a priori* information about them in chosen test regions. It is necessary to find the best way to express the classes *M* of one classification system in terms of *N* classes of the second classification system. One possible heuristic approach to this problem was reported in some publications (Aronoff, 1989; Belchansky et al., 1994a). (*M* image classes are associated with the exact ground data on *N* classes in terms of pixels of image classes found in each *N* class; each image class is attributed to one *N* class according to the maximum pixel number at a certain threshold of corresponding image classes found in *N* classes; the principle of classification using a loss function is considered.) By its maximum pixel number at a certain threshold of corresponding classes of test fields, the classification technique (Aronoff, 1989) corresponds to the idea of an *a posteriori* analysis (attribution of image classes to corresponding ground classes on the basis of the maximum *a posteriori* probability).

To develop the structure and logics of algorithms and software for classification of boreal forest vegetation according to data obtained by OKEAN-01 satellite (MSU-M, RAR, RM-08) and at test fields, we partially used basic concepts and results described in these publications, which allowed us to obtain more efficient algorithms, and software for classification using methods of numerical optimization.

6.2 STUDY AREA

To test the efficiency of classification and mapping of boreal forest types, we chose a north-east region of Yakutia, Russia. Field data collection was spread over 63 degrees north latitude, 144.5 degrees east longitude; 65 degrees north latitude, 139.5 degrees east longitude; 65.5 degrees north latitude, 145 degrees east longitude; 67 degrees north latitude and 142.5 degrees east longitude. The study area covered Yana's and Indigirka's river basins. There are two forest types in the region: the north-taiga type and middle-taiga type. Besides, it has three vertical zones: the forest zone, highland-tundra zone and Arctic desert zone. This region was chosen as the study area because it includes the most varied mix of the boreal species of interest.

The boreal forest of the study area is characterized by several features of climate, structure and biogeochemistry. The climate is continental, with extreme daily, seasonal and annual variation in air temperature. Boreal trees grow in a unique solar radiation mode characterized by long growing-season day lengths but low solar incidence angles. Consequently, boreal forest have very narrow canopies. Mosses and lichens are structural components of boreal forest, and are important in determining the physical environment and biogeochemical cycles. A seasonal snow pack role in hydrology and land surface climatology are essential. Snow melt occurs over a short period of time, but then cycles through the system very slowly because of permafrost and high water tables. Permafrost is important in the circumpolar boreal forest, especially in Siberia. Saturated, boggy soils and peat are also common, especially on permafrost areas. The boreal forest vegetation mapped in this report included the cover types shown below.

6.3 DATA

OKEAN-01 N6 satellite images were acquired for the study area. The OKEAN data (RAR, MSU-M, RM-0.8) (Figure 6.1) included the scene #853 (August 1,1991), the data structure was exposed in mode 1 (RAR, RM-08, MSU-M (0.8–1.1 μm)). The OKEAN-01 images were georegistered to a UTM (Universal Transverse Mercator) map projection. A subimage of this scene (1301 rows by 591 columns) was used for boreal forest classification study.

Ground changes embraced the territory of the following coordinates: 63 degrees north latitude, 145 degrees east longitude; 65 degrees north latitude, 139 degrees east longitude; 65.5 degrees north latitude, 145 degrees east longitude; 67 degrees north latitude, 142.5 degrees east longitude (Yakutia). For each cover type included in the classification scheme, different topographic locations were used to define training and test areas. The field survey goal was to use different sites with different slopes, slope positions and aspects that would give a wide spectral variability range to characterize each of the cover types.

The ground database intended for the test regions made it possible to construct a raster file of ground data in the system of forest classification aimed specifically at studying the problems of global change (e.g., climate, biodiversity, etc.).

In general the structure and content of the concrete topic-oriented classification systems should be unambiguously connected with the structure of models of corresponding topical regions.

The boreal forest classification scheme is presented in Table 6.1.

Table 6.1 Boreal Forest Vegetation Classification Scheme

CLASS	DESCRIPTION
Cl. 1	Dominant — daurskaya larch and juvenile with middle density 1950/1 hect; lower tier cover: ledum with mosses cover; mesorelief character — plateau, the upper floodlands
Cl. 2	Dominant — daurskaya larch and juvenile with middle density — 2065/1 hect; lower tier cover: cowberries-mosses; mesorelief character — plateau, the upper floodlands
Cl. 3	Dominant — daurskaya larch and juvenile with middle density — 1500/1 hect; lower tier cover: lichen cover; mesorelief character — hill summit, plateau
Cl. 4	Dominant — daurskaya larch and juvenile with middle density — 1500/1 hect; lower tier cover: tundra birch (ernik); mesorelief character — hill summit, plateau
Cl. 5	Dominant — daurskaya larch and juvenile with middle density — 2100/1 hect; lower tier cover: motley grass
Cl. 6	Dominant — daurskaya larch and juvenile with middle density 2200/1 hect; lower tier cover: cowberries; mesorelief character — plateau
Cl. 7	Dominant — daurskaya larch and juvenile with middle density — 50/1 hect; lower tier cover: ledum, alder; mesorelief character — gentle slope, plateau
Cl. 8	Lake
Cl. 9	5 years burnt
Cl. 10	Mountain tundra (lichen tundra with grasses and shrubs)
Cl. 11	Dominant — daurskaya larch and juvenile with middle density — 2330/1 hect; lower tier cover: dod-rose, lichen; mesorelief character — gently slope, plateau
Cl. 12	Dominant — daurskaya larch; lower tier cover: cowberry, alder; mesorelief character — upper floodlands
Cl. 13	Dominant — daurskaya larch and juvenile with middle density — 1000/1 hect; lower tier cover: ledum cowberry
Cl. 14	Cedar
Cl. 15	Conifer
Cl. 16	Alder
Cl. 17	Pebbly bank
Cl. 18	Sea

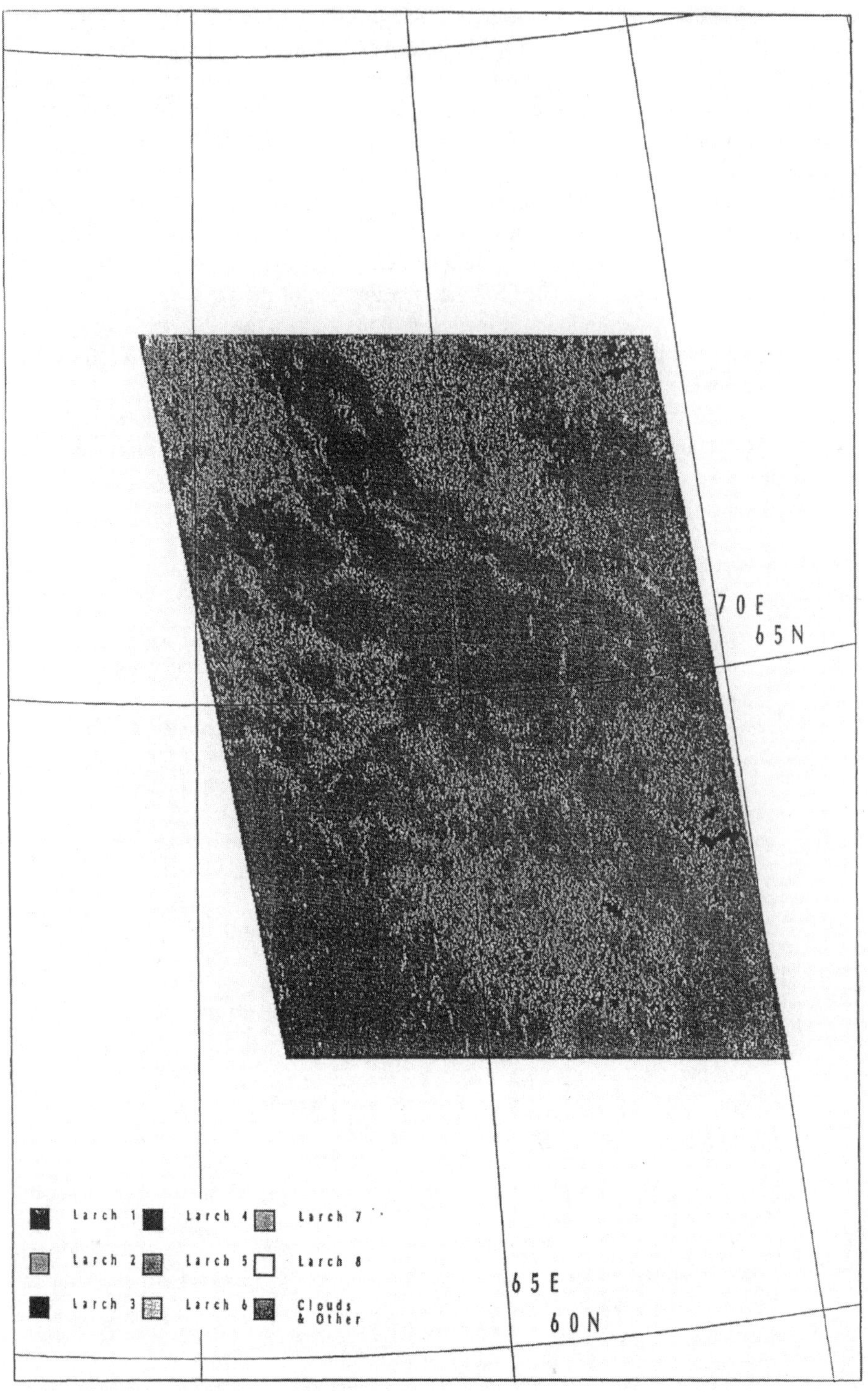

Figure 6.1 OKEAN data for Jakutsk District (forest classification).

6.4 MAPPING BOREAL FOREST VEGETATION

The data processing methodology used in the first step of our study of boreal forest vegetation classification includes: OKEAN-01 data (RAR, RM-08, MSU-M) preliminary processing (filtering, georegistration, radiometric and geometric distortion correction); generating additional band combinations from the original remotely sensed data; selecting training areas; generating statistics; spectral pattern analysis; supervised, unsupervised or combined classification; assessing the classification accuracy. The data processing methodology is illustrated in Figure 6.2.

Before classification, the exploratory phase was fulfilled. It included generating additional bands from the original remotely sensed data, selecting training areas, generating training statistics and spectral pattern analysis. Generating additional bands was important to define beforehand if there are any bands, band combinations or other derivatives that might provide additional information. Once all the bands, ratios and band derivatives have been generated, bands providing the most information for delineating ground classes with the aid of spectral pattern analysis can be defined.

Data processing and exploration were performed according to a previous methodology (Figure 6.2). Supervised, unsupervised and combined classifications were used for boreal forest vegetation classification. The software and methodology for integrated processing of data acquired by OKEAN satellite and simultaneous satellite land observations for boreal forest studies using a database of forest GIS are presented in Belchansky et al. (1994b). A rigorous accuracy assessment was performed for each classification.

6.5 SOME RESULTS

The boreal forest classification scheme included 18 classes (Table 6.1). The number of representative training areas for each class was selected within the OKEAN-01 satellite scene. Training statistics were generated and spectral pattern analysis was conducted using RAR, RM-08, MSU-M bands, band combinations and derivatives for solution of the boreal forest classification problem. The spectral pattern analysis for classification of cover type from OKEAN-01 satellite data included 18 classes.

Five variants — variant 1 (RAR, PC1, PC2), variant 2 (RAR, PC2), variant 3 (RAR), variant 4 (PC2) and variant 5 (RM-08, RAR, MSU-M) — were selected as source data for the cover type classification process. (PC1 and PC2 are the first and second principal components of the original remotely sensed data.)

PC1 accounted for 60.82% of the variability found in RAR, RM-08, MSU bands; PC2 — 29.86%.

The results of supervised classifying the OKEAN-01 satellite data for cover type using the selected algorithms are as follows: overall accuracy of boreal forest types classification for eight classes (Cl. 1–Cl. 7, Cl. 18) was for variants I, II, III, IV, V — 43.62%, 44.68%, 42.55%, 43.62%, and 53.19%, respectively, for classification scheme of Table 6.1. Each classification result was assessed by comparing it with ground data to form an error matrix for different values of ED, but classification accuracy was computed based on class labeling of the pixels belonging to the test sites.

Every ground data pixel was about 100 hectares but we did not have a reliable digital vegetation map of the whole area to estimate the overall classification accuracy of the scene.

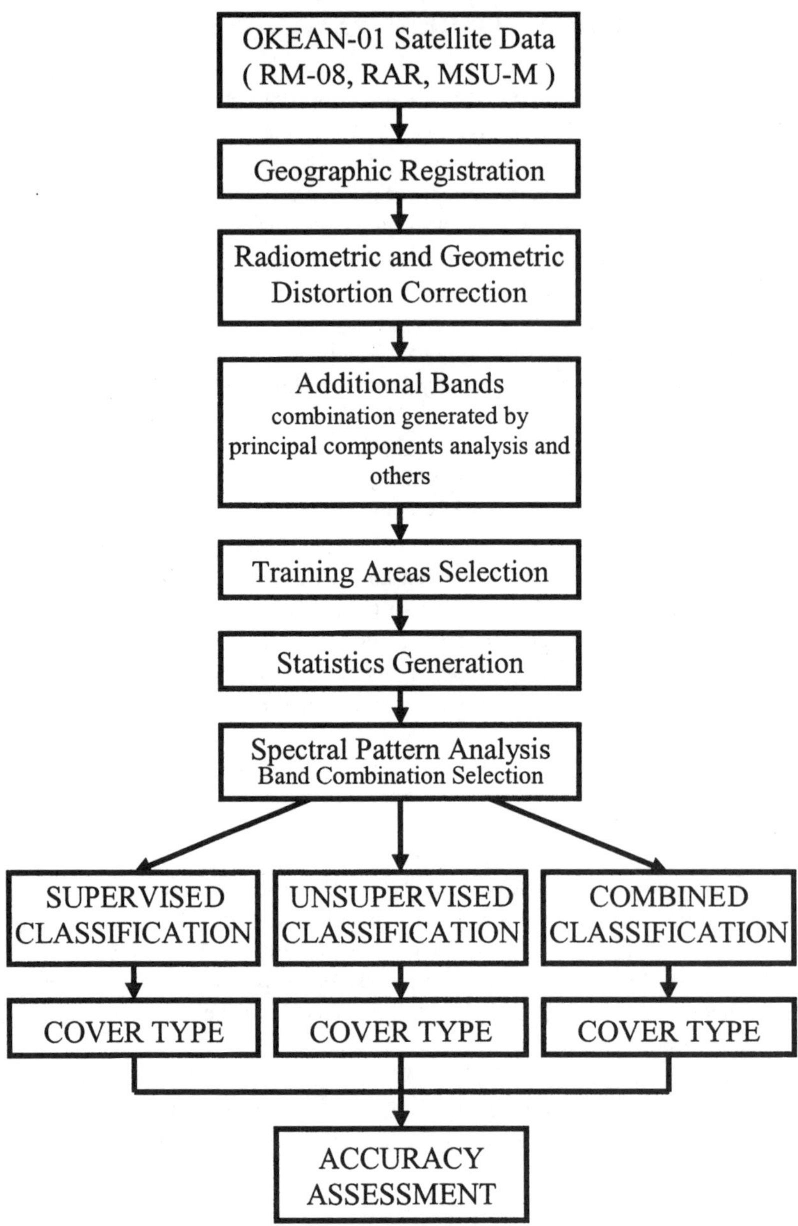

Figure 6.2 OKEAN-01 satellite data processing flow chart.

Classification accuracy for land classes Cl. 1–Cl. 7 and Cl. 18 was assessed because species of boreal forest Cl. 1–Cl. 7 have a significant contribution to global change. The best boreal forest cover types classification for Cl. 1–Cl. 7 and Cl. 18 classes included processing RM-08, RAR, MSU-M bands with with minimum distance algorithm classification (overall accuracy = 53.19%). Pixel distribution, overall, producer's and user's accuracy for these classifications are presented in Table 6.2.

Table 6.2 Pixels Distribution, Overall, Producer's and User's Accuracies

CLASS	CL1	CL2	CL3	CL4	CL5	CL6	CL7	CL18	Total
CL1	10	12	1	1	2	0	0	0	26
CL2	4	9	1	1	0	0	1	0	16
CL3	8	2	4	1	0	1	0	0	16
CL4	1	0	0	1	0	0	0	0	2
CL5	9	7	1	4	4	0	1	0	26
CL6	6	11	0	1	3	8	4	0	33
CL7	1	1	0	1	0	2	4	0	9
CL18	0	0	0	0	0	0	0	60	60
Total	39	42	7	10	9	11	10	60	188

OVERALL ACCURACY = 100/180 = 53.19%

CLASS	PRODUCER'S ACCURACY	USER'S ACCURACY
CL1	10/39 = 25.641%	10/26 = 38.462%
CL2	9/42 = 21.429%	9/16 = 56.250%
CL3	4/7 = 57.143%	4/16 = 25.000%
CL4	1/10 = 10.000%	1/2 = 50.000%
CL5	4/9 = 44.444%	4/26 = 15.385%
CL6	8/11 = 72.727%	8/33 = 24.242%
CL7	4/10 = 40.000%	4/9 = 44.444%
CL18	60/60 = 100.000%	60/60 = 100.000%

The relatively low agreement between satellite-derived classes and the ground data should not be interpreted as an exact measure of OKEAN-01 satellite data proficiency. The results of this study demonstrate comparative abilities between the satellite instruments only in the context of the analytical methodologies and volume of ground data that were used. Let us consider another methodology of classification.

6.6 IMPROVEMENT OF BOREAL FOREST CLASSIFICATION USING MINIMUM LOSS FUNCTION CRITERIA

6.6.1 General Description of the Classification Goal

Let us assume that a certain preliminary processing of the satellite OKEAN-01 (MSU-M, RAR, RM-08) yielded data with M classes in a certain many-dimensional spectral space. It is necessary to carry out the best (in the sense of the minimum of a certain loss function) classification of satellite data in terms of classes using ground data on N classes of a sounded surface.

The problem is solved primarily by determination of physically interpreted loss functions and the subsequent realization of the optimization procedure on a computer. (The logics of the many-dimensional space synthesis and selection of the initial number of classes M will be considered in due course.)

Since vegetation types are usually recorded in the form of relevant maps, a starting point for the loss function construction is a more exact definition of each type.

Usually, satellite data are classified by a limited selection of data of ground test fields. The most convenient method of defining the map precision as a whole and for each type is to evaluate the minimum possible thresholds of accuracy. Let us designate them as $Q_{i\min}$ and P_i, respectively, i being a type (class) number in the map. P_i in the general case is taken *a priori* (e.g., 95%, etc.) and its value is determined first of all by a necessary value of *a priori* data and the aim of the map usage (e.g., evaluation of mineral resources, bio-diversity, etc.).

Value $Q_{i\min}$ is calculated by the following approach. Let us assume that a vegetation map resulting from the classification is tested for the exactness of each type of vegetation by the data of ground test fields. Evidently, R_i (risk) probability that each class of unacceptable precision will pass the testing is determined by the expression:

$$R_i = 1 - P_i = \sum_{Y=0}^{X_i} \frac{N_i!}{(Y! - Y)!} Q_i^{(N_i - Y)} (1 - Q_{(i)})^Y, \tag{6.1}$$

where Q_i is an acceptable precision of the evaluation of classification quality, X_i is the number of acceptable incorrectly classified pixels, N_i is total pixels of class i of tested fields, Y_i is the number of incorrectly classified pixels.

Then at taken P_i value of a reliability level, one can easily calculate from (6.1) the values of $Q_{i\min}$ determined by condition $X_i = Y_i$, i.e. from the following relation:

$$R_i = \sum_{Y=0}^{X_i} \frac{N_i!}{(Y! - Y)!} Q_{i\min}^{(N_i - Y)} (1 - Q_{i\min})^Y. \tag{6.2}$$

Given value i for each considered class $Q_{i\min}$ at taken R_i (value R_i is assumed to be the same for all classes), it is easy to determine the loss function for classification in the simplest case in the following equation:

$$F_i = (1 - Q_{i\min}) W_i n_i, \tag{6.3}$$

where n_i is the total number of pixels of class i on the test territory, W_i is the weight of i-class or all the classes:

$$F = \sum_i F_i. \tag{6.4}$$

Taking into account the entire picture of the distribution of test pixels in the initial classes of the image, the loss function can be presented in a general case as:

$$F = \sum_{i=1}^{N} \sum_{\substack{i=1 \\ j \neq 1}} \left(\frac{A_{ij}}{\sum_{\substack{i=1 \\ j \neq 1}} A_{ij}} \right) W_{ij} (1 - Q_{i\min}) n_i, \tag{6.5}$$

where N is the taken number of classes of the sounded surface, A_{ij} is the number of pixels attributed to class j from class i, W_{ij} is the weight of incorrectly classified pixels.

Thus, having taken N of surface classes and M of image classes, as well as a known number of pixels of all test fields corresponding to relevant classes of the sounded surface, one can find the best marking of image classes by corresponding surface classes (taken ground data) on the basis of the minimization of the loss function by the integer method.

We have been of the opinion until now that the number of image classes M is known. Nevertheless the logics of the formation of initial multiband satellite data, the size and structure of the initial spectral space and class number M in the image used for classification may be several orders higher than the number of sought-for classes. The algorithm was constructed from the logics of the enumeration of sought-for image classes on the basis of quantization level measurements by the taken value of a spectral distance and ground data. The specificity of the logics of the classification algorithm functioning is considered below.

6.6.2 The Structural Scheme of Data Processing and Classification System

A generalized scheme of blocks for data processing and classification is shown in Figure 6.3. The scheme consists of the following basic elements: space images, topical systems for classification of a sounded surface and ground data; preliminary data processing and GIS and data classification.

Let us consider more thoroughly the logics of the data classification block functioning. In accordance with the constructed algorithms and software, the following components are formed at the first step:

- The ground data raster file (information about a sounded surface in the taken topical system of classification)
- A set of M images ($M = K + L$, where K is the total number of bands, L is the number of bands synthesized on the basis of the method of main components, texture analysis, etc.).

To reduce the data volume and to make the optimal marking of image classes by the corresponding topical systems of the surface classes in the module of cluster analysis for a taken ED of spectral scattering thresholds values in the Euclidean space more exact, we form a file of image classes on the basis of M bands of satellite data and raster file of ground data (Figure 6.3).

The resulting file of image classes goes to the module of marking of image classes, in which, by ground data and risk criterion taken for the raster file, one can solve the problem of optimal marking and obtain a table of correspondence between image classes and classes for ground data on the sounded surface. Further, in the file of image classes and in the table of correspondence, a digital table is formed in the module of image classification that contains the results of classification in terms of classes of ground data.

Some peculiarities of functioning of the data processing and classification systems are exemplified below.

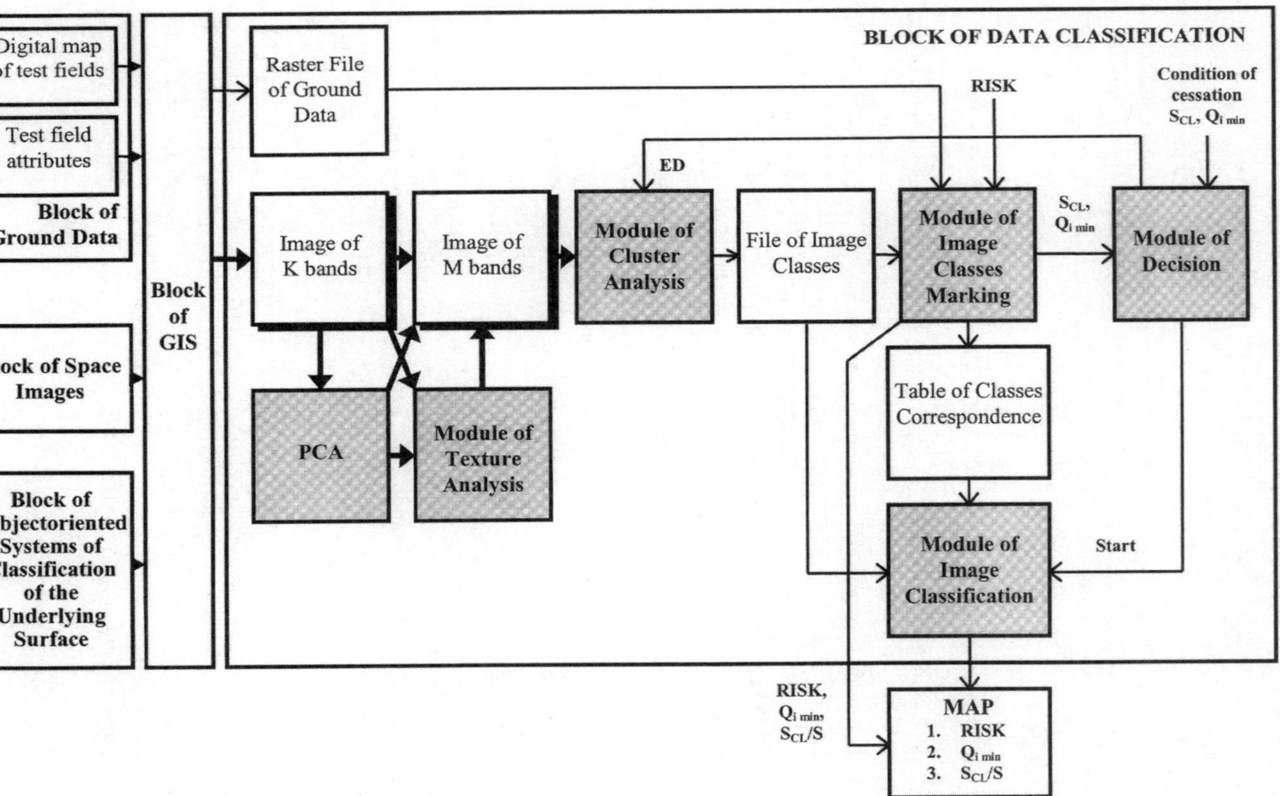

Figure 6.3 Structural scheme of data processing and classification system.

6.6.3 Classification of Boreal Forest Vegetation Using Okean-01 Satellite Data

To illustrate the logics of algorithm functioning, let us consider the basic steps of processing for data classification and subject mapping using three data channels (RM-08, RAR, MSU-M). In this case, the cluster analysis module (Figure 6.3.) receives three-dimensional data ($M = 3$).

Earlier, we studied the efficiency of differentiation of the following classes: Cl. 1, Cl. 2, Cl. 3,..., Cl. 7, Cl. 18, on the basis of the preliminary spectral analysis, synthesis of additional bands and different techniques of cluster analysis (forest types Cl. 1 to Cl. 7 are the most important for climate studies). Therefore, to compare the results, let us assume that classes Cl. 1, Cl. 2, Cl. 3, ..., Cl. 7, Cl. 18 are sought for.

Table 6.3 Matrix of Pixel Distribution (ED = 8)

Classes of the sounded surface	Image Classes																					
	1	2	3	4	5	6	7	8	9	11	12	13	14	16	17	18	19	20	21	23	24	25	Total
1	0	0	0	8	1	3	3	4	0	0	1	0	0	1	1	1	1	1	1	0	0	0	27
2	0	0	4	2	1	2	2	2	0	0	2	1	0	2	0	4	1	0	0	0	0	0	23
3	0	0	0	2	1	1	0	0	0	0	1	0	0	0	0	0	0	0	0	0	0	0	5
4	0	0	1	4	0	1	0	0	0	0	0	0	0	1	0	0	0	0	0	0	0	0	7
5	0	0	1	1	1	0	0	0	0	0	0	0	0	2	0	0	0	0	0	0	0	0	6
6	0	0	1	1	0	0	1	2	0	0	0	0	1	0	0	0	0	0	0	0	0	0	6
7	0	0	0	0	0	1	1	0	1	1	1	1	0	0	0	0	0	0	0	0	0	0	6
18	3	3	0	0	0	0	0	0	0	0	0	0	0	0	0	0	0	0	0	1	1	1	9
Total	3	3	7	18	4	8	7	8	1	1	5	2	1	6	1	5	2	1	1	1	1	1	89

Evidently, the most complete scheme of classes by three-dimensional satellite data is the case of clusterization at Euclidean distance ED = 0. Taking into account a large number of illustrating tables, let us consider the case when ED = 8, and then give the generalized results for other variants.

For variant ED = 8, the distribution of pixels of image classes in ground data classes (Cl. 1, Cl. 2, Cl. 3, ..., Cl. 7, Cl. 18) is given in Table 6.3 (Table 6.3 was made using the files of image classes and sounded surface, see Figure 6.3). Each pixel of ground data

Table 6.4 Table of Classes Correspondence (ED = 8)

Classes of sounded surface	Image classes
1	4 6 7 8 17 20 21
2	3 12 18
3	
4	
5	
6	14
7	9 11
18	1 2 23 24 25
Out	5 10 13 15 16 19 22

Table 6.5 Matrix of Classification Accuracy (ED = 8)

Classes of sounded surface	Number of correctly classified pixels	Total pixels in a class	Correctly classified pixels %	Risk	Minimum exactness
1	21	27	77.78	0.001	0.4672
2	10	23	43.48	0.001	0.1501
3	0	5	0.00	0.001	0.0000
4	0	7	0.00	0.001	0.0000
5	0	6	0.00	0.001	0.0000
6	1	6	16.67	0.001	0.0002
7	2	6	33.33	0.001	0.0083
18	9	9	100.00	0.001	0.4642
For the entire matrix	43	89	48.31		

covered 100 hectares, which corresponded to RAR resolution which received the data of all the channels. It is evident that the amount of data on the sounded surface was much less than that of the data on the territory covered by one OKEAN-01 satellite image. This affected significantly the exactness of final results.

In the module of marking (see Figure 6.3) at minimization of the loss function with single weighting and risk value $R = 0.001$, the optimum table of class correspondence for the variant under study takes the form of Table 6.4.

As seen from Table 6.5, the minimum loss function when image classes 4, 6, 7, 8, 17, 20, 21 obtain the name of the class — 1; 3, 12, 8, become class 2, etc.

The results of evaluation of classification accuracy are listed in Table 6.5. At ED = 8, the mean percentage of correctly classified pixels is 48.31, minimum accuracy of classification in the subject map is (%): 46.72 for Cl. 1, 15.01 for Cl. 2, 0.0 for Cl. 3, 0.0 for l.4, 0.05 for Cl. 5, 0.02 for Cl. 6, 0.083 for Cl. 7, 46.42 for Cl. 8. For the variant under consideration, area S_{CL} of the classified territory is 32.72% (at ED = 8).

The obtained minimum accuracy of classification of the territory under consideration and its value (32.72% of the picture area) are ambiguously connected with ED and amount of ground data. At the same ground data, the decrease in ED changes the efficiency of classification.

Estimates of classification accuracy for some ground data, but at ED = 0 and 4, are given in Tables 6.6 and 6.7. A share of classified territory is 0.3 and 18.3%, respectively.

Tables 6.6 and 6.7 show that the percentage of correctly classified pixels and minimum possible values of classification accuracy for each class of the subject map sharply increases at decreasing ED, however, the share of the classified territory in the entire picture diminishes.

The attainment of the necessary efficiency in classification (minimum admissible accuracy at a chosen risk, a share of classified territory) of boreal forest vegetation on the basis of OKEAN-01 satellite data is only possible at the adequate volume of ground data. However, the "taken efficiency level" should be determined by a concrete subject (e.g., structure of models for studies of the carbon balance in global changes and admissible exactness of initial data: vegetation types, accuracy of their classification, etc.).

The developed approach, algorithms, and software make it possible to rapidly process data for classification of boreal forest vegetation and to evaluate the results obtained from different points of view.

Table 6.6 Matrix of Classification Accuracy (ED = 0)

Classes of sounded surface	Number of correctly classified pixels	Total pixels in a class	Correctly classified pixels %	Risk	Minimum accuracy
1	27	27	100.00	0.001	0.7743
2	22	23	95.65	0.001	0.6634
3	5	5	100.00	0.001	0.2512
4	7	7	100.00	0.001	0.3728
5	6	6	100.00	0.001	0.3162
6	4	6	66.67	0.001	0.0942
7	6	6	100.00	0.001	0.3162
18	9	9	100.00	0.001	0.4642
For the entire matrix	86	89	96.63		

Table 6.7 Matrix of Classification Accuracy (ED = 4)

Classes of sounded surface	Number of correctly classified pixels	Total pixels in a class	Correctly classified pixels %	Risk	Minimum accuracy
1	20	27	74.07	0.001	0.4282
2	17	23	73.91	0.001	0.3991
3	2	5	40.00	0.001	0.0101
4	5	7	71.00	0.001	0.1437
5	3	6	50.00	0.001	0.0378
6	5	6	83.33	0.001	0.1815
7	4	6	66.67	0.001	0.0942
18	9	9	100.00	0.001	0.4642
For the entire matrix	65	89	73.03		

For instance, the use of spectral analysis, synthesis of additional bands, etc. to isolate the above mentioned eight classes within the frames of the same ground data allowed us to obtain an average value of correctly classified pixels which only made up 53.19%. The use of classification algorithms for the same ground data and classes increased the accuracy up to 96.63% as well as made it possible to evaluate a classified territory share and the minimal possible accuracy of classification for each type of vegetation. This is very important for the solution of various problems in concrete subject regions.

6.7 CONCLUSION

Obviously, the most complete system of image classes corresponds to ED = 0. In this case the overall accuracy is 96.63% and the area of classified territory is 0.3% of the whole subimage. When ED = 4, the overall accuracy is 73.03% and the area of classified territory is 18.3%; when ED = 8, these indices are 48.31% and 32,72%, respectively.

On the contrary, the overall accuracy was 53.19% for the same forest cover classes when spectral pattern analysis with minimum distance algorithm classification was used.

Thus, application of our algorithms allows one to raise the overall accuracy of classification and to evaluate the part of classified territory and the value of minimum accuracy level. For example, when ED = 0, the values of minimum accuracy level for different forest cover classes are: for class Cl. 1 — 77.43%, for class Cl. 2 — 66.34%, for class Cl. 3 — 25.12%, for class Cl. 4 — 37.28%, for class Cl. 5 — 31.62%, for classes Cl. 6–Cl. 9 — 42%, for class Cl. 7 — 31.62% and for class Cl. 18 — 46.42%.

Results indicated that OKEAN satellite data have a potential for discriminating up to seven or eight boreal forest cover types. The relatively low agreement between satellite-derived classes and the ground data should not be interpreted as an exact measure of OKEAN-01 satellite data proficiency. The results of this study demonstrate comparative abilities between the satellite instruments only in the context of the analytical methodologies and ground data that were used.

The redundancy of geobotanical description for radar data gives rise to significant problems in constructing ground-derived classification systems for boreal forest vegetation. Hence statistical methods elaboration for quantitative estimation of training area numbers and for sample sites elimination whose spectral characteristics are strongly visible is necessary. Alternative classification methods may also provide more efficient ways to extract SAR information, such as second-order or neural network classifiers (Hepner et al., 1990) that consider textural properties of the image data. A good understanding or a more satisfactory interpretation of remotely sensed image data should include the description of both the spectral and textural aspects.

REFERENCES

Aronoff, S. A., 1989.Approach to optimized labelling of image classes. *Photogram. Eng. and Rem. Sens.* 50(6), pp. 719–727.

Belchansky, G. I. and Ovchinnikov, G. K., 1994a. Improvement of Boreal Forest Classification According to Satellite "OKEAN" Data. *Soviet Journal of Remote Sensing*, Vol. 11(6), pp. 960–970.

Belchansky, G. I., Mordvintsev, I. N., Ovchinnikov, G. K., Petrosyan, V. G., Douglas, D. C. and Pank, L., 1994b. Classification of the vegetation of boreal forests using OKEAN satellite data. *Soviet Journal of Remote Sensing*, Vol. 11 (2), pp. 226–244.

Hepner, G. F., Logan, T., Ritter, N. and Bryant, N., 1990. Artificial neural network classification using a minimal training set-comparison to conventional supervised classification. *Photogrammetric Engineering and Remote Sensing*, 56 (4), pp. 469–473.

Rosenfield, G.H. and Fitzpatrick-Lins, K., 1986. A coefficient of agreement as a measure of thematic classification accuracy. *Photogrammetric Engineering and Remote Sensing*, 52(2), pp. 223–227.

Way, J.et al., 1990. The effect of changing environment conditions on microwave signatures of forest ecosystems: preliminary results of the March 1988 Alaskan aircraft SAR. *International Journal of Remote Sensing*, 11:1119–1144.

Werle, D., 1986. The role of spaceborne Radar (SIR-A) image interpretation in monitoring tropical forest conversion: Report, Ottawa, On: RADARSAT Project Office, pp. 28–45.

7

Comparative Assessment of ALMAZ-1 SAR, ERS-1 SAR, JERS-1 SAR and Landsat-TM Satellite Data for Tundra Habitat Studies

7.1 INTRODUCTION

Multispectral optical satellite scanners, such as Landsat-TM, are commonly used for tundra habitat studies. However, periods of prolonged darkness or persistent cloud cover often preclude data acquisition on desired dates. Appearance of microwave satellite systems essentially increased potential monitoring environments by affording all-weather data collection. Compared with multispectral images, active microwave measurements have unique sensitivity to terrain surface, soil composition and moisture regime (Curlander and McDonough, 1991). Some aspects associated with classifying tundra vegetation using SAR data have been identified (Way et al., 1990). However, more studies are required to evaluate the diversity of methods for processing and analyzing radar data in the context of specific ecological applications and system characteristics.

Each SAR system investigated by this study (ERS-1, JERS-1 and ALMAZ-1) used a different combination of wavelength and polarization, which resulted in different responses to different landscape features (Curlander and McDonough, 1991). Other variables that influenced this evaluation of satellite detection capabilities included procedures of raw data analysis, methods for optimal focusing raw SAR data and filtering, defining "ground-truth" systems, classifying satellite data and associating satellite and ground data.

This chapter demonstrates consistent analytical methodologies to better elucidate differences within the information content of the satellite data. The relative capabilities of ALMAZ-l SAR, JERS-1 SAR, ERS-1 SAR data (standard SAR processing products) and the Landsat-TM multispectral satellite were studied based on the results of classifying land-cover and terrain types in Northern Alaska. A variety of adaptive filters (Median, Lee, Frost and Sigma) were used for suppressing speckle noise. The primary objectives of this study were to develop the database of the satellite and ground polygon measurements and a data processing system for tundra vegetation classification. The objectives included designing a modified ground-derived classification system, evaluating the utility of the satellite data for tundra vegetation classification and assessing the capability of integrating ALMAZ-1, JERS-1 and ERS-1 SAR data with Landsat-TM data for mapping tundra habitats.

7.2 STUDY AREA

Areas encompassing portions of the Prudhoe Bay region on the Arctic coastal plain of northern Alaska were selected for study (70°12′–70°20′ N, 148°20′–148°40′ W) (Figure 7.1). The region is wet tundra, with numerous shallow lakes. The vegetation is characterized by highly intersperse mosaics of sedges, mosses, lichens and low shrubs. The climate is arctic, with cool short summers and very cold winters. Permafrost underlies the entire region, and the active layer thaws annually between June and August to 10–100 cm depths. Annual precipitation averages around 200 mm, with approximately two thirds falling as snow. Petroleum extraction since the late 1960s has directly and indirectly altered the landscape with roads, pipelines, gravel pads, facilities, excavations and redistributed hydrologic regimes. Detailed geobotanical descriptions of the study area are presented in Walker et al. (1986). Within the Prudhoe Bay region, three integrated geobotanical and historical disturbance test maps were chosen. These squares (maps W22, W32 and W34) contain three basic types of data: geobotanical, historical natural disturbances and historical anthropogenic disturbances.

7.3 SATELLITE DATA

ALMAZ-1 SAR, ERS-1 SAR, JERS-1 SAR, and Landsat-TM satellite images were acquired for the study area. All SAR data were received as full-resolution, geometrically and radiometrically corrected images (Level 1B) from their respective distribution centers: NPO Mashinostroenia in Moscow, Russia (ALMAZ-1, Figure 7.2), Alaska SAR Facility in Fairbanks, Alaska (ERS-1 and JERS-1), Earth Observation Satellite Data Company in Lanham, Maryland (Landsat-TM, Figure 7.3). ALMAZ-1, ERS-1, JERS-1 and Landsat-TM data, respectively, included scenes #7753 (July 31, 1992), #5911100 (July 13, 1992), #4131 (August 9, 1992) and #Y5051921175X0 (August 2, 1985). Some characteristics of satellite instruments are presented in Table 7.1. Landsat-TM image was georegistered to UTN coordinates, and three SAR images were subsequently co-registered to the Landsat-TM image. Georegistration used a second-order polynomial equation, and the data were concurrently resampled to a 10-m pixel size using cubic convolution interpolation.

7.4 GROUND DATA

Three previously published geobotanical maps (W22, W32, W34) were used as models of "ground-truth" (Figure 7.1). The maps were published by Walker et al. (1986) as part of a cumulative impact study of the region's oil development activities. Each map covered approximately 21 km and was produced by interpretation of aerial photographs (1:6000 scale color infrared) and ground reconnaissance data. The maps were selected as models of "ground truth" because of their detailed spatial resolution and comprehensive habitat characterization. More than 11000 polygons delineated the geobotanical features and anthropogenic disturbances across the three mapped areas. Each polygon was characterized with respect to several criteria: vegetation, landscape, surface form, soil, disturbance type.

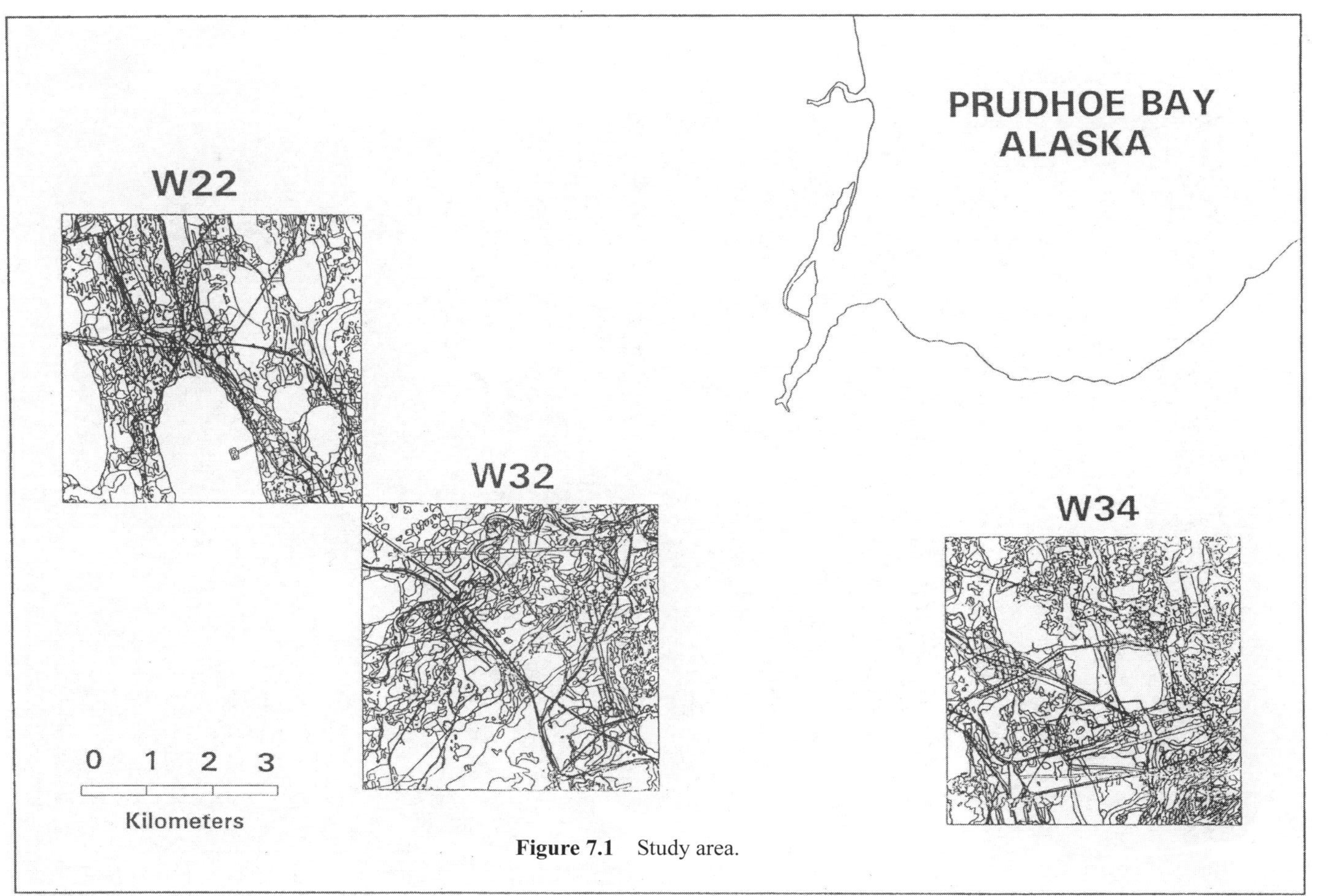

Figure 7.1 Study area.

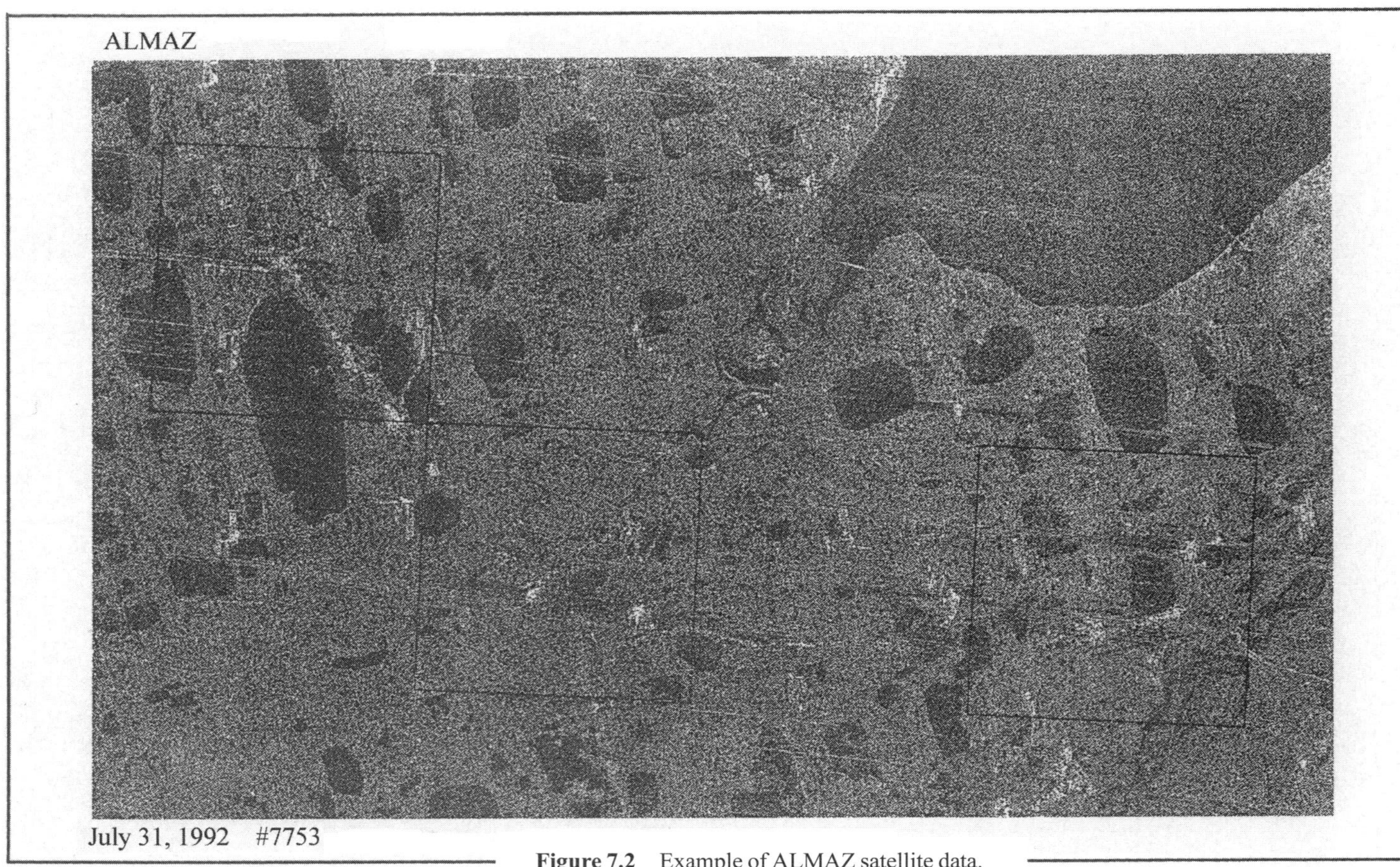

Figure 7.2 Example of ALMAZ satellite data.

Figure 7.3 Example of raw Landsat-TM satellite data.

Table 7.1 Some Characteristics of Satellite Data

SATELLITE	INSTRUMENT	WAVELENGTH	RESOLUTION	POLARIZATION
ALMAZ-1	SAR	9.60 cm	15 m/1 look	HH
ERS-1	SAR	5.66 cm	25 m/3 looks	VV
JERS-1	SAR	23.50 cm	18 m/3 looks	HH
Landsat	TM			
	Visible	0.45–0.52 μm	30 m	
		0.52–0.60 μm	30 m	
		0.63–0.69 μm	30 m	
	Near-IR	0.76–0.90 μm	30 m	
	Mid-IR	1.55–1.75 μm	30 m	
		2.08–2.35 μm	120 m	

7.5 SATELLITE-DERIVED MAP

The structure of the data processing system is presented in Figure 7.4. It outlines the data processing methodology used in this chapter for tundra vegetation classification and comparative assessment of satellite data efficiency. The data processing system includes: SAR data preliminary processing (raw data analysis, optimal focusing, SAR data filtering, geo-registration); selecting training areas; polygon data to raster data conversion; classification procedures; merger signatures procedure; optimized classification using minimum loss function criteria; assessing the classification accuracy. A methodology of classification using minimum loss function criteria was discussed in Chapter 6.

ALMAZ-1, ERS-1, JERS-1 and Landsat-TM satellite images and ground-truth maps were co-registered to UTM Zone-6 map-coordinates. For comparison with satellite data, it was necessary to reduce the complexity of the geobotanical maps by grouping the polygons into a generalized classification scheme. All generalized classes were required to have at least 9 ha of total map area. Less extensive classes were not considered for analysis due to their small representation. Five generalized classification schemes were defined. Four schemes were based primarily on vegetation composition: A with 23 classes (Table 7.2), B with 17 classes (Table 7.2), C a more general scheme with eight classes (Table 7.3). E with 20 classes (Table 7.2), and D with 17 classes based primarily on surface form characteristics (Table 7.4). All schemes included human disturbance classes.

Master ground-truth maps were generated for each classification scheme by merging the original polygons according to the generalized class definitions. The polygon master maps were converted to a raster format with 10 m × 10 m pixel size. All mixed pixels that occurred along polygon boundaries were excluded. Systematic methods were used to evaluate the information content of the satellite data with respect to the ground classification schemes.

Six satellite image databases were analyzed: the four original images ALMAZ-1, ERS-1, JERS-1 and Landsat-TM (Figure 7.5); and two image combinations: ALMAZ-1 + ERS-1 + JERS-1, ERS-1 + JERS-1.

For each of the image databases, one analysis was conducted for each of the five ground classification schemes. A consistent methodology was used to evaluate each image database with each ground map (Figure 7.4). The first step generated a maximum number of statistically distinct image-class signatures for each image database. The signatures were

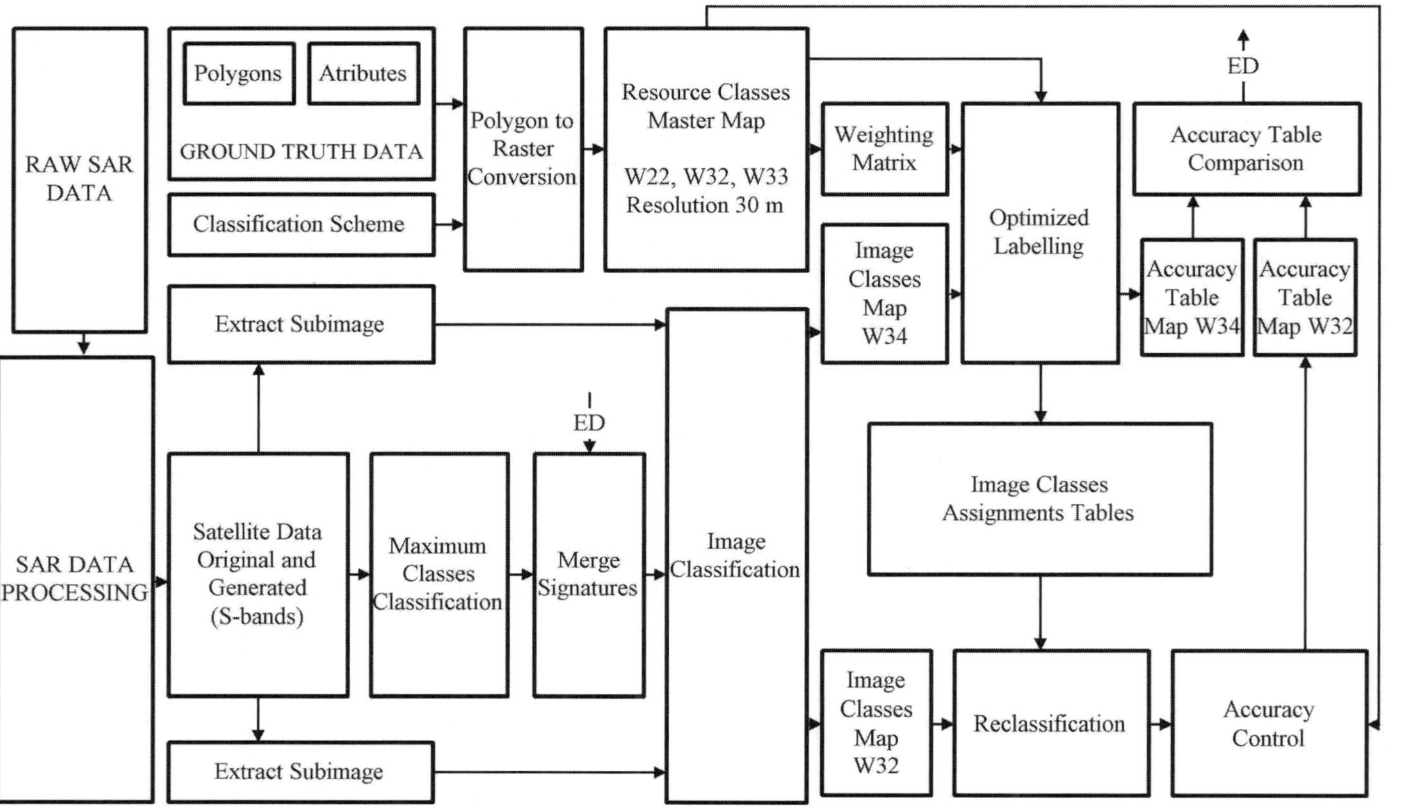

Figure 7.4 Data collection and processing system.

Table 7.2 Ground Classification Schemes A, B and E

CLASS	CLASS	CLASS	VEG1	VEG2	VEG3	HD831	HD832	HD833	ND83
E1	B1	A1	1	0	0	0	0	0	0
	—	—	1	2	0	0	0	0	0
	—	—	1	3	0	0	0	0	0
	—	—	1	3	2	0	0	0	0
	—	—	1	5	0	0	0	0	0
	—	—	1	41	0	0	0	0	0
E2	B2	A2	2	0	0	0	0	0	0
	—	—	2	1	0	0	0	0	0
	—	—	2	3	0	0	0	0	0
	—	—	2	21	0	0	0	0	0
	—	—	2	41	0	0	0	0	0
E3	B3	A3	3	0	0	0	0	0	0
	—	—	3	1	0	0	0	0	0
	—	—	3	2	0	0	0	0	0
	—	—	3	5	0	0	0	0	0
	B4	A4	3	21	0	0	0	0	0
	—	—	3	21	5	0	0	0	0
	—	—	3	21	41	0	0	0	0
	—	—	3	41	0	0	0	0	0
	—	—	3	41	21	0	0	0	0
E4	—	—	5	0	0	0	0	0	0
	—	—	5	1	0	0	0	0	0
	—	—	5	1	2	0	0	0	0
	—	—	5	2	0	0	0	0	0
	—	—	5	2	41	0	0	0	0
	—	—	5	21	0	0	0	0	0
	B5	A5	5	21	41	0	0	0	0
	—	—	5	41	0	0	0	0	0
E5	B6	A6	21	0	0	0	0	0	0
	B7	A7	21	3	0	0	0	0	0
	—	—	21	3	41	0	0	0	0
	—	—	21	3	41	0	0	0	0
	—	—	21	3	62	0	0	0	0
	B8	A8	21	41	0	0	0	0	0
		A9	21	41	3	0	0	0	0
		A10	21	41	99	0	0	0	0
	—	—	21	41	48	0	0	0	0
	—	—	21	41	62	0	0	0	0
	—	—	21	62	0	0	0	0	0
	—	—	21	99	0	0	0	0	0
E6	B9	A11	41	0	0	0	0	0	0
	—	—	41	3	0	0	0	0	0
	B10	A12	41	21	0	0	0	0	0

Table 7.2 (continued)

CLASS	CLASS	CLASS	VEG1	VEG2	VEG3	HD831	HD832	HD833	ND83
E6		A13	41	21	99	0	0	0	0
	—	—	41	21	3	0	0	0	0
	—	—	41	21	62	0	0	0	0
	B11	A14	41	62	0	0	0	0	0
	—	—	41	62	21	0	0	0	0
	—	—	41	63	0	0	0	0	0
	—	—	41	63	21	0	0	0	0
	B12	A15	41	99	0	0	0	0	0
E7	—	—	62	0	0	0	0	0	0
	—	—	62	5	0	0	0	0	0
	—	—	62	21	0	0	0	0	0
	—	—	62	21	41	0	0	0	0
	B13	A16	62	41	0	0	0	0	0
	—	—	62	41	21	0	0	0	0
E8	—	—	—	—	—	1	0	0	0
	B14	A17	—	—	—	3	0	0	0
E9	B15	A18	—	—	—	4	0	0	0
	—	—	—	—	—	4	6	0	0
	—	—	—	—	—	4	7	0	0
	—	—	—	—	—	4	8	0	0
E10	B16	A19	—	—	—	5	0	0	0
	—	—	—	—	—	5	6	0	0
	—	—	—	—	—	5	6	11	0
	—	—	—	—	—	5	7	0	0
	—	—	—	—	—	5	8	0	0
	—	—	—	—	—	5	10	0	0
	—	—	—	—	—	5	11	0	0
E11	—	A20	—	—	—		0	0	0
	—	—	—	—	—	6	0	11	0
	—	—	—	—	—	6	5	0	0
	—	—	—	—	—	6	5	10	0
	—	—	—	—	—	6	5	11	0
	—	—	—	—	—	6	8	0	0
	—	—	—	—	—	6	10	0	0
	—	—	—	—	—	6	10	5	0
	—	—	—	—	—	6	11	0	0
	—	—	—	—	—	6	11	5	0
	—	—	—	—	—	6	15	0	0
E12	—	—	—	—	—	7	0	0	0
	—	—	—	—	—	7	5	0	0
	—	—	—	—	—	7	10	0	0
	—	—	—	—	—	7	11	0	0
E13	—	A21	—	—	—	8	0	0	0
	—	—	—	—	—	8	6	0	0
	—	—	—	—	—	8	11	0	0

Table 7.2 (continued)

E14	—	—	—	—	—	10	0	0	0
	—	—	—	—	—	10	0	6	0
	—	—	—	—	—	10	5	0	0
	—	—	—	—	—	10	5	6	0
	—	—	—	—	—	10	6	0	0
	—	—	—	—	—	10	6	5	0
	—	—	—	—	—	10	7	0	0
	—	—	—	—	—	10	8	0	0
E15	—	—	—	—	—	11	0	0	0
	—	—	—	—	—	11	5	0	0
	—	—	—	—	—	11	5	6	0
	—	—	—	—	—	11	6	0	0
	—	—	—	—	—	11	6	5	0
	—	—	—	—	—	11	7	0	0
	—	—	—	—	—	11	8	0	0
	—	—	—	—	—	11	12	0	0
E16	B17	A22	—	—	—	13	0	0	0
	—	—	—	—	—	13	4	0	0
	—	—	—	—	—	13	5	0	0
	—	—	—	—	—	13	6	0	0
	—	—	—	—	—	13	16	0	0
E17	—	A23	—	—	—	15	0	0	0
E18	—	—	—	—	—	0	0	0	52
E19	—	—	—	—	—	0	0	0	64
E20	—	—	—	—	—	0	0	0	99

Table 7.3 Ground Classification Schemes A, B and C

CLASS	CLASS	CLASS	VEG1	VEG2	VEG3	HD831	HD832	HD833	ND83
C1	B1	A1	1	0	0	0	0	0	0
C2	B2	A2	2	0	0	0	0	0	0
	B3	A3	3	0	0	0	0	0	0
C3	B4	A4	3	21	0	0	0	0	0
	B7	A7	21	3	0	0	0	0	0
C4	B6	A6	21	0	0	0	0	0	0
C5	B8	A8	21	41	0	0	0	0	0
		A9	21	41	3	0	0	0	0
		A10	21	41	99	0	0	0	0
	B10	A12	41	21	0	0	0	0	0
		A13	41	21	99	0	0	0	0
C6	B9	A11	41	0	0	0	0	0	0
	B12	A15	41	99	0	0	0	0	0
C7	B11	A14	41	62	0	0	0	0	0
	B13	A16	62	41	0	0	0	0	0
C8	B14	A17	—	—	—	3	0	0	0

Table 7.4 Ground Classification Scheme D

CLASS	CLASS	SURF1	SURF2	HD831	HD832	HD833
D1	—	2	0	0	0	0
D2	—	3	0	0	0	0
D3	—	4	0	0	0	0
D4	—	5	0	0	0	0
D5	—	6	0	0	0	0
D6	—	7	0	0	0	0
D7	—	10	0	0	0	0
D8	—	21	0	0	0	0
D9	—	2	14	0	0	0
D10	—	3	14	0	0	0
D11	—	4	14	0	0	0
D12	—	7	14	0	0	0
D13	—	7	6	0	0	0
D14	A17	—	—	3	0	0
D15	A18	—	—	4	0	0
D16	A19	—	—	5	0	0
D17	A22	—	—	13	0	0

Legend for Tables 7.2, 7.3 and 7.4

VEG1 Primary vegetation
VEG2 Secondary vegetation
VEG3 Tertiary vegetation
SURF1 Primary surface form
SURF2 Secondary surface form

HD831 Primary Human Disturbance 1983
HD832 Secondary Human Disturbance 1983
HD833 Tertiary Human Disturbance 1983
ND83 Natural Disturbance 1983

CODE VEGETATION CODES

1. Water
2. Aquatic grass marsh
3. Aquatic sedge marsh
5. Aquatic moss marsh
21. Wet sedge tundra
41. Moist, nontussock-sedge, dwarf-shrub tundra
62. Dry, dwarf-shrub, crustose-lichen tundra
63. Dry, dwarf-shrub, forb, lichen tundra (Dryas river terraces)
99. Barren

SURFACE FORM CODES

2. High-centered polygons, center-trough relief less than 0.5 m
3. Low-centered polygons, center-rim relief greater than 0.5 m
4. Low-centered polygons, center-rim relief less than 0.5 m
5. Mixed high-centered and low-centered polygons
6. Frost scars
7. Strangmoor and/or disjunct polygon rims (generally well-defined features visible on 1:6000 scale photographs)

10. Nonpatterned ground
14. Termokarst pits (density greater than 4 pits per 3/8 in. circle on 1:6000 scale photographs)
21. Water

HUMAN DISTURBANCE CODES

1. Gravel roads
2. Peat roads
3. Gravel pads (includes reverse and flaring pits)
4. Continuous flooding, more than 75% open water
5. Discontinuous flooding, less than 75% open water
6. Constructions-induced termokarst
7. Vehicle track-deeply rutted and/or with thermokarst
8. Vehicle track-not deeply rutted
9. Winter road
10. Gravel and construction debris (more than 75% cover)
11. Gravel and construction debris (less than 75% cover)
12. Heavy dust or dust-killed tundra
13. Excavation of river gravels or other gravel sources, roadcuts or construction excavations
14. Barren tundra caused by oil-spills, burns, blading, etc.
15. Barren tundra caused by previous flooding
16. Constructions-induced eolian deposits

NATURAL DISTURBANCE CODES

52. River channel change, flooding previously dry area
64. River channel change, exposing previously flooded vegetation
99. River channel change, exposing previously flooded gravel bar

merged using Euclidean distance (ED) thresholds. An optimal ED was determined by creating iterative groups of merged signatures, each group being derived with a different ED value (ED = 0, 1, 2, 3…).

The ground-map areas W34 (training) and W32 (validation) were then classified with each ED signature group. For area W34, the image classes were assigned to ground classes using minimum loss function criteria (Chapter 6). During the labeling procedure, the number of pixels representing each ground class was adjusted such that all classes had equal weighting (common habitat types did not have precedence over less common types).

The minimum loss function identified the optimal labeling table for attaining maximum agreement statistics between satellite image classes and the respective ground classification scheme.

The labeling table developed for W34 area was applied to W32 area. If agreement statistics with the ground data for W34 and W32 areas were dissimilar (poor stability between training and validation data), the ED was rejected. Otherwise, the ED value was used to generate image classes for all three maps combined, and the minimum loss labeling method was applied again. Multistep ED processing was used to define an optimum ED with respect to the minimum loss function criteria when classification results were compared between training sites and test sites.

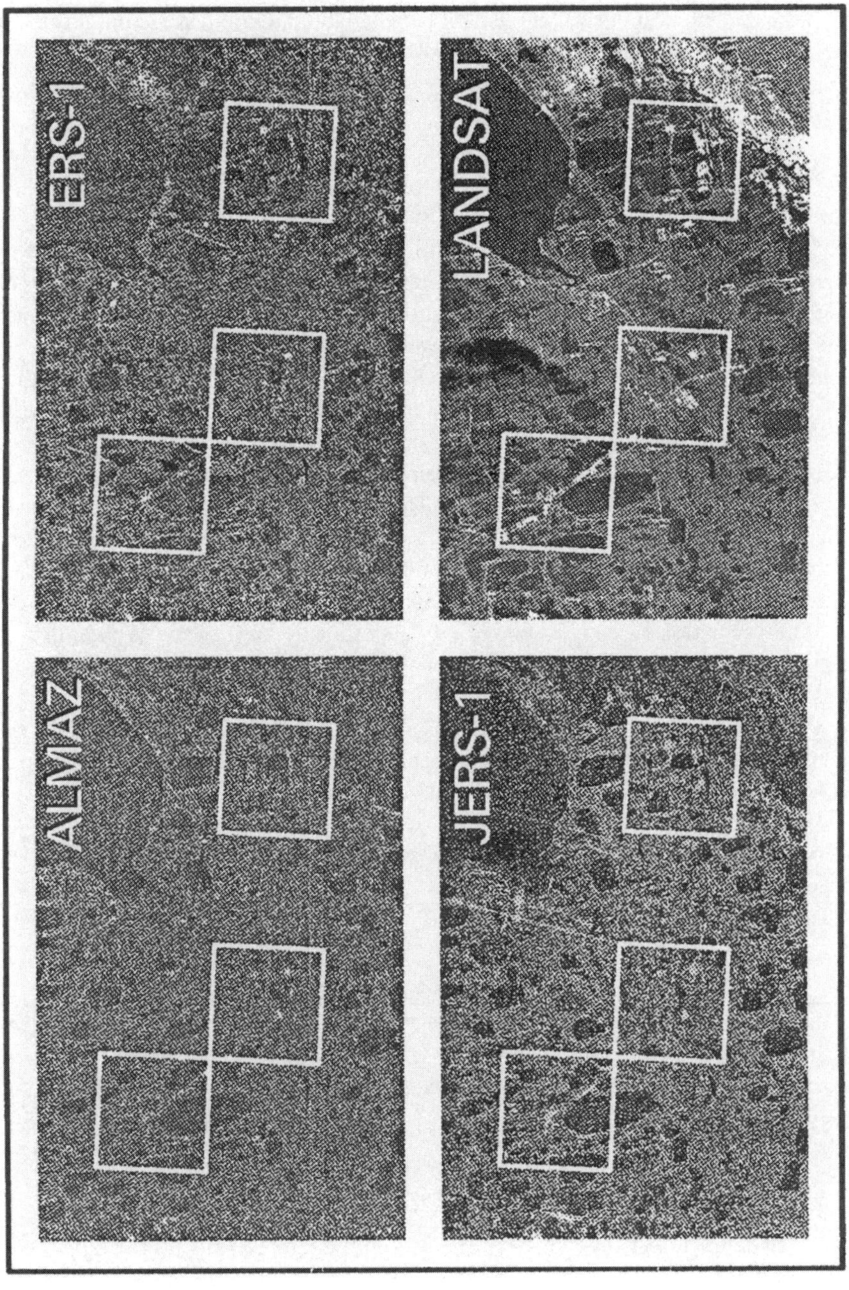

Figure 7.5 Satellite data.

Confusion matrices and Kappa statistics (Rosenfield and Fitzpatrick-Lins, 1986) were derived.

For each image database and ground-classification scheme, the comprehensive agreement statistics were compared for all accepted ED values. Results are reported for the ED value that produced the best overall agreement between the satellite-derived classification and the respective ground map.

7.6 RESULTS

The making of a modified ground-derived classification system to use with remote sensing techniques is an important problem. Geobotanical description of every tundra polygon embraces the following attributes: vegetation, surface forms, landforms, soils, percentage of open water (secondary geobotanical features were mapped when they covered more than 30% of a given map polygon). Three maps have 8584 polygons whose area is not less than nine Landsat-TM pixels. There are 50 classes having area greater than 15 ha. Each class has descriptions of vegetation, soils, etc. Our next step was to test the utility of ALMAZ-1, ERS-1 and JERS-1 data for vegetation classification. For this examination we constructed different ground-derived classification systems. The first one was the vegetation/landcover based scheme. This scheme included 23 classes and took into account the following attributes: primary, secondary and tertiary vegetation (if it covered more than 10% of a given map-polygon) and anthropogenic disturbance. Each class has a total area at three maps more than 9 ha. The vegetation legend is a modified version of Level C of the Walker (1983) hierarchical vegetation mapping classification system. The anthropogenic disturbance legend was taken from Walker (1986).

The ground-derived classification system is shown in Table 7.2. As stated above, the logic of classification algorithm allows one to change the value of ED from 0 with step 1. Preliminary analysis has shown that the optimal values are: ED = 3 (ALMAZ-1), ED = 7

Table 7.5 Accuracy Assessment of Satellite Data Classification (Scheme A, Pixel Size 10 m × 10 m)

Class	Satellite data	ALMAZ ED = 3		ERS-1 ED = 7		JERS-1 ED = 7		ERS-1+ JERS-1 ED = 5		ALMAZ+ ERS-1 + JERS-1 ED = 21		LANDSAT-TM ED = 11	
	Map pixels number	agree. %	comm. %	agree. %	comm. %	agree. %	comm. %	agree. %	comm. %	agree. %	comm. %	agree. %	comm. %
A1	117188	42.75	46.19	28.62	45.93	46.05	45.93	53.22	38.62	67.69	34.30	85.58	13.08
A6	34183	13.73	92.24	25.64	91.66	19.83	93.73	24.39	91.05	14.03	91.09	51.85	84.72
A7	23156	0.00	—	0.00	—	0.00	—	3.89	90.16	3.04	89.92	10.16	85.02
A8	89270	14.46	84.38	19.41	84.38	0.00	—	7.16	86.48	3.54	84.53	0.00	—
A11	23761	0.00	—	0.00	—	0.00	—	1.65	95.86	3.37	95.43	27.10	89.33
A17	31317	44.42	89.47	23.68	91.34	69.21	91.22	46.34	88.38	53.60	87.58	78.42	57.60
A18	29197	21.70	92.66	17.80	94.28	0.00	—	14.72	94.04	27.37	92.91	59.09	79.02
A22	16581	0.00	—	0.00	—	0.00	—	0.00	—	0.00	—	24.96	48.43
Total	454741	19.34	80.66	15.83	84.17	18.12	81.88	20.68	79.31	22.52	77.47	37.93	62.07
Kappa, %		7.22		3.67		7.81		9.39		11.42		29.92	

Table 7.6 Accuracy Assessment of Satellite Data Classification (Scheme B, Pixel Size 10 m × 10 m)

Class	Satellite data	ALMAZ ED = 3		ERS-1 ED = 8		JERS-1 ED = 7		ERS-1 + JERS-1 ED = 5		ALMAZ + ERS-1 + JERS-1 ED = 21		LANDSAT-TM ED = 12	
	Map pixels number	agree. %	comm. %	agree. %	comm. %	agree. %	comm. %	agree. %	comm. %	agree. %	comm. %	agree. %	comm. %
B1	117188	42.75	46.13	19.05	43.36	42.85	45.17	53.22	38.74	67.69	34.45	78.78	8.30
B6	34183	20.88	91.85	42.75	91.38	28.72	90.53	42.73	90.01	29.17	90.02	68.33	86.53
B7	24010	0.00	—	0.00	—	0.00	—	3.83	88.99	3.07	89.03	4.20	80.14
B8	94808	14.45	83.41	19.09	84.84	25.30	83.12	7.91	85.30	3.55	83.44	0.00	—
B9	23761	0.00	—	0.00	—	0.00	—	0.00	—	0.00	—	4.06	78.55
B14	31317	60.64	90.02	33.67	90.25	49.14	86.75	43.49	87.70	54.02	87.07	68.73	37.06
B15	29489	0.00	—	3.24	93.73	0.00	—	8.25	93.25	20.96	92.64	59.43	85.35
B17	17362	0.00	—	0.00	—	0.00	—	0.00	—	0.00	—	17.68	49.46
Total	453417	19.83	80.17	14.65	85.35	21.92	78.08	21.73	78.25	23.33	76.66	35.27	64.37
Kappa, %		7.36		2.53		7.85		10.15		12.10		27.25	

(ERS-1), ED = 7 (JERS-1), ED = 21 (ALMAZ-1 + ERS-1 + JERS-1), ED = 5 (ERS-1 + JERS-1) (pixel size for all satellite data was 10 m × 10 m).

The optimum assignment of image classes to classification scheme A (Table 7.2) is presented in Table 7.5 for each satellite database. It is initially obvious that only a fraction of the 23 ground classes received image class assignments. Because the labeling methodology served to minimize errors with respect to the ground data, results indicated that the information content of the satellite data can be optimally used to simultaneously discriminate only a subset of the 23 classes in classification scheme A. As one can see from Table 7.5, three classes A1 (water), A6 (wet sedge tundra) and A17 (gravel pads) prevail when JERS-1 data are processed, and five classes A1, A6, A8, A17 and A18 (area of continuous flooding) prevail when ALMAZ-1 data are processed. The classification accuracy for every class is: 46.05% (A1); 19.83% (A6); 69.21% (A17), and 42.75% (A1); 13.73% (A6); 14.46% (A8), 44.42% (A17), 21.70% (A18), respectively.

The same five classes prevail when ERS-1 data are processed (A1, A6, A8, A17, A18). The classification accuracy for these classes is: 28.62%, 25.64%, 19.41%, 23.68%, 17.80%, respectively. It should be noted that classification accuracy for classes A1, A17, A18 is larger for ALMAZ-1 data than for ERS-1 data, and for classes A1 and A17 it is larger for JERS-1 data than for ALMAZ-1 and ERS-1 data.

Seven classes prevail when ALMAZ-1, JERS-1 and ERS-1 data are jointly processed (A1, A6, A7, A8, A11, A17, A18) but classification accuracy for classes A7, A8, A11 is not large.

Not surprisingly, the information content of Landsat-TM was able to achieve the closest representation of scheme A.

Combining two or more SAR images improved the discrimination of water (A1) compared to the individual SAR database. The number of correctly classified pixels for ground-derived classes A9, A10, A13 is 0 for any SAR data. Hence, at the next step of analysis we do not consider classes with tertiary vegetation as a single class. Among classes of anthropogenic disturbances (A17–A23) we left only those which have an area more than 100 pixels both on map W34 and on map W32 (classes A17, A18, A19, A22).

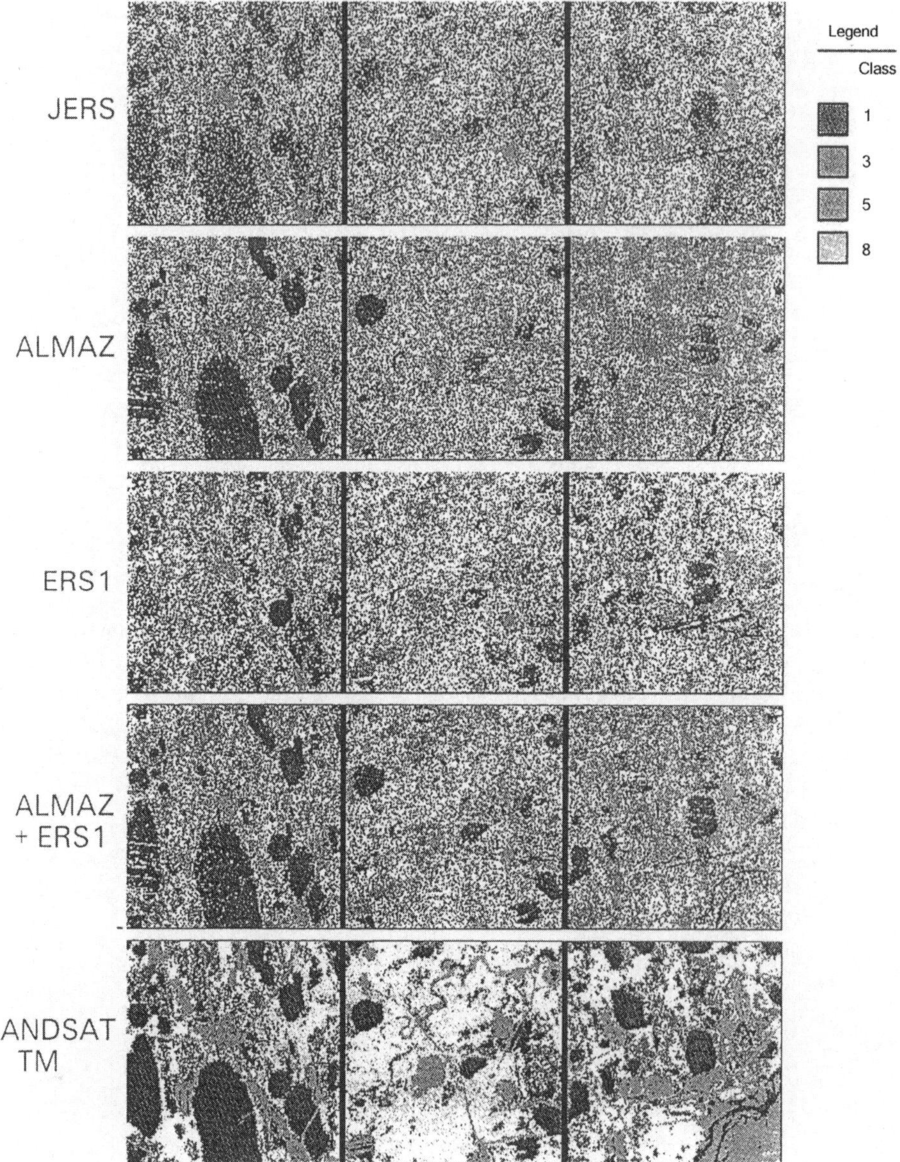

Figure 7.6 Classification scheme C. Example of classification results.

Thus, we have generated a new variant of the vegetation/landcover-based classification system — scheme B (Table 7.2). For this classification system the optimal values of ED are: for JERS-1 data ED = 7, for ALMAZ-1 data ED = 3, for ERS-1 data ED = 8, for ERS-1 + JERS-1 data ED = 5, for ALMAZ-1 + ERS-1 + JERS-1 data ED = 21. The results of efficiency assessment of JERS-1, ALMAZ-1 and ERS-1 satellite data for optimal ED are given in Table 7.6.

The class numbers corresponding to the classification system for 23 classes are given in Table 7.2. Comparing Table 7.6 with Table 7.5 exposes the growth of classification accuracy for vegetation class A6 (wet sedge tundra). This increase is considerable for ERS-1 data, about 17%. Distinguishing classes for radar data (B1, B6, B7, B8, B9, B14, B15) have the greatest areas on three maps plots. The classification accuracy of class B1 (water) is the same as in Table 7.5 for ALMAZ-1, ERS-1 + JERS-1 and ALMAZ-1 + ERS-1 + JERS-1 data.

Analysis of Table 7.6 gives rise to an assumption that two classes do not distinguish if the first one has type A primary vegetation and type B secondary vegetation and the second one vice versa. In this case, we combined similar classes and got Table 7.3 where class C1 — water, class C2 — aquatic graminoid march, class C3 — aquatic/wet sedge complex, class C4 — wet sedge tundra, class C5 — wet/moist tundra complex, class C6 — moist sedge tundra, class C7 — moist/dry tundra complex and class C8 — gravel roads and pads (Figure 7.6). For this classification system the optimal values of ED are: for JERS-1 data ED = 7, for ALMAZ-1 data ED = 3, for ERS-1 data ED = 4, for ALMAZ-1 + ERS-1 + JERS-1 data ED = 21, for ERS-1 + JERS-1 data ED = 4. The results of the efficiency assessment of ALMAZ-1, ERS-1 and JERS-1 satellite data for optimal ED are given in Table 7.7.

In the framework of classification scheme C, optimal separation was observed for a subset of four classes (Table 7.3): water (C1), wet/aquatic sedge complex (C3), wet sedge and moist-sedge shrub tundra complexes (5) and gravel pads (C8).

Comparative analysis of Table 7.6 and Table 7.7 showed a classification possibility for three vegetation classes: wet sedge tundra (class B6, Table 7.6), wet/aquatic sedge complex (class C3, Table 7.7), moist/wet tundra complex (class C5, Table 7.7). Overall accuracies for Table 7.7 are considerably higher than for Tables 7.5 and 7.6: 44.82% for JERS-1 data classification; 38.93% for ALMAZ-1 data classification; 40.59% for ERS-1

Table 7.7 Accuracy Assessment of Satellite Data Classification (Scheme C, Pixel Size 10 m × 10 m)

Class	Satellite data	ALMAZ ED = 3		ERS-1 ED = 4		JERS-1 ED = 7		ERS-1+ JERS-1 ED = 4		ALMAZ + ERS-1 + JERS-1 ED = 21		LANDSAT-TM ED = 11	
	Map pixels number	agree. %	comm. %	agree. %	comm. %	agree. %	comm. %	agree. %	comm. %	agree. %	comm. %	agree. %	comm. %
C1	117188	78.46	48.74	71.02	49.81	78.25	47.01	78.57	39.34	83.33	36.88	92.25	13.23
C3	40202	32.64	84.61	29.84	79.52	9.74	86.94	34.90	80.79	40.09	81.24	39.14	73.10
C4	34186	1.26	90.52	0.44	91.29	0.64	90.63	0.91	89.12	1.34	89.54	3.21	89.49
C5	117666	35.63	62.63	49.75	62.26	60.03	59.07	53.90	57.78	51.16	55.72	75.71	45.19
C8	31317	10.23	40.83	10.24	43.85	22.20	21.51	20.73	20.41	26.53	21.26	74.58	15.55
Total	386908	38.93	61.06	40.59	59.41	44.82	55.17	45.06	54.73	44.71	55.26	61.36	38.14
Kappa, %		18,17		18,72		23.20		25.87		25.86		48.54	

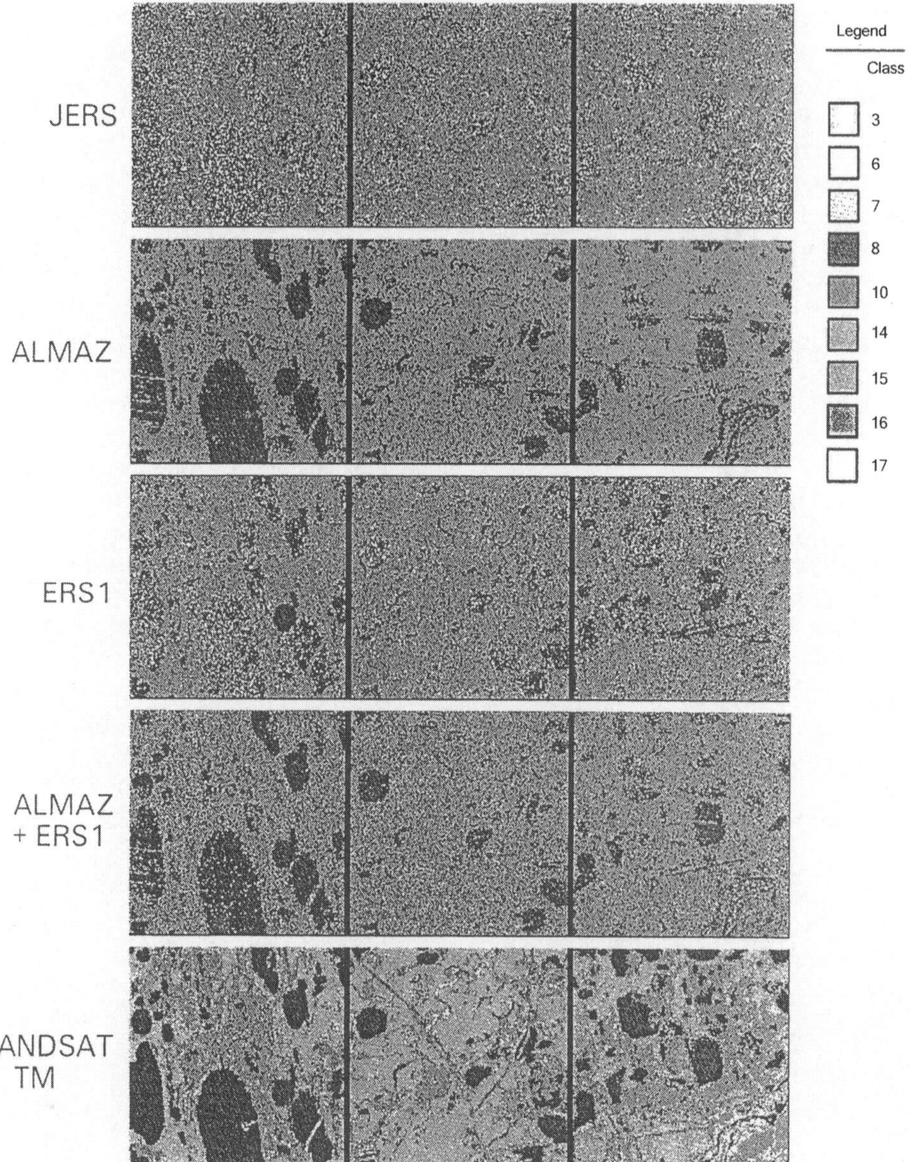

Figure 7.7 Classification scheme D. Example of classification results.

data classification, 44.71% for ALMAZ-1 + ERS-1 + JERS-1 data classification and 45.06% for ERS-1 + JERS-1 data classification. These accuracies have increased by 20%–25% in comparison with 17 class classifications. One can note an increase in the classification accuracies of water class (22%–43%). The classification accuracy for class A1 (water) is highest for ALMAZ-1 + ERS-1 + JERS-1 data processing, for class C3 (wet/aquatic sedge complex) — for the same data processing and for class C5 (moist/wet tundra complex) — for JERS-1 data.

Scheme C contained more generalized ground classes (compared with scheme A and B) that appeared to have a closer correspondence with the information content of the satellite database. Kappa statistics indicated that the improved agreements were greater than those expected by chance due to the smaller number of classes in scheme C. The subset of four classes with optimal separability occurred along a moisture gradient, ranging from water to gravel pads. Between these two endpoints, the satellite data showed a potential for separating vegetation communities associated with very wet areas from those occurring in drier moisture regimes. Results of classification for optimal ED are presented in Figure 7.6.

In order to estimate the utility of SAR data for tundra vegetation classification we considered an additional classification scheme based on the surface form (Table 7.4).

Surface forms are described according to the legends used in the North Slope Borough's geographic information system (GIS) (Walker, 1986). The results of efficiency assessment of ALMAZ-1, ERS-1 and JERS-1 satellite data for optimal ED for the classification system based on the surface form are given in Table 7.8 and are shown in Figure 7.7.

Analysis of the results in Table 7.8 shows that SAR data classification accuracy for classes D14 and D15 is higher than for these classes in the vegetation-based classification system. Among the surface form classes in scheme D, the SAR data were again best suited for discriminating water (class D8) and gravel pads (D14). Less successful SAR separations were aligned at detection of non-textured surfaces (D7), strangmoor (D6), or areas of continuous flooding (D17). Results of classification for optimal ED are presented in Figure 7.7.

Table 7.8 Accuracy Assessment of Satellite Data Classification (Scheme D, Pixel Size 10 m × 10 m)

Class	Satellite data	ALMAZ ED = 3		ERS-1 ED = 0		JERS-1 ED = 7		ERS-1 + JERS-1 ED = 4		ALMAZ + ERS-1 + JERS-1 ED = 21		LANDSAT-TM ED = 11	
	Map pixels number	agree. %	comm. %	agree. %	comm. %	agree. %	comm. %	agree. %	comm. %	agree. %	comm. %	agree. %	comm. %
D3	27325	0.00	—	0.00	—	0.00	—	4.26	89.69	10.28	89.35	71.93	84.13
D6	69769	13.37	87.95	8.44	91.41	9.30	87.83	4.48	89.60	1.70	87.47	2.36	60.65
D7	45086	13.71	90.07	27.51	89.47	21.18	90.33	14.10	89.91	6.96	90.52	40.23	85.58
D8	127776	42.31	44.60	30.00	41.78	44.62	43.40	56.90	36.86	65.51	31.60	80.90	11.21
D10	25996	0.00	—	0.00	—	0.00	—	0.00	—	0.00	—	3.81	91.83
D14	31317	62.35	90.47	47.45	91.44	41.28	86.75	44.56	88.02	49.91	86.66	72.62	43.16
D15	29489	3.40	92.77	6.10	94.03	18.17	94.93	23.10	94.58	35.46	93.69	19.93	74.30
D17	17362	0.00	—	0.00	—	0.00	—	0.00	—	0.00	—	17.90	46.61
Total	456048	19.76	80.24	16.07	83.93	20.03	79.97	22.18	77.80	23.03	76.95	38.50	61.08
Kappa, %		7.53		4.69		7.99		10.42		11.51		29.49	

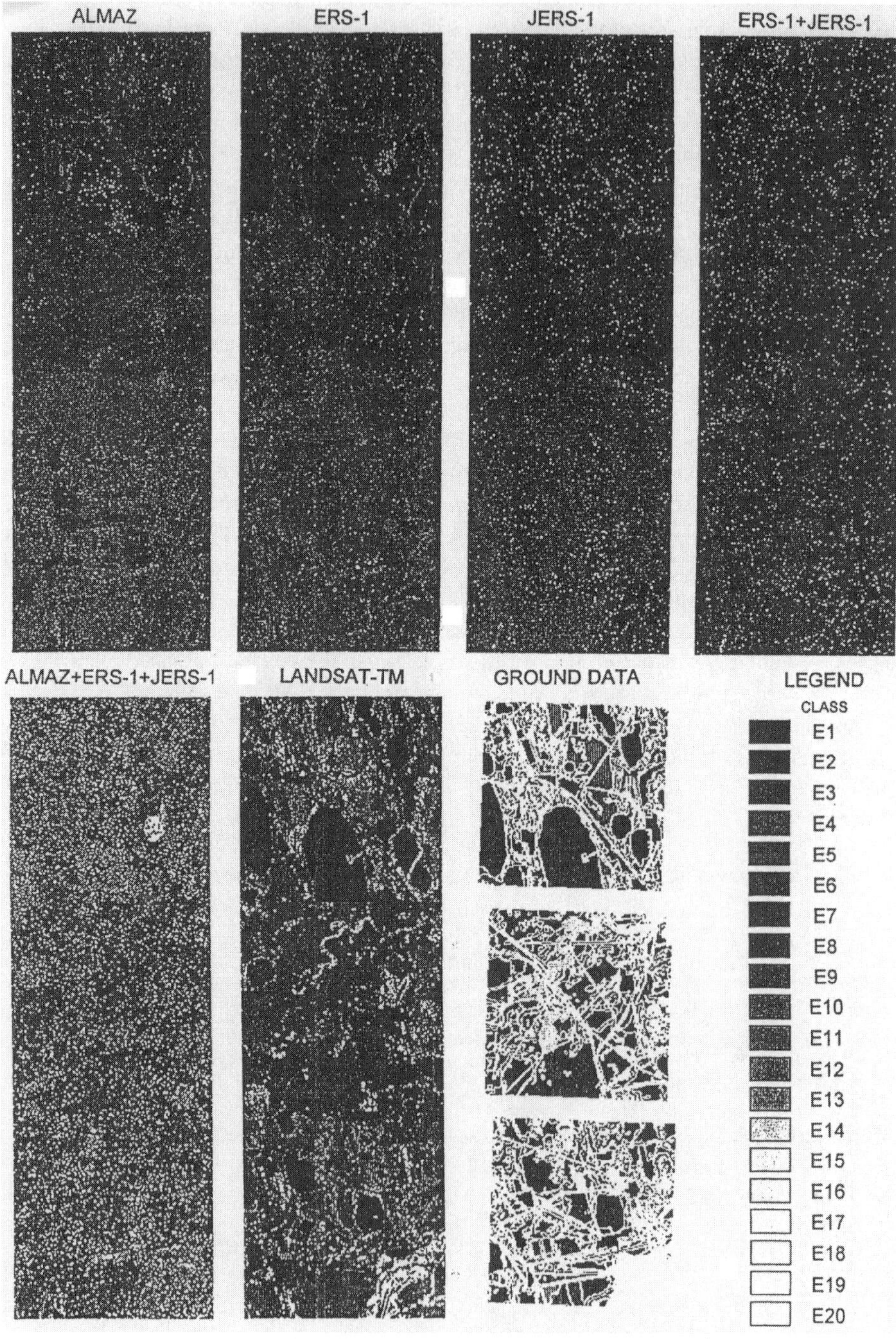

Figure 7.8 Classification scheme E. Example of classification results.

Table 7.9 Accuracy Assessment of Satellite Data Classification (Scheme A, Pixel Size 30 m×30 m)

Class	Satellite data	ALMAZ ED = 3		ERS-1 ED = 6		JERS-1 ED = 6		ERS-1 + JERS-1 ED = 5		ALMAZ + ERS-1 + JERS-1 ED = 24		LANDSAT-TM ED = 12	
	Map pixels number	agree. %	comm. %	agree. %	comm. %	agree. %	comm. %	agree. %	comm. %	agree. %	comm. %	agree. %	comm. %
A1	10157	79.23	23.05	67.05	38.70	63.44	42.58	72.45	29.73	74.45	24.49	93.00	6.11
A4	836	0.0	—	0.0	—	0.0	—	2.27	81.90	0.72	82.35	1.32	66.67
A6	2046	0.0	—	29.57	91.59	0.0	—	23.26	88.73	32.31	90.08	34.65	86.99
A7	1608	0.0	—	0.0	—	0.0	—	5.91	85.25	4.42	85.39	25.93	81.48
A8	6534	8.71	81.00	13.09	81.03	19.94	82.03	10.76	81.54	0.0	—	59.55	49.23
A11	1142	0.0	—	0.0	—	0.0	—	0.26	93.18	0.18	95.12	3.24	87.11
A12	834	0.0	—	0.0	—	0.0	—	0.24	50.00	0.0	—	2.16	53.85
A17	3407	35.10	42.86	21.43	83.87	37.60	58.04	37.80	73.69	49.25	70.61	73.08	39.50
A18	2369	63.91	91.65		93.32	34.70	93.01	22.96	92.57	24.69	90.95	40.48	48.41
A22	2038	0.0	—	0.0	—	0.0	—	8.64	87.79	0.0	—	48.14	41.29
Total	33750	33.56	66.36	27.89	71.93	29.19	70.40	31.62	67.63	31.48	64.52	56.19	43.29
Kappa, %		21.43		15.03		14.14		18.95		21.01		47.46	

Table 7.10 Accuracy Assessment of Satellite Data Classification (Scheme E, Pixel Size 30 m×30 m)

Class	Satellite data	ALMAZ ED = 4		ERS-1 ED = 6		JERS-1 ED = 7		ERS-1 + JERS-1 ED = 7		ALMAZ + ERS-1 + JERS-1 ED = 19		LANDSAT-TM ED = 10	
	Map pixels number	agree. %	comm. %	agree. %	comm. %	agree. %	comm. %	agree. %	comm. %	agree. %	comm. %	agree. %	comm. %
E1	10164	94.78	58.83	95.53	61.52	91.87	61.42	93.83	51.85	72.45	42.39	95.69	8.69
E5	12419	42.84	48.98	31.82	49.31	46.59	46.02	59.10	43.67	64.46	43.72	90.72	32.85
E6	3387	0.0	—	0.0	—	0.0	—	0.59	93.36	0.38	86.02	13.14	63.94
E8	3425	41.28	58.79	30.42	75.58	30.60	45.07	34.13	62.98	47.39	71.82	71.82	32.99
E9	2380	0.0	—	0.0	—	0.0	—	1.64	89.87	0.97	91.51	50.84	55.00
E13	30	0.0	—	0.0	—	0.0	—	0.0	—	3.45	33.33	0.0	—
E16	2252	0.0	—	0.0	—	0.0	—	1.38	78.32	0.98	89.16	51.69	41.57
Total	37515	43.63	56.07	39.19	60.58	43.11	56.09	48.38	50.78	45.46	48.86	70.03	29.03
Kappa, %		22.85		17.67		21.64		29.55		28.91		60.32	

Landsat-TM data discriminated a similar subset of surface form classes with higher overall agreement.

The optimum assignment of image classes to classification schemes A and E (Table 7.2) for a variant of pixel size 30 m×30 m is given in Table 7.9 for scheme A and in Table 7.10 for scheme E for each satellite database. Reanalyses of the combined ERS-1, JERS-1 and ALMAZ-1 database, using a 30 m×30 m pixel size, showed an essential improvement for all classification schemes, indicating that superior results of merging SAR data can be attained if speckle noise is reduced (Tables 7.9, 7.10). The most effective classification was for classification scheme E for image data sets with pixel size 30 m×30 m. Results of classification for optimal ED are shown in Figure 7.8. The information content of Landsat-TM was able to achieve the closest representation of schemes A and E (Tables 7.9 and 7.10).

7.7 CONCLUSION

This chapter demonstrates comparative abilities between the satellite systems only in the context of the analytical methodologies that were applied. Alternative classification methods may also provide more efficient ways to extract SAR information, such as second-order (Barber et al., 1993) or neural network classifiers (Hepner et al., 1990) which consider textural properties of image data.

The results of this chapter indicate that SAR data have a potential for discriminating up to four or five geobotanical classes in wet tundra regions. The relatively low agreement between satellite-derived classes and the ground data should not be interpreted as an exact measure of SAR proficiency. It is important to note that the "ground-truth model" contained an unknown degree of map error as well as temporal disparity with the satellite images.

The methods used to select an optimum ED for merging the maximum number of image signatures was based on subsets of the full study area (W34, W32). This strategy identified an ED that was better suited for classifying "unknown" areas (W22 or elsewhere). Had all ground data from all three maps been used to optimize the ED, without validating the results with independent data, agreements statistics would have improved but the results would have relevance only within the region of ground data. The total area of each ground class was weighted so all ground classes had equivalent consideration with respect to image-class assignment. Without the weighting, more extensive map classes would tend to dominate the assignments, which would tend to improve agreement statistics at the expense of higher commission errors.

The redundancy of geobotanical description for SAR data gives rise to significant problems in constructing ground-derived classification systems for tundra vegetation. Hence statistical methods elaboration for quantitative estimation of training area numbers and for sample sites elimination whose spectral characteristics distinguish strongly is necessary.

Methods used to "focus" the raw SAR signal data (image synthesis) can also influence the detection capabilities of various environmental targets (Curlander and McDonough, 1991). A full-resolution (Level 1B) SAR image is synthesized at its respective processing center using a fixed set of focusing parameters. "Standard" SAR products may or may not represent the optimal image synthesis for a given application. Research is needed to evaluate the effects of different SAR focusing method with respect to the detection capabilities of various environmental targets. The topic of SAR focusing is addressed in the following chapter.

ALMAZ-1, ERS-1 and JERS-1 satellite systems have lower efficiency in comparison with Landsat-TM data for tundra vegetation classification but they provide systematic all-weather observation of vegetation components that are significant when examining tundra habitat in global change conditions.

REFERENCES

Aronoff, S. A., 1989. Approach to optimized labeling of image classes. *Photogram. Eng. and Rem. Sensing*, 50 (6), pp. 719–727.

Barber, D. G., Shokr, M. E., Fernandes, R. A., Soulis, E. D., Flett, D. G. and LeDrew, E. F., 1993.

A comparison of second-order classifiers for SAR sea ice discrimination. *Photogram. Eng. and Rem. Sensing*, 59 (9), pp. 1397–1408.

Belchansky, G. I., Ovchinnikov, G. K. and Douglas, D. C., 1995. Comparative evaluation of ALMAZ, ERS-1, JERS-1, and Landsat-TM for discriminating wet tundra habitats. *Polar Record*, 31 (177), pp. 161–168.

Curlander, J. C. and McDonough, R. N., 1991. Synthetic aperture radar: systems and signal processing. John Wiley & Sons, INC. New York.

Durand, J. M., Gimonet, B. J., Perbos, J. R., 1987. SAR data filtering for classification. IEEE Transactions on geoscience and remote sensing, V. GE-25, No. 5.

Jorgenson, J. C., Douglas, D. C., and Raynolds, M. K., 1991. Comparison and implementation of classified vegetation maps derived from Landsat -TM and Spot imagery data based for delineating wildlife habitat availability and distribution. AFWRC INTERIM REPORT – 1988–1990. 1002 terrestrial research. Alaska Fish and Wildlife Research Center. Anchorage, pp. 119–129.

Hepner, G. F., Logan, T., Ritter, N. and Bryant, N., 1990. Artificial neural network classification using a minimal training set-comparison to conventional supervised classification. *Photogram. Eng. and Rem. Sensing*, 56 (4), pp. 469–473.

Rosenfield, G. H. and Fitzpatrick-Lins, K., 1986. A coefficient of agreement as a measure of thematic classification accuracy. *Photogram. Eng. and Rem. Sensing*, 52 (2), pp. 223–227.

Stenback, J., and Congalton, R., 1990. Using Thematic Mapper imagery to examine forest understory. *Photogram. Eng. and Rem. Sensing*, 56(9), pp. 1285–1290.

Walker, D. A., 1983. A hierarchical tundra vegetation classification especially designed for mapping in northern Alaska. Fourth International Conference Proceedings, University of Alaska. Fairbanks. Alaska. National Academy Press. Washington, DC, USA, pp. 1332–1337.

Walker, D. A., Webber, P. J., Walker, M. D., Lederer, N. D. et al., 1986. Use of geobotanical maps and automated mapping techniques to examine cumulative impacts in the Prudhoe Bay Oil field. Alaska. Environmental Conservation 13 (2), pp. 149–160.

Way, J. et al., 1990. The effect of changing environment conditions on microwave signatures of forest ecosystems: preliminary results of the March 1988 Alaskan aircraft SAR. *Int. J. Remote Sensing*, Vol. 11, pp. 1119–1144.

8

Dependence Between SAR Data Focusing Parameters and Efficiency of Discriminating Tundra Habitat

8.1 INTRODUCTION

One problem that arises in the context of environmental research is selection of an appropriate scale of measure and accuracy to detect, quantify or monitor a topic of interest. For satellite remote sensing studies, the problem becomes a matter of applying a temporal and spatial resolution that is compatible with the dynamics and extent of the environmental parameters under investigation. Problems associated with attaining suitable temporal resolution have been reduced by the all-weather data collection capabilities of SAR systems. For SAR, the question of spatial resolution and accuracy can be addressed, in part, by the methodology used to process raw data.

A first SAR processing parameter, called a "number of looks", specifies a number (N) of subbeams combined to form the SAR image, resulting in coarser resolution (by a factor of N) and lower speckle variance (by a factor of $1/N$) (Curlander and McDonough, 1991).

A second SAR processing parameter, called a "window function", is used to compensate ambiguous information. Several window functions that use different weightings have been developed to address the ambiguity problem (Harris, 1978).

The ground-based SAR processing facilities of ALMAZ-1, ERS-1, JERS-1 and RADARSAT satellite data use a fixed set of data-focusing parameters (number of looks and window function) for synthesizing radar images. These parameters were chosen to provide an image product with robust qualities for the general multidisciplinary scientific community. However, for a specific ecological application, these parameters may not be optimum from the viewpoint of appropriate spatial resolution and accuracy of classification of the landscape types.

Therefore, one of the objectives of this chapter was to demonstrate an example of developed IBM PC/AT-compatible SAR processing software that could be used to evaluate the efficiency of different data-focusing parameters in the context of a researcher's specific ecological application.

8.2 BASIC FOCUSING ALGORITHM

The primary objective of this chapter is to demonstrate PC/AT-compatible software for focusing SAR data. The basic focusing algorithm includes two blocks of compression: range compression

$$d(s,x) = F_s^{-1}[H_2^{-1}(f_s)F_s[e(s,x)]], \tag{8.1}$$

and azimuth compression

$$\sigma(s,x) = F_x^{-1}[H_1^{-1}(f_x|s)F_x[d(s,x)]], \tag{8.2}$$

where F_s, F_x (F_s^{-1}, F_x^{-1}) denote a linear direct (inverse) Fourier transform with respect to variables s (range) and x (azimuth); $e(s, x)$ is a complex reflected signal (hologram); $d(s, x)$ is the result of range compression; $\sigma(s, x)$ is the complex image; $H_1^{-1}(f_x|s)$ and $H_2^{-1}(f_s)$ are the frequency transfer functions of two matched filters, azimuth and range (Wu and Jin, 1982; Fitch, 1988; Curlander and McDonough, 1991).

Resolution of a matched filter and, consequently, spatial resolution of synthesized gain pattern over the respective coordinate (s or x) is defined by its bandwidth. By changing the bandwidth, the user can control the SAR spatial resolution. The use in operations (8.1)–(8.2) of matched filters with a maximal frequency band allows the user to obtain the best spatial resolution ($\delta_{S\min} \times \delta_{X\min}$).

Apart from spatial resolution, an important characteristic of $H_1^{-1}(f_x|s)$ and $H_2^{-1}(f_s)$ filters is the structure and magnitude of synthesized gain pattern side lobes, which can be controlled by real weighting functions (Harris, 1978) repeatedly incorporated into the frequency transfer functions of the matched filters.

The Doppler parameters $f_c(s)$ (Doppler centroid) and $f_r(s)$ (rate of Doppler frequency change) entering into the expression for the frequency transfer function of azimuth filter $H_1^{-1}(f_x|s)$ are usually defined through the known parameters of orbital movement and orientation of the platform, however, these parameters may be unavailable or insufficiently correct. That is why the algorithm includes two procedures for estimating $f_c(s)$ and $f_r(s)$ from the reflected signal $e(s, x)$.

The block scheme of the basic algorithm is presented in Figure 8.1. Each block realizes the following function:

- "range compression with $\delta_{S\min}$ resolution" is operation of range compression with a best resolution, $d(s,x) = F_s^{-1}[H_2^{-1}(f_s)F_s[e(s,x)]]$
- "estimation of $f_c(s)$" is Madsen procedure for estimating the Doppler centroid (Madsen, 1989)
- "estimation of $f_r(s)$ and azimuth compression with $\delta_{S\min}$ resolution" are operations of estimation of the Doppler frequency change rate and azimuth compression with a best resolution, $\sigma(s,x) = F_x^{-1}[H_1^{-1}(f_x|s)F_x[d(s,x)]]$

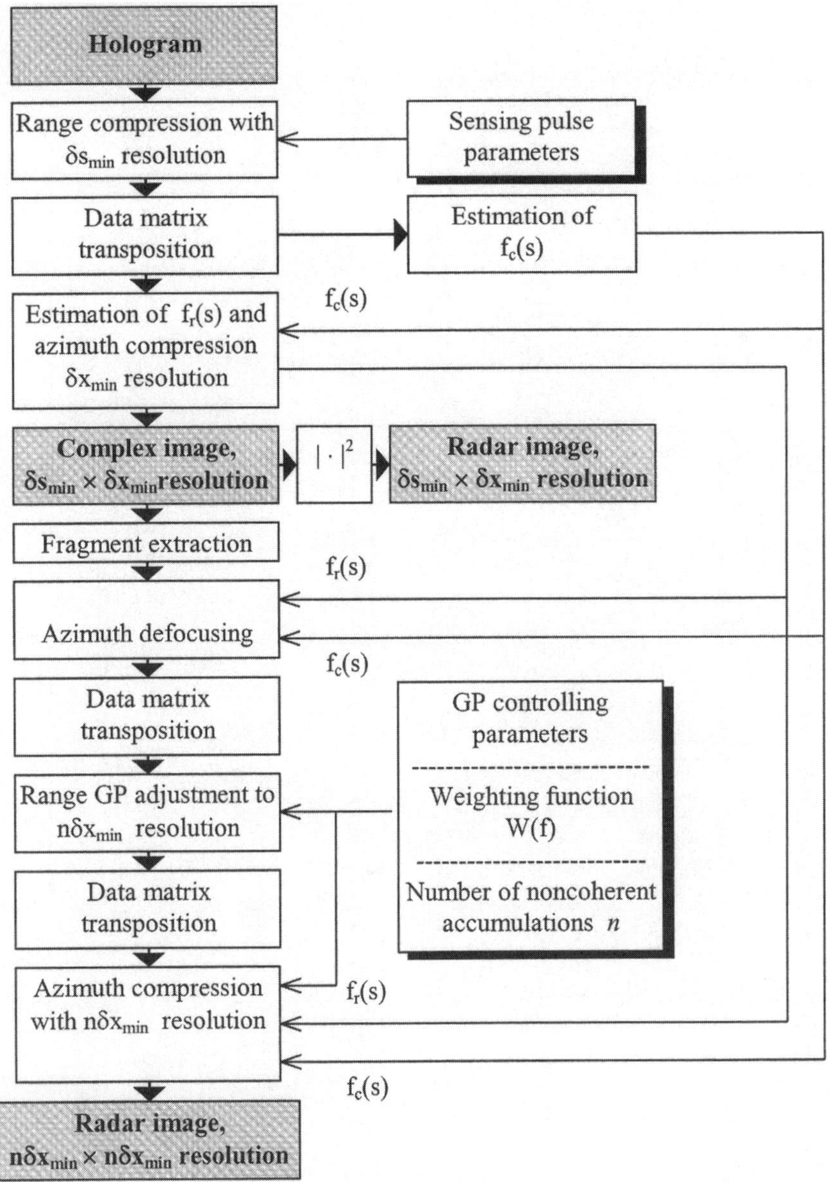

Figure 8.1 Block scheme of the basic algorithm for SAR data focusing.

- "$|\cdot|^2$" is transformation of the complex image into RI, $\sigma^o(s, x) = |\sigma(s, x)|^2$, where $\sigma^o(s, x)$ is RI
- "fragment extraction" is an extraction of the complex image fragment for its controlled focusing
- "azimuth defocusing" is an operation inverse (with an accuracy to Fourier transform) to azimuth compression with a best resolution, $d'(s, f_x) =$

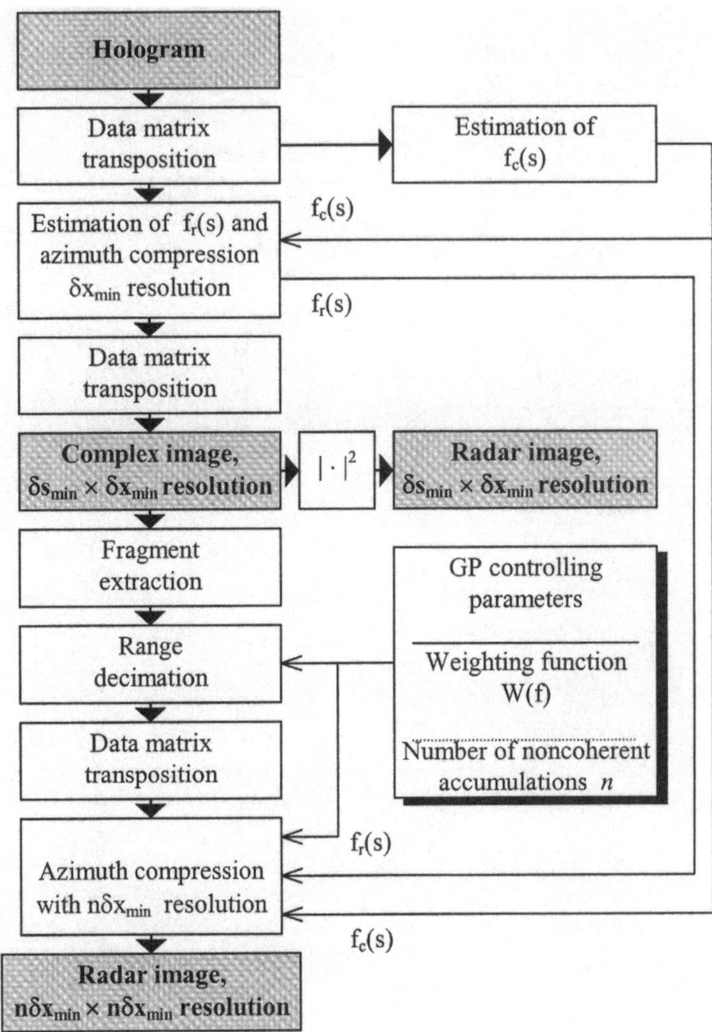

Figure 8.2 Block scheme of algorithm for ALMAZ-1 SAR data focusing.

$[H_1^{-1}(f_x|s)]^* F_x[\sigma(s,x)]$, where * is the sign of the complex matching

- "range synthesized gain pattern (GP) adjustment providing $n\delta x_{\min}$ resolution" is range GP adjustment by multiple incorporation into the transfer function $H_2^{-1}(f_s)$ of the weighting function $W^n(f_s)$ tuned to the frequency band of the width B_x/n, $d''(s,f_x) = F_s^{-1}[W^n(f_s)F_s[d'(s,f_x)]]$, where B_x is the frequency bandwidth of the Doppler spectrum
- "azimuth compression with $n\delta x_{\min}$ resolution" is azimuth compression with a flexible GP adjustment using the method of noncoherent accumulations (Curlander and McDonough, 1991) for suppression of speckle noise and multiple incorporation of weighting functions $W_i^n(f_x^n)$ tuned to respective frequency bands

of the width B_x/n, into the transfer functions of partial azimuth filters $H_{1i}^{-1}(f_x|s)$ $(i = 1,..., n)$

$$\sigma^0(s,x) = (1/n)\sum_{i=1}^{n}\left|F_x^{-1}[W_i^n(f_x)H_1^{-1}(f_x|s)d''(s,f_x)]\right|^2;$$

- "data matrix transposing" is an operation of two-dimensional matrix transposing in accordance with the multipass Eklundth algorithm (Eklundth, 1972) modified by adding one more passage to enable transposing of a rectangular matrix (the classic Eklundth algorithm transposes only square matrices).

It should be noted that the processed matrix size must be powers of two. This restriction allows the discrete Fourier transform and transposition to be efficiently realized.

8.3 FOCUSING ALGORITHM FOR ALMAZ-1 SAR DATA

A block scheme of the focusing algorithm for the ALMAZ-1 SAR data is presented in Figure 8.2. This algorithm differs from the basic one (Figure 8.1) in that it lacks range focusing, while including an additional block of "range decimation" that matches the range size of SAR pixel to its azimuth size by means of complex image interpolation. This difference is due to the absence of intrapulse modulation in the ALMAZ-1 SAR. The other feature of the ALMAZ-1 SAR is that the reflected signal counts $e_a(s_i, x_j)$ are, in this case, the real values rather than complex ones, since the phase of reception after sending a consecutive sensing pulse increases by $\pi/2$ and the recurrent rate of sensing pulses is twice that of the nominal width of the Doppler band. For restoration of the complex counts from the real ones, the user can employ the following relation:

$$e(s_k, x_l) = \begin{cases} e_a(s_k, x_{2l}) - je_a(s_k, x_{2l+1}), & l \text{ is even} \\ -e_a(s_k, x_{2l}) + je_a(s_k, x_{2l+1}), & l \text{ is odd} \end{cases}. \tag{8.3}$$

8.4 FOCUSING ALGORITHM FOR ERS-1, JERS-1, RADARSAT SAR DATA

The focusing algorithm for the ERS-1, JERS-1, RADARSAT SAR data uses, as initial data, a complex image synthesized by focusing with the best spatial resolution (standard product of the Alaska SAR Facility, Olmsted 1993), which substantially reduces computations. The algorithm adjusts the synthesized gain pattern (GP) to a new azimuth and range resolution, the method of noncoherent accumulations being used for suppressing speckle noise during azimuthal adjustment. The block scheme of this algorithm is shown in Figure 8.3.

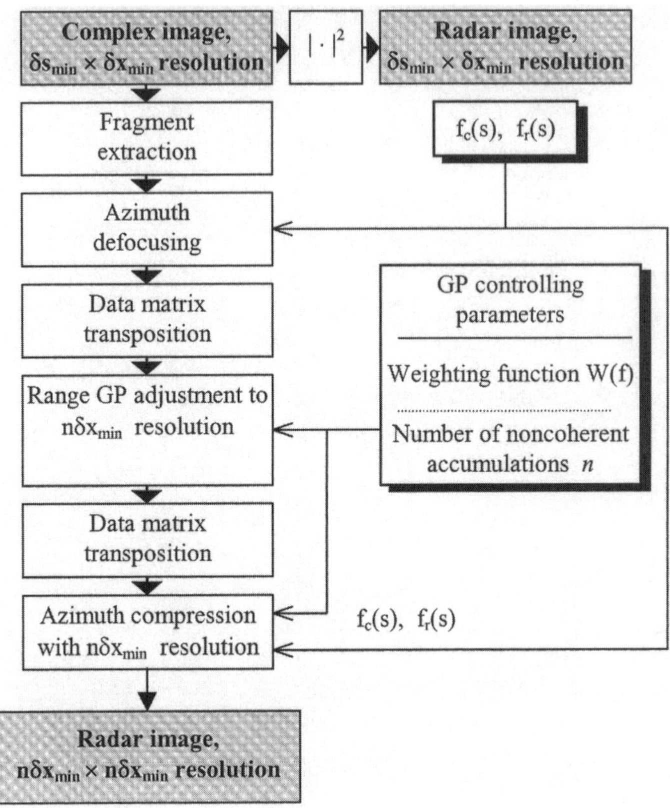

Figure 8.3 Block scheme of algorithm for ERS-1, JERS-1, RADARSAT SAR data focusing.

8.5 STRUCTURE OF FOCUSING SOFTWARE

The software for the SAR data focusing is realized in a C++ language and consists of two modules: one for focusing ALMAZ-1 SAR data and one for focusing ERS-1, JERS-1, RADARSAT SAR data. Both modules include a set of program modules realizing the functional blocks of respective block schemes (Figures 8.2 and 8.3). The program modules of data-focusing software are presented in Table 8.1.

The program modules (Table 8.1) are realized using two preliminary developed libraries: one for digital processing of signals (DPS) and one for data access. The DPS library maintains a basic set of DPS operations with linear real and complex data, while the access library maintains a basic set of operations of access to two-dimensional data matrices.

The sizes of DPS and access libraries are 1870 and 990 rows of the initial code, respectively. The sizes of focusing modules for ALMAZ-1 SAR and ERS-1, JERS-1, RADARSAT SAR data are 1380 and 1000 rows of initial code, respectively.

Firmware Medium. Software for SAR data focusing is IBM PC/AT-compatible and operates in a WINDOWS 3.X medium. All computations are floating-point of double

Table 8.1 Program Modules of Data-focusing Software

Module
Module for Doppler centroid estimation
Module for Doppler frequency change estimation and azimuth compression with δx_{min} resolution
Module for range compression with δs_{min} resolution
Module for fragment extraction from data matrix
Azimuth defocusing module
Module for range GP adjustment to $n\delta x_{\mathrm{min}}$ resolution
Module for azimuth compression with $n\delta x_{\mathrm{min}}$ resolution
Range decimation module
Module for complex image conversion to radar image

accuracy. The working memory must be not less than 2 Mb, while the external memory can be from 100 to 1000 Mb depending on data size. The recommended processor is not smaller than 386/387.

8.6 ASSESSMENT OF DEPENDENCE BETWEEN SAR DATA FOCUSING PARAMETERS AND TUNDRA HABITAT CLASSIFICATION

The primary goals of this chapter were to elaborate the integrated database of satellite and ground polygon data, to construct modified ground-derived classification schemes and to investigate the dependence between SAR data focusing parameters and tundra landcover and terrain-type classification accuracy for different ground-derived classification schemes.

8.6.1 PC/AT SAR Processor

PC/AT-compatible software was developed to process ALMAZ-1 raw data and ERS-1 complex images with user-defined options for data-focusing. The focusing algorithm based on a matched filtering approach includes range compression and azimuth compression blocks but does not perform any compensation for range migration (it does not affect the image quality because the ALMAZ-1 range migration is negligible and compensation for the ERS-1 range migration is performed by the ASF SAR processor when a complex image is generated). The clutterlock and autofocuse methods (Curlander and McDonough, 1991) are used for specifying the Doppler parameters precisely. The algorithm is able to change spatial resolution by means of the multilook technique. It also provides a set of window functions for flexible adjustment of the synthesized gain pattern (Harris, 1978). Range compression was eliminated for ALMAZ-1 data-focusing because the ALMAZ-1 radar pulse is not frequency modulated.

A post-processing software module was developed that optionally applies a variety of adaptive filters (Median, Frost, Lee and Sigma) for suppressing speckle noise (Durand et al., 1987).

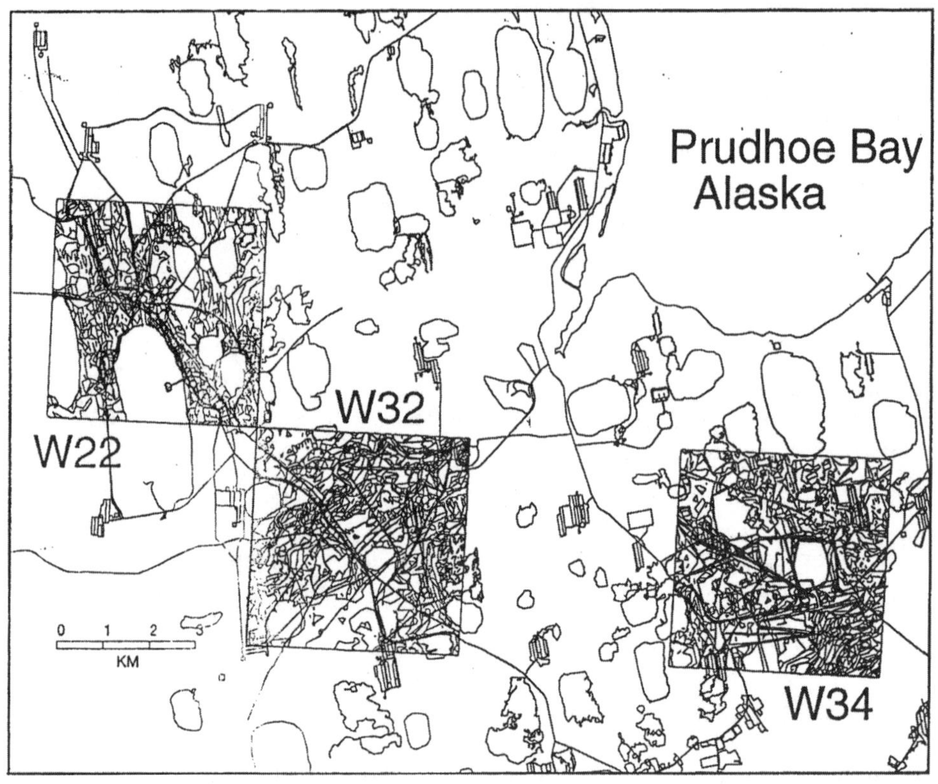

Figure 8.4 Study area.

8.6.2 Study Area

A 21 km^2 area of the Prudhoe Bay region on the northern coastal plain of Alaska was used to evaluate the effects of SAR data-focusing parameters on the efficiency of discriminating landcover and terrain classes (Figure 8.4). The study area W34 was mapped at 1:6000 scale to a geobotanical classification scheme by Walker et al. (1986). Map W34 contains three basic types of data: geobotanical, historical natural disturbances and historical anthropogenic disturbances. Four vegetation/terrain classification schemes, adapted from previously published (Walker et al., 1986) 1:6000 scale geobotanical maps, were used as "ground truth" for evaluating the image classifications.

8.6.3 Data Processing

Data from one ALMAZ-1 satellite overpass (July 31,1992, orbit # 7753) and one ERS-1 satellite overpass (July 13,1992, scene #25911100) were used as source data. Four standard data products were acquired for the ALMAZ-1 overpass from the NPO Mashinostroienia SAR processing facility in Moscow:

Table 8.2 Agreement and Commission Statistics Comparing a Generalized Landcover and Surface Form "Ground-truth" Map to Classifications of ALMAZ-1 SAR Images That Were Synthesized Using Different Numbers of "Looks" and Different Windowing Functions

Class	10 m×10 m pixels number	ALMAZ-1 1 look standard product		ALMAZ-1 1 look Rectangle		ALMAZ-1 1 look Hamming		ALMAZ-1 1 look Kaiser–Bessel	
		agree, %	comm., %	agree, %	comm., %	agree, %	comm., %	agree, %	comm., %
A1	23270	52.60	64.26	32.58	66.63	52.73	65.86	62.84	65.16
A7	6392	0.00	—	0.64	89.10	0.22	88.14	0.00	—
A17	13125	71.92	86.98	85.34	87.48	69.61	87.14	52.75	86.44
A22	11112	4.14	93.52	0.37	96.75	2.59	95.72	12.27	93.46
Total	113853	19.44	80.56	16.57	83.43	19.07	80.93	20.12	79.88
Kappa, %		6.22		3.81		5.65		6.58	
B1	23270	60.38	65.18	53.64	68.56	71.02	67.49	62.84	64.96
B14	13125	73.54	86.69	75.58	86.37	65.65	86.11	71.16	86.85
B17	11118	0.00	—	0.13	95.92	0.00	—	0.03	97.30
Total	112849	21.00	79.00	19.86	80.14	22.28	77.72	21.24	78.76
Kappa, %		7.23		5.95		7.82		7.39	
C1	23270	97.90	72.01	98.22	71.70	98.99	72.09	98.96	71.87
C5	32212	0.00	—	0.00	—	0.00	—	1.79	48.80
C8	13125	24.33	70.23	26.90	69.55	24.12	67.76	21.82	67.75
Total	92351	28.13	71.81	28.57	71.43	28.37	71.63	28.66	71.19
Kappa, %		5.61		6.24		5.71		5.97	
D2	13619	2.12	90.94	0.00	—	0.00	—	0.00	—
D7	11788	73.11	87.58	73.91	88.15	64.46	87.74	63.92	87.87
D14	13125	34.21	88.83	32.52	89.28	39.95	88.80	38.68	88.59
D17	11118	1.09	94.17	0.82	94.03	3.26	94.01	4.87	93.44
Total	114860	11.77	88.23	11.38	88.62	11.50	88.50	11.45	88.55
Kappa, %		1.19		0.81		0.89		0.87	

Class	20 m×20 m pixels number	ALMAZ-1 2 looks Rectangle		ALMAZ-1 2 looks Hamming		ALMAZ-1 2 looks Kaiser–Bessel	
		agree, %	comm., %	agree, %	comm., %	agree, %	comm., %
A1	5406	64.39	51.17	69.63	44.31	74.20	46.10
A7	1505	5.05	83.51	2.26	88.15	0.00	—
A17	3217	80.23	84.56	58.87	81.65	60.30	82.21
A22	2974	6.49	87.96	24.24	91.30	25.42	89.98
Total	25914	24.43	75.57	24.75	75.01	25.88	74.10
Kappa, %		11.55		12.38		13.25	
B1	5406	64.39	51.14	69.63	44.32	74.20	46.09
B7	1525	2.69	82.55	0.00	—	0.00	—
B14	3217	86.07	84.73	87.22	85.18	74.04	83.94
B17	2975	4.40	45.87	0.00	—	13.95	88.02
Total	25740	24.95	75.05	25.52	74.43	26.45	73.55
Kappa, %		11.93		12.68		13.65	
C1	5406	96.80	66.00	97.54	65.49	97.52	65.44
C5	7244	12.13	48.11	21.49	48.58	18.37	48.59

Table 8.2 (continued)

Class	20 m×20 m pixels number	ALMAZ-1 2 looks Rectangle		ALMAZ-1 2 looks Hamming		ALMAZ-1 2 looks Kaiser–Bessel	
		agree, %	comm., %	agree, %	comm., %	agree, %	comm., %
C8	3217	42.62	63.63	34.85	57.81	32.86	60.40
Total	20973	35.68	64.12	37.91	62.07	36.52	62.66
Kappa, %		14.85		16.42		15.40	
D2	3099	4.52	90.65	1.26	87.00	4.84	92.42
D7	2731	71.18	84.43	68.91	84.39	72.24	84.83
D8	6148	43.25	45.48	49.24	38.26	57.27	39.73
D10	2229	2.69	87.50	0.00	—	0.00	—
D14	3217	22.01	64.39	20.42	74.96	14.77	25.20
D17	2975	26.25	85.02	33.61	84.98	27.93	83.58
Total	26543	23.70	76.30	24.88	75.12	26.18	73.79
Kappa, %		12.28		13.56		14.72	
Class	30 m×30 m pixels number	ALMAZ-1 4 looks Rectangle		ALMAZ-1 4 looks Hamming		ALMAZ-1 4 looks Kaiser–Bessel	
		agree, %	comm., %	agree, %	comm., %	agree, %	comm., %
A1	2227	71.44	42.89	69.82	33.18	70.59	34.28
A7	629	6.52	80.84	2.86	85.60	0.00	—
A17	1440	82.08	82.31	62.22	76.09	66.46	78.45
A22	1371	5.11	90.13	45.51	85.66	44.06	83.77
Total	10555	27.32	72.24	29.30	70.69	29.68	70.31
Kappa, %		14.29		16.90		17.17	
B1	2227	71.44	43.00	69.82	33.23	70.59	34.36
B7	631	3.65	75.00	0.00	—	0.00	—
B8	2620	5.34	50.00	0.00	—	0.00	—
B14	1440	88.19	82.54	79.38	81.64	73.96	80.98
B17	1371	2.04	28.21	30.49	78.27	35.52	80.14
Total	10515	29.03	70.87	29.63	70.26	29.71	70.09
Kappa, %		15.65		17.03		17.15	
C1	2227	94.97	53.25	96.14	51.40	97.08	57.37
C5	2911	43.18	44.87	47.41	47.39	32.88	45.96
C8	1440	53.96	55.57	44.17	57.94	40.76	62.44
Total	8553	48.51	51.49	48.60	51.32	43.33	55.91
Kappa, %		30.15		29.76		23.87	
D2	1237	13.82	86.33	8.89	87.41	12.05	88.70
D7	1138	61.25	83.05	80.67	83.75	74.96	82.98
D8	2520	54.80	39.11	63.73	32.75	64.64	33.78
D10	902	21.95	83.90	0.00	—	0.00	—
D14	1440	30.76	59.84	21.39	21.43	19.58	23.78
D17	1371	17.43	75.13	29.32	74.46	31.29	75.67
Total	10924	28.64	71.36	30.61	69.25	30.59	69.40
Kappa, %		17.63		19.67		19.47 (continued)	

Note: Agree = agreement; comm. = commission
Class definitions and descriptions are given in Chapter 7 of this book.

- Full resolution image (10 m resolution with 5 m pixel spacing) which was derived using 1 look and the Rectangle window function;
- raw data (5I × 5Q complex format). Two standard data products were acquired for the ERS-1 overpass from the Alaska SAR facility in Fairbanks:
- full resolution image (25 m resolution and 12.5 m pixel spacing) which was derived with 4 looks and the Hamming window function;
- complex image (10 m resolution, 8 m pixel spacing and 16I×16Q format) which was derived with 1 look and the Hamming window function.

Different image variants were synthesized from the raw ALMAZ-1 data and the complex ERS-1 image using the PC/AT software developed. Nine image variants were generated from the ALMAZ-1 data: (1, 2, and 4 looks together with each of three window functions: Rectangular, Hamming and Kaiser–Bessel). Two image variants were generated from the complex ERS-1 image: (2 looks and 4 looks with the Hamming window). One more ERS-1 image variant (1 look, the Hamming window) was derived from complex

Table 8.3 Agreement and Commission Statistics Comparing a Generalized Landcover and Surface Form "Ground-truth" Map to Classifications of ERS-1 SAR Images

Class	10 m × 10 m pixels number	ERS-1 1 look Hamming		20 m × 20 m pixels number	ERS-1 2 looks Hamming		30 m × 30 m pixels number	ERS-1 4 looks Hamming		ERS-1 4 looks ASF product	
		agree, %	comm., %		agree, %	comm., %		agree, %	comm., %	agree, %	comm., %
A1	23270	0.00	—	5406	42.88	55.10	2227	49.17	46.09	64.57	49.03
A17	13125	94.40	88.42	3217	50.23	86.33	1440	53.61	84.49	42.99	83.00
A22	11112	6.88	88.85	2974	38.90	87.04	1371	38.15	85.20	43.11	85.36
Total	113853	11.55	88.45	25914	19.65	80.35	10555	22.64	77.33	25.09	74.78
Kappa, %		0.15			6.82			9.16		11.54	
B1	23270	17.31	68.42	5406	54.20	60.49	2227	75.44	58.19	64.57	49.26
B14	13125	88.53	88.39	3217	67.21	87.61	1440	57.01	84.34	59.24	85.24
B17	11118	0.00	—	2975	3.73	86.30	1371	14.15	83.63	20.86	84.80
Total	112849	13.87	86.13	25740	20.21	79.73	10515	25.92	74.06	24.51	75.45
Kappa, %		1.40			6.25			11.40		10.59	
C1	23270	99.40	73.95	5406	96.89	69.22	2227	96.41	64.25	97.53	65.10
C5	32212	3.89	48.86	7244	24.17	47.24	2911	29.03	45.38	28.00	45.01
C8	13125	5.44	34.31	3217	12.19	36.05	1440	25.00	63.15	19.65	64.58
Total	92351	27.18	72.82	20973	35.19	64.76	8553	39.19	60.70	38.23	61.55
Kappa, %		2.49			11.43			17.42		16.05	
D2	13619	0.00	—	3099	5.20	91.99	1237	44.14	86.20	35.97	85.55
D7	11788	55.67	88.69	2731	52.18	87.03	1138	41.39	84.88	47.19	85.64
D8	26684	0.00	—	6148	19.00	45.19	2520	37.10	39.17	42.26	38.37
D14	13125	41.66	87.27	3217	21.20	81.08	1440	16.94	69.91	19.86	80.03
D17	11118	13.94	88.83	2975	29.61	88.72	1371	13.35	87.73	8.61	86.87
Total	114860	11.82	88.18	26543	16.26	83.73	10924	21.78	78.20	22.44	77.47
Kappa, %		1.33			4.90			10.08		10.68	

Note: All images were synthesized using the Hamming window function (class definitions and descriptions are given in Chapter 7 of this book); agree = agreement; comm. = commission

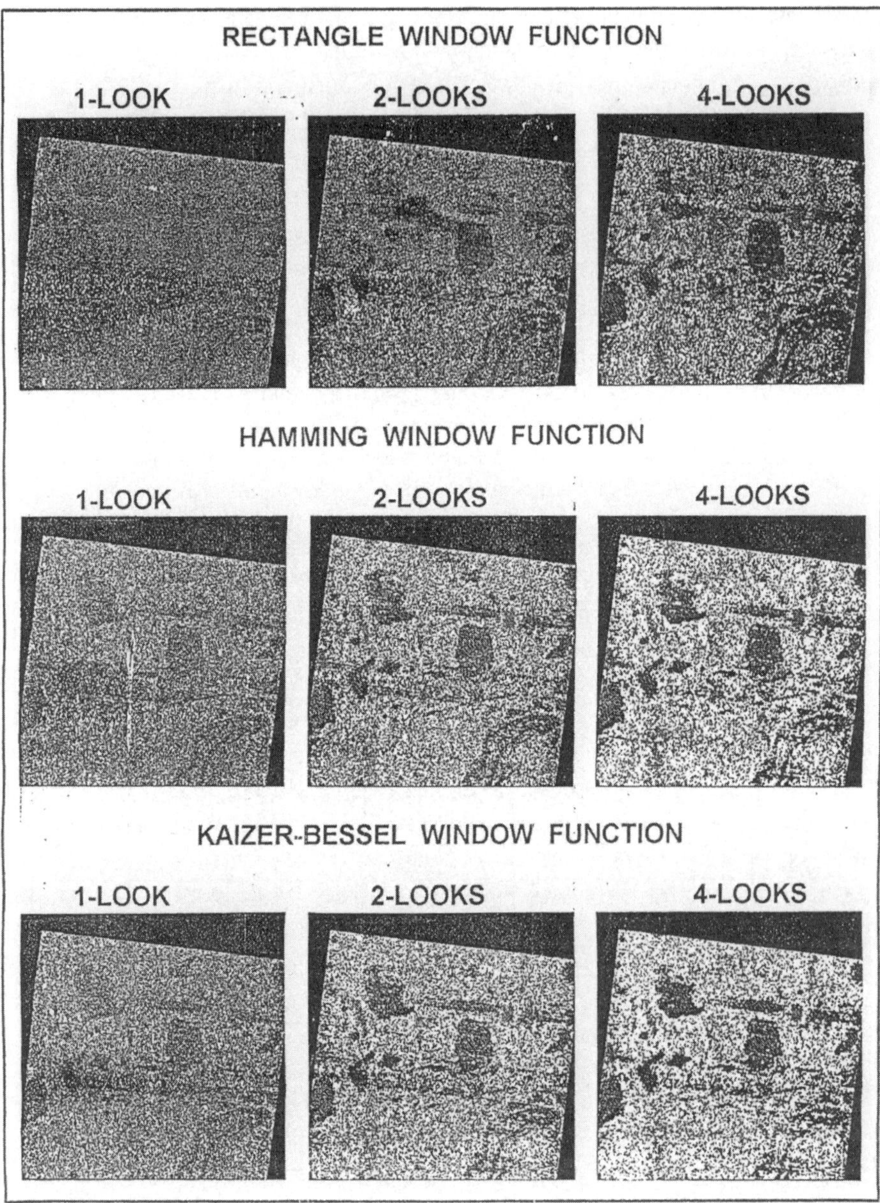

Figure 8.5 Classification of ALMAZ image processing variants (scheme A).

image with pixel-by-pixel intensity evaluation. Speckle-suppression filters were not applied to the derived images. Including the ALMAZ-1 and ERS-1 standard full resolution products, 14 images (10 ALMAZ-1 and 4 ERS-1) of site W34 were subsequently classified and compared to the "ground-truth" map.

Standard ALMAZ-1 and ERS-1 SAR-processing products, synthesized images and ground truth maps were co-registered to UTM Zone-6 map-coordinates.

The geobotanical ground classification scheme was remapped into four, more general, classification schemes. Three schemes were based primarily on vegetation complexes (A with 23 classes, B with 17 classes and C with eight classes) and one scheme (D with 17 classes) was based primarily on surface form characteristics. Scheme class definitions and descriptions are given in Chapter 6 of this book.

8.7 RESULTS

All classification results are presented in Tables 8.2, 8.3 and Figure 8.5 for ALMAZ-1 image processing variants for classification scheme A. Let us consider the results for classification schemes C and D.

The ALMAZ-1 image classes were optimally assigned with minimum loss function criteria (see Chapters 6, 7) to three ground classes in scheme C: Class C1 (water), C5 (wet tundra) and C8 (gravel pads) (Table 8.2). Increasing the number of "looks" during image synthesis improved the capability of ALMAZ-1 to discriminate among the three classes. Optimal efficiency of 1-look ALMAZ-1 separated only two classes, while 4-look processing had close to 50% agreement across all three classes and lower commission errors. The Hamming window function resulted in better discrimination of wet tundra (C5), while the most effective detection of gravel pads (C8) was observed with the Rectangle window function (Table 8.3).

With respect to classification scheme D, which was based on surface form characteristics, increasing the number of "looks" for ALMAZ-1 increased the ability to discriminate between a greater number of ground classes (Table 8.2). Classification agreements were also slightly higher for the images that were processed using consecutively higher numbers of "looks". The effects of window functions were less consistent across the surface from classes of scheme D (Table 8.2). For 2-look and 4-look ALMAZ-1 processing, the Rectangle window was more efficient for detecting Class D14 (gravel pads) and less efficient than the Hamming or Kaiser–Bessel window for discriminating water (Class D8).

Classification results for ERS-1 SAR images that were derived with the PC/AT software using 2 looks and 4 looks (Hamming window) are presented together with the corresponding results for the 1-look and 4-look (Hamming window) standard image products from the Alaska SAR Facility (Table 8.2). Similar to the results for ALMAZ-1, increasing the number of "looks" improved ERS-1 discriminatory capabilities for both classification schemes, however, the magnitude of improvement for ERS-1 (Table 8.2) was less than that observed for ALMAZ-1 (Table 8.2). With regard to ERS-1 4-look data-focusing, slight differences were observed between the results from the PC/AT software processor versus the Alaska SAR Facility processor (Table 8.2). Although the methodological principles of the two processors were identical, minor differences could be attributable to the different computer hardware platforms or subtle differences in parameter approximations and software compilers.

8.8 CONCLUSION

This chapter contributes additional knowledge toward the goal to more effectively apply SAR data to ecological questions. For any specific SAR application, it is important to understand the optimal combination of data-focusing parameters for extracting maximum information from the raw SAR data or complex SAR images. In the context of the classification schemes and methods of this study, the results showed that different focusing parameters have a substantial influence on the proficiency of extracting information from the SAR data. Increasing the number of "looks" improved classification capabilities of general tundra landcover types. Improvement was likely due to the proportional reduction in speckle noise that accompanied the degraded resolution. The choice of window function appeared to have a secondary role that was more sensitive to discrimination of individual classes.

Some effects of processing SAR data with different "looks" are visually detectable. For example, Figure 8.6 illustrates ALMAZ-1 images, over a portion of site W34 that were synthesized using 1, 2 and 4 looks and the Rectangle window function. The dark linear feature near the image center is an airport runway (1800 m long) which is bounded by infrastructures that appear bright. A gravel road traverses the upper right corner, and the dark features on the left and the upper center are lakes. Note the degraded resolution and suppressed speckle noise of the 4-look image compared with the 1-look image. This "smoothing" effect generally improved agreement results between classified image pixels and the W34 ground map. Speckle suppression filters can also be used to "smooth" radar images. The lower illustration in Figure 8.6 shows the 1-look ALMAZ-1 image after application of the Lee filter (Durand et al., 1987); the image resulted from the Lee filtering of full ALMAZ-1 scene is presented in Figure 8.7. Filtering also degrades effective resolution, however, the Lee filter visually appears to have good potential for smoothing speckle noise while maintaining spatial and textural integrity.

The PC/AT SAR processing software presented in this chapter provides a tool for future research of SAR analytical techniques. The preliminary conclusions for tundra established a baseline for developing more specific hypotheses and expanding methodologies, and stimulated the development of additional post-image processing software that will use a Fourier analysis for separating texture features and for providing independent supplementary information about image properties.

The Fourier module should improve the discrimination of targets with unique textures as well as provide methods for studying the effects of window functions on the characterization of textural attributes.

Researchers who have a defined remote sensing application and compatible ground-truth data can more thoroughly evaluate SAR with respect to their project goals using the PC/AT SAR processor. Regardless of whether the environment is sea ice, desert, tundra, tropical forest or otherwise, expanded research is needed to better utilize the information collected by existing SAR systems and to design future system improvements.

For further improvement of SAR processing, it is necessary to enhance the data-focusing software in the direction of creating informational support for an appropriate choice of data-focusing parameters oriented to the best classification of image pixels.

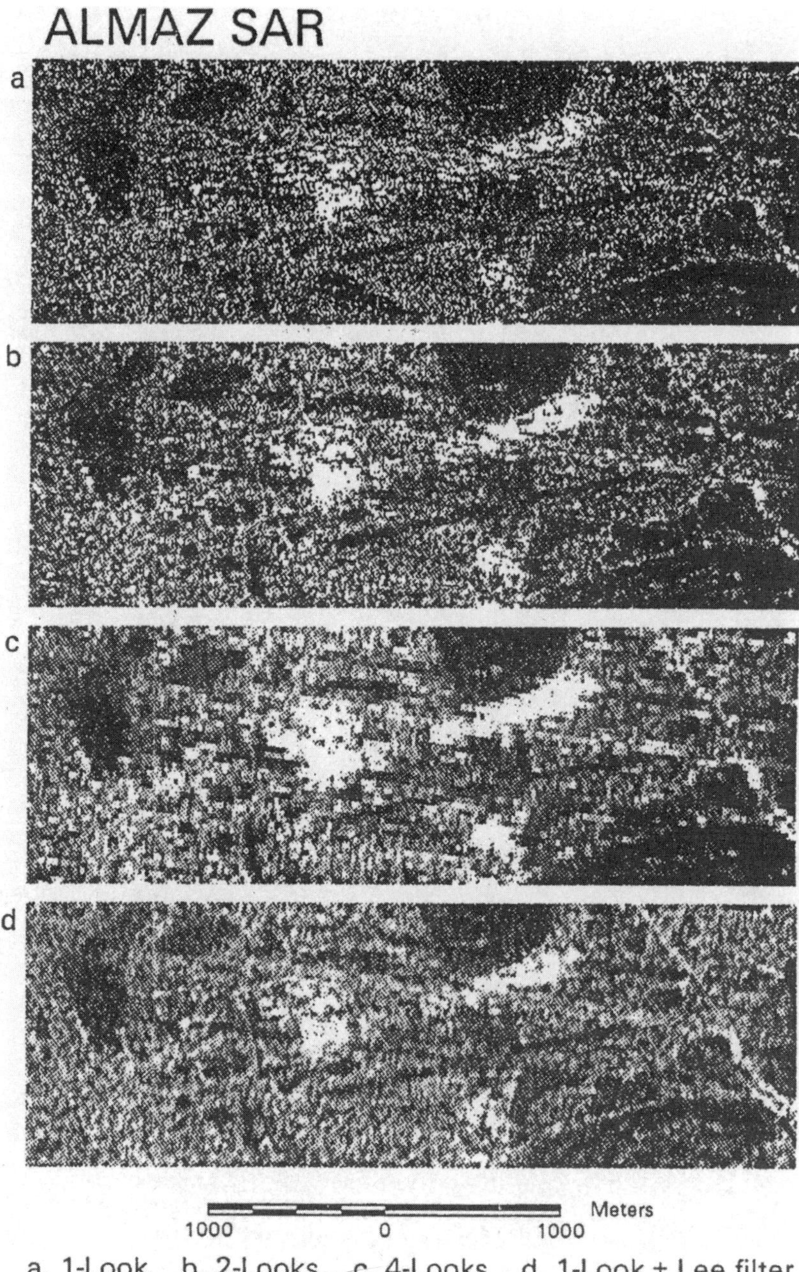

Figure 8.6 Examples of ALMAZ SAR image synthesis using 1, 2, and 4 looks and 1 look with Lee filter applied. The area depicted includes infrastructures of the Prudhoe Bay oil field located in wet tundra on the northern coastal plain of Alaska.

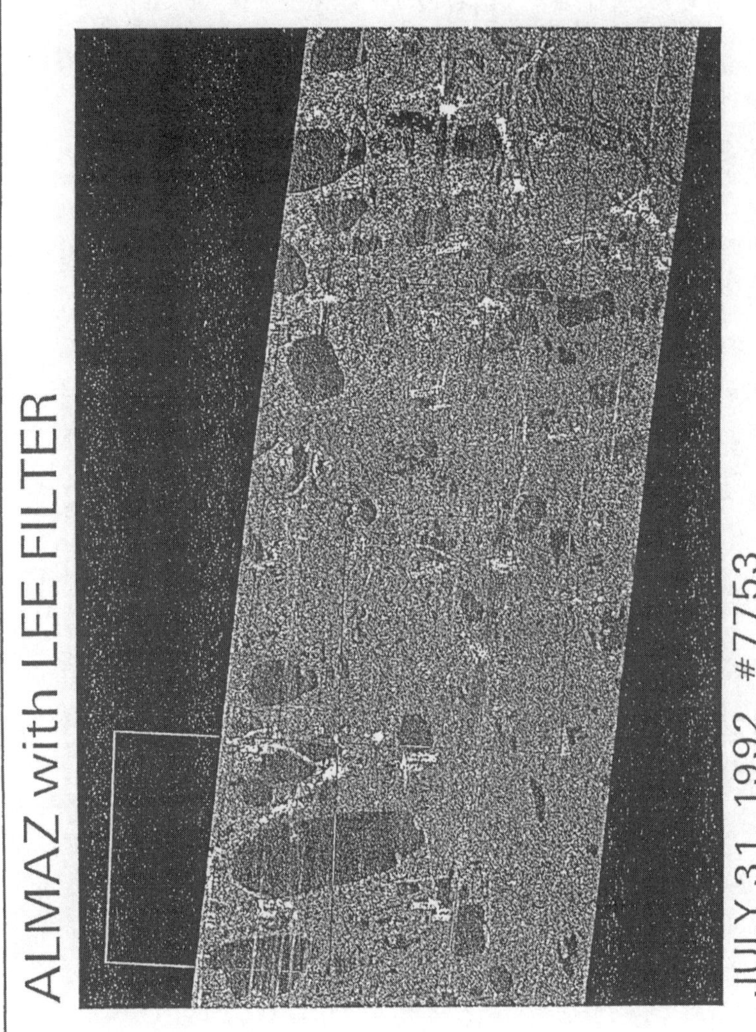

Figure 8.7 The result of Lee filtering of the full ALMAZ scene with the test square frames overlaid.

REFERENCES

Aronoff, S. A., 1984. An approach to optimized labeling of image classes. *Photogram. Eng. and Rem. Sensing*, 50(6), pp. 719–727.

Belchansky, G. I., Ovchinnikov, G. K. and Douglas, D. C., 1995. Comparative evaluation of ALMAZ-1, ERS-1, JERS-1 and Landsat-TM for discriminating wet tundra habitats. *Polar Record*, 31(177), pp. 161–168.

Curlander, J. C. and McDonough R. N., 1991. Synthetic aperture radar: systems and signal processing. New York: John Wiley & Sons, Inc.

Durand, J. M., Gimonet, B. J. and Perbos, J. R., 1987. SAR data filtering for classification. IEEE *Transactions on Geoscience and Remote Sensing*, 25(5), pp. 629–637.

Eklundth, J. O., 1972. *IEEE Transactions on Computers*, C-21, 801.

Fitch, J. P., 1988. Synthetic Aperture Radar, N.Y.: Springer-Verlag.

Harris, F. J., 1978. On the use of windows for harmonic analysis with the discrete Fourier transform. *Proceedings of the IEEE*, 66(1), pp. 51–83.

Li, F. K., Held, D. N., Curlander, J. C. and Wu, C., 1985. IEEE Transactions on Geoscience and Remote Sensing, GE-23, 1, 47 p.

Madsen, S.N., 1989. IEEE Transactions on Aerospace and Electronic Systems, AES-25, 2, 134 p.

Olmsted, C., 1993. Alaska SAR Facility Scientific SAR User's Guide. Geophysical Institute, University of Alaska, Fairbanks. (ASF-SD003).

Walker, D. A., Binnian, E. F., Lederer, N. D., Nordstrand, E. A., Meehan, R. H., Walker, M. D. and Webber, P. J., 1986. Cumulative landscape impacts in the Prudhoe Bay Oil Field 1949–1983. In: Meehan, H. and P. J. Webber (editors). Tundra Development Review: Toward a Cumulative Impact Assessment Method, U. S. Fish and Wildlife Service, Anchorage, Alaska. Section 2B.

Wu, C. and Jin, M., 1982. IEEE Transactions on Aerospace and Electronic Systems, AES-18, 5: 563.

9

Studying Regional Aspects of Polar Bear Ecology in the Russian Arctic Using Satellite Data

9.1 INTRODUCTION

This chapter presents an example of polar bear regional ecology studies using microwave satellite data. The main aspects of these studies include regional and seasonal changes in habitat parameters; daily, seasonal and annual variability of movement rates; individual and group specific direction and migration patterns, and the character of using a particular type of habitat. Dr. Gerald Garner (the Alaska Biological Science Center) made many contributions to this work.

Polar bear (*Ursus Maritimus*) population takes the top place in the Arctic ecosystem food chain and is divided into 15 ecogeographical populations (Amstrup et al., 1986, 1994; Arthur et al., 1996; Belikov et al., 1995; Calvert et al., 1995; Demaster et al., 1980; Derocher and Stirling, 1990; Garner et al., 1990, 1994, 1995; Wiig, 1995). Total size of the population size is about 21470–28310 and its density is characterized by the value of one individual per 141–269 km^2 (Derocher and Stirling, 1990; Polar Bears, 1995). The principles of determining 15 populations are based on the general ecological/geographical criteria (Uspensky, 1989).

At present the degree of ecology studies of different populations is characterized by great discontinuity. The populations of the Chukchi, Laptev, Franz-Joseph and Novaya Zemlya, Spitsbergen and East Greenland regions are least studied (Polar Bears, 1995) because other groups (USA and Canada Arctic sea regions) were studied more intensively during the last decade using ground, airborne and spaceborne observation systems. For example, the movement data were obtained through Argos data collection and location system and have been used for USA and Canadian population studies since 1984 (Pank et al., 1985; Fancy et al., 1988; Douglas et al., 1992).

A joint American–Russian study of polar bear ecology was performed during 1991 in the Severnaya Zemlya Islands of the east-central Russian Arctic in the framework of American–Russian scientific cooperation on environment protection. The multidisciplinary studies of several Russian polar bear populations were activated in 1992–1998 based on the satellite monitoring data (telemetry and remote sensing) (Garner et al., 1991, 1995;

Belchansky et al., 1992; Belikov et al., 1995). The preliminary estimations were provided for the five female polar bears in the region of the Chukchi and Bering Seas. For the region of Severnaya Zemlya Islands the sample size was small, therefore no efforts were made to determine the statistical significance of the variation in total movement. Some problems of complex studies of the polar bear regional ecology based on multiyear synchronous telemetry and multispectral microwave space monitoring data was discussed in White and Carrott (1990). In particular, it was mentioned that space monitoring provided high accuracy in obtaining georegistered locations and movements of both individual female polar bears and regional groups in their natural habitat conditions.

The distribution of polar bears is circumpolar in the Northern Hemisphere, but there are gaps between key ecological parameters for different populations. Russian Arctic populations lay in the largest one. Among the poorly studied characteristics of polar bear population we can mention the seasonal and annual activity levels, the routes and individual movement strategies, seasonal and annual orientations of the individual home range, relative use and availability of resources. The resources usually include habitat types and categories — land, islands, constant multiyear sea ice and mixed sea ice types.

This chapter describes how satellite telemetry and remote sensing microwave data are used for studying the animal movement patterns, estimating the rate of daily, seasonal and annual movement variability and understanding the use of resources to their availability.

9.2 METHODOLOGY

9.2.1 Study Area

The study area includes the Barents, Kara, Laptev, East-Siberian and Chukchi sea regions and the adjacent part of the Arctic Ocean (Figure 9.1). This area is limited by the Eurasia coastline on the southern part, wide and relatively deep strait between Spitsbergen and Scandinavia in the west, the narrow shallow Bering Strait on the east and freely communicated with the Arctic Ocean on the northern part. Arctic seas are characterized by the lack of warm (mean winter and mean summer Celsius temperatures are –19.4, –24.3, –27.5, –29, –25 and 7.5, 3.5, 1.5, 1.5, 2.5 accordingly) and weak radiation heating (Dobrovolsky and Zaligin, 1982).

9.2.2 Data Description

Radio-collared bears were traced by satellites based on the Argos Data Collection and Location System (Argos 1984; Fancy et al., 1988). The satellite telemetry data were collected by G. W. Garner and D. C. Douglas (Alaska Biological Science Center, USGS). To minimize the degree of dependence among data, they retained for each animal only one set of the data and location per transmission period (i.e. the 8-hour period during which signals were received during 5 or 6 days). Locations provided by the Argos system were coded with the index of accuracy (range 1–3), determined by the number of messages received during an overpass transmitter stability, and geometry of the location (Argos, 1984). During each satellite overpass, the locations of the radio-collars were calculated from the Doppler shift of the frequency of the signals (Fancy et al., 1988). The lowest value

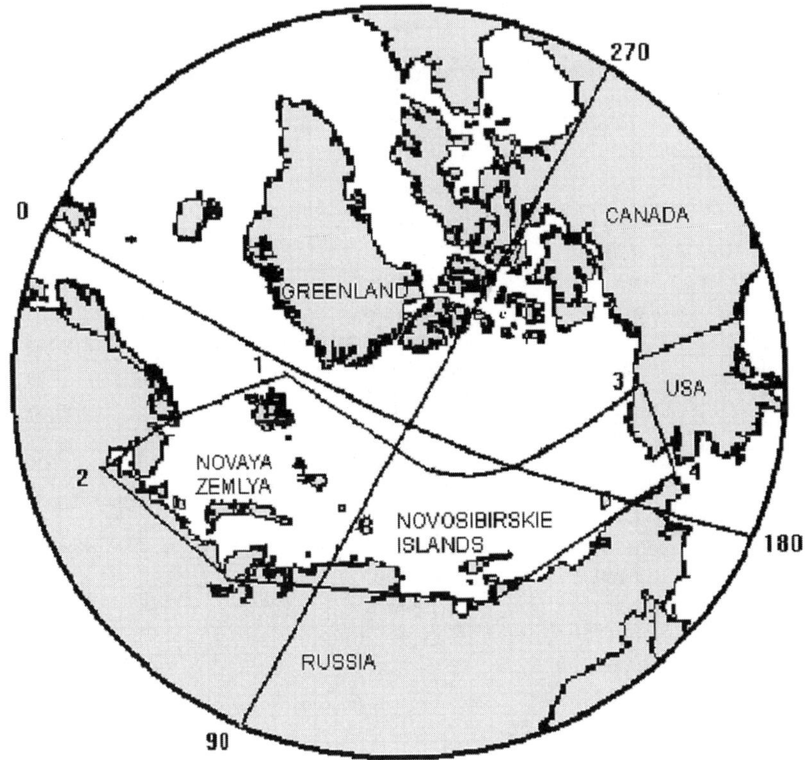

Figure 9.1 Location of the study area in the Barents, Kara, Laptev, East-Siberian and Chukchi Sea regions (point coordinates: 1 – 84.03N, 0.27E; 2 – 65.86N, 33.52E; 3 – 69.82N, 145.32W, 66.05N, 169.9W).

of the quality factor we used was associated with the location error of <1 km as estimated by Argos Service. The general description of radio-collared bears is shown in Table 9.1.

The digital images of the study area were obtained from the Scientific Research Centers for Natural Resources Studies (NITS IPR), Moscow Region, Dolgoprudny (Russian satellites OKEAN-01 N7 and N8, passive-radiometer and active-radar microwave data) and from the National Snow and Ice Data Center, Colorado Univercity (DMSP F-13 satellite, SSM/I–7-channel brightness temperature data). Corresponding weekly sea ice condition maps were obtained from the NOAA Ice Center, USA.

9.2.2.1 Registration of Radio-collared Polar Bears to Different Populations

Registration was performed taking into account the annual individual areas and according to the potential population home range intersection measure. While performing the registration procedure the potential population home range and annual individual area contours were generated in polar stereographic projection using literature and radio-tracking data, accordingly (Polar Bears, 1995). The main principle of polar bear registration to the concrete populations included 95% intersection between female contour local-individual annual areas and corresponding home range populations. If the mentioned condition was not observed, animals were registered to the independent populations.

Table 9.1 General Descriptions of Polar Bear Radio-tracking for the Period between 1 April 1995 and 30 December 1997

Number	Identification number of animals (#)	Observation period	Habitat	Number of locations
1	6341	1995–1997	1,2	95
2	6342	1995–1997	2	119
3	6348	1995–1997	1,2	68
4	6352	1995–1996	4,5	43
5	6356	1995	5	20
6	6357	1995–1996	5	42
7	6358	1995–1997	2	103
8	6359	1995	2	9
9	6361	1995–1997	2,3	54
10	6363	1995	2	20
11	6365	1995–1996	2,3	18
12	6403	1995–1997	1,2	109
13	6404	1995–1997	1,2	49
14	6407	1995	3	44
15	6408	1995–1997	1,2	49
16	6409	1995–1997	1,2	58
17	6410	1995	1	21
18	6411	1995	2	19
19	7602	1995–1997	1	80
20	7622	1995	3,4	19
21	7892	1995–1996	3	62
22	7894	1995–1997	2	99
23	7898	1995–1997	1	115
24	7899	1995–1997	1	47
25	7906	1995–1997	1,2	46
26	7913	1995–1997	1,2	110
27	10931	1995	1,2	69
28	10932	1995, 1997	1,2	4
29	10935	1995, 1996	1	60
30	10936	1995, 1996	2	65
31	20057	1995, 1996	1,2	43
32	20060	1995, 1996	1,2	58
33	20073	1995, 1996	1,2	71
34	20080	1995, 1996	1	36

Note: Regions were defined as follows: 1 — Barents, 2 — Kara, 3 — Laptev, 4 — East-Siberian, 5 — Chukchi Seas

9.2.2.2 Definition of the Major Periods of Sea Ice Activity

Polar bear movements in the Arctic seas are greatly influenced by sea ice distribution and types (Garner et al., 1995). For correct estimating and comparative analysis (in- and inter-regional, seasonal etc.) of polar bear activity (annual, seasonal, daily) parameters one needs to know the key parameters of seasonal ice condition changes. The major periods of sea ice activity were determined using Arctic sea-ice area and population home range intersection based on the threshold of sea ice extent values (minimal and maximal). Four seasons of ice condition changes were defined relative to the thresholds mentioned and using remote sensing data: daily ice concentrations and physical surface temperature from the OKEAN-01 N7, N8 and SSM/I satellite data, and Navy/NOAA weekly sea ice maps.

9.2.2.3 *Comparative Analysis of Movement Rates and Migration Patterns for Different Groups of Individuals*

Daily movement rate was calculated from the movement data by dividing the distance between consecutive locations for a bear by the time interval in hours depending on the habitat region and research season. The estimations were considered to be lower threshold level because daily activity was determined on the least distance between neighboring 6 d. locations (Amstrup and Durner, 1995; Messier et al., 1992). We used the Student's *t*-test with the significance level of 0.05 to compare the movement rates for different groups.

The seasonal activity of different population groups was calculated as a mean distance between maximum distances traveled by animals during each season (minimum, advancing, maximum and receding sea ice periods).

Annual activity of an *i*-individual ($i \subset I$) from *k*-populations ($k \subset K$, $K = \{1, 2, ..., 5\}$) in the period of an *m*-year ($m \in M$) was determined by the formula

$$d_{ikm} = \underset{j_1, j_2 \in J_{ikm}}{MAX} D(X_{ikmj_1}, X_{ikmj_2}), \; i \in I, k \in K, m \in M,$$

where $I = I_1 \cup I_2 \cup I_3 \cup I_4 \cup I_5$ is the set of all radio-collared animals, I_k is the set of *k* group of animals, $(X_{ikmj_1}, X_{ikmj_2}) - j_1$ and j_2 are the locations of *i* radio-collared females from *k* group in the studied year *m*, J_{ikm} is the set of geographical locations of *i* individual from the *k* group in the studied year *m* and *D* is the Euclidean distance between respective locations. That is, d_{ikm} is the maximum distance between two most remote locations during the year for every individual. Population annual activity was calculated as a mean of d_{ikm}.

Also the mean total distance traveled by animals during the year from different groups and the navigation efficiency coefficient were also calculated. The navigation efficiency coefficient was defined as the ratio of annual activity to the total distance traveled by the animals in one year.

To determine the factors affecting sea ice distribution or types on the radio-collared individual or group movement direction, complex analysis of biotelemetry data and sea ice conditions was provided using satellite remote sensing data. Animal movement directions were characterized by the vectors defining the movements in the latitude-longitude directions based on the individual locations during every year. Sea-ice condition changes were evaluated with vectors showing the main drifting pack ice seasonal directions (melting/freezing) (Walsh et al., 1985). Two-year mean annual directions of ice condition changes were determined with daily sea ice type mosaics (new ice, young ice, FY ice and MY ice) for the period 1996–1997. Kruskal–Wallis (K–W) tests were used for most comparisons. Depending on the data source the paired and unpaired *t*-test were used for comparative analyses of polar bear activity parameters (Conover, 1980).

9.2.2.4 *Studying Habitat Preferences*

To determine animal habitat use it is necessary to estimate the available area of each well-defined habitat category. The method of mapping the study area for analysis of the habitat preferences was described in a series of publications (Alldredge and Ratti, 1986, 1992; Johnson et al., 1980; Marcum and Loftsgaarden, 1980; Neu et al., 1974; Thomas and Taylor, 1990). The inefficiency of this approach is caused by the very high fluctuation of polar bear habitat parameters (ice concentration and types, and others), so even weekly or decade maps data are not enough. Therefore correct habitat preference estimation requires operative methods of monitoring that correspond with a real polar bear movement dynamic and a variation of their environment parameters (Manly et al., 1993; Arthur et al., 1996). In all cases the sea ice types, concentrations and other habitat categories such as mixed sea ice types, islands, coastal area and others are useful for examining the animal habitat preference. In these studies we use a nonmapping procedure for estimating the category proportions of several habitat parameters simultaneously (Marcum et al., 1980). The technique was provided during a study of habitat preferences in the Barents and Kara Seas of a control set of 19 polar bears. The preference parameters were estimated for islands, continent coastline, multiyear sea ice and mixed sea ice types (1. Spitsbergen, 2. Franz-Josef, 3. Novaya Zemlya, 4. Severnaya Zemlya, 5. Taimir and Kolsky Peninsula coast, 6. Constant MY sea ice of the Arctic Ocean, 7. Open water, drifting FY and MY sea ice). The estimation procedure of habitat preference includes:

1. Telemetry data analysis and determination of the sample population home range, estimating the frequency of the animal locations in every category type. Home range was defined using the convex area such as:

$$A_1{*}X_1 + B_1{*}Y_1 > 0 \text{ (or} < 0),$$
$$\ldots\ldots\ldots\ldots\ldots\ldots\ldots\ldots\ldots\ldots,$$
$$A_j{*}X_j + B_j{*}Y_j > 0 \text{ (or} < 0),$$
$$\ldots\ldots\ldots\ldots\ldots\ldots\ldots\ldots\ldots\ldots,$$
$$A_J{*}X_J + B_J{*}Y_J > 0 \text{ (or} < 0),$$

where A_j, B_j (j = 1, ..., J) are the coefficients defining j side of home range, X_j, Y_j are longitude and latitude of a point inside the contour home range, J is sides number of the convex contour of the population home range.

2. Generation of a set of uniformly distributed points indicating polar bear hypothetical locations inside the convex area based on the Monte-Carlo method and determination of the habitat type utilization frequency assuming the uniform animal distribution.

3. Comparing the proportion of the random points falling in each habitat type with the proportion of polar bear locations occurring in each habitat type using the chi-square test of homogeneity. The *Ho* hypothesis detailed check was performed by the following technique. Let us define P_{i1} — the proportion of points randomly distributed that fall in the category i ($i \in I$, $I = \{1, 2, \ldots, 7\}$); P_{i2} — the proportion of polar bear locations that fall in category i; n_1 — the number of total points randomly distributed over the study area; n_2 — the total number of animal locations, I — the set of habitat types used; ∞ — the significance level of testing *Ho* using the Bonferroni approach (Miller, 1966, Wright, 1992). Than the lower (L_i) and upper (U_i) bounds of the confidence interval are defined as follows:

$$L_i = P_{i1} - P_{i2} - Z_{1-\infty/14}{*}\sqrt{\frac{(1-P_{i1})P_{i1}}{n_1} + \frac{(1-P_{i2})P_{i2}}{n_2}},\ i \in I,$$

$$U_i = P_{i1} - P_{i2} + Z_{1-\infty/14*}\sqrt{\frac{(1-P_{i1})P_{i1}}{n_1} + \frac{(1-P_{i2})P_{i2}}{n_2}}, \; i \in I.$$

If the hypothesis of homogeneity of 2 multinomial distributions is rejected (the chi-square test) then the conclusion is that a bear does not use each habitat category in proportion to its occurrence. The next step is to determine which categories the animals prefer. In this test, the 100% simultaneous confidence interval (L_i, U_i) for every category types was obtained. The *Ho* hypothesis test was carried out on the basis of the following alternative variants in terms of these values L_i, U_i, P_{i1}, and P_{i2}:

a) If the confidence interval (L_i, U_i) includes 0, then the decision is that $P_{i1} = P_{i2}$ and that category *i* is used in proportion to its availability.

b) If the confidence interval (L_i, U_i) doesn't contain zero and $L_i > 0$, $U_i > 0$, the decision is that category *i* is used significantly less than in proportion to its availability.

c) If the confidence interval (L_i, U_i) doesn't include 0 and $L_i < 0, U_i < 0$, the decision is that category *i* is used significantly more than in proportion to its availability.

9.3 ANALYSIS OF THE RESULTS

9.3.1 Radio-collared Bear Registration

Figure 9.2(a, b, c, d) shows the movement trajectories of 15 polar bear females from April 1995 to December 1997. Detailed analysis of polar bear movements suggested that these females might be referred to four known ecogeographical groups (Spitsbergen — 15 females, Franz-Josef/Novaya Zemlya — 12 females, Laptev — 4 females, Chukchi — 3 females). However, because of the strong intersection (over 95%) between individual annual areas for eight animals from Spitsbergen populations with the home range of Franz-Josef/Novaya Zemlya population, it would be useful to separate them to the transitive Barents–Kara group of animals. This doesn't contradict the conclusion about the strict position of annual individual areas from the Spitsbergen populations in the Barents Sea and adjacent parts of the Arctic Ocean shown by Wiig (1995) using telemetry monitoring of 69 radio-marked polar bears in 1988–1989.

9.3.2 Definition of the Major Periods of Sea Ice Activity

Table 9.2 shows four periods relating to the major periods of sea ice activity in the region of the Arctic seas. The periods are defined as follows: 1 — minimum sea ice, 2 — advancing sea ice, 3 — maximum sea ice, 4 — receding sea ice. Results show that for every period the length and beginning of seasonal changes of ice activity vary depending on the year and region. Particularly, the Laptev Sea differs from the Barents Sea in the aspects of quicker new ice freezing, long maximum extent period and short melting period. For example, the Barents and Laptev Seas are characterized by the following periods: freezing 12.10.95–06.01.96, 10.10.95–24.11.95, maximum sea ice extent 07.01.1996–02.06.1996, 25.11.95–04.07.1996, melting 03.06.1996–19.09.1996, 05.07.1996–28.08.1996, accordingly. For estimating the Chukchi Sea major periods we included the historical data (Garner et al., 1995).

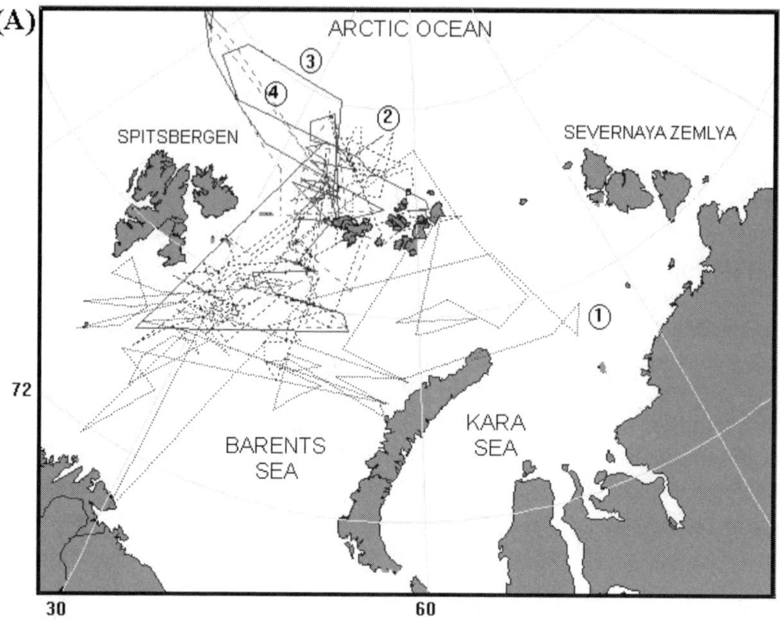

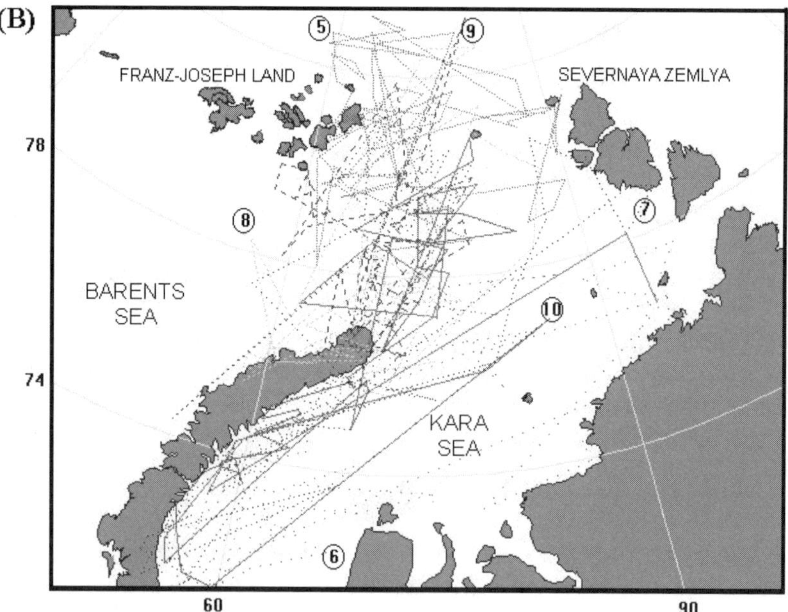

Figure 9.2 Movement trajectories of 15 radio-collared polar bears during the period from April 1995 to December 1997 (females: a: 1 – 6408, 2 – 10935, 3 – 7898, 4 – 20080; b: 5 – 6341, 6 – 6342, 7 – 6403, 8 – 6348, 9 – 6405, 10 – 20060).

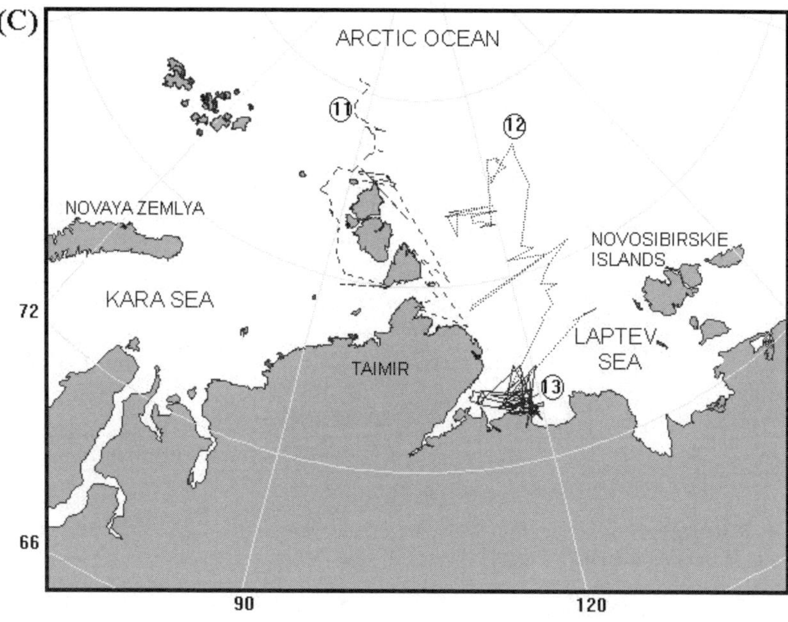

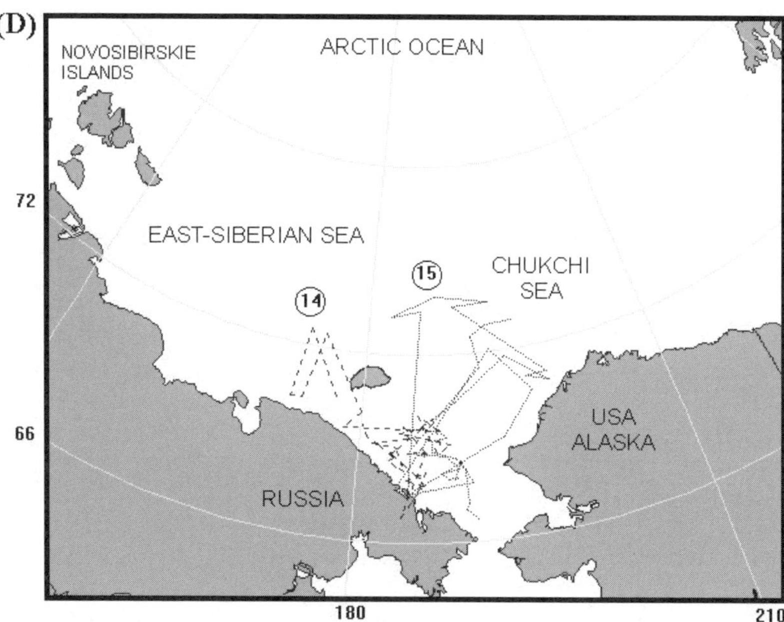

Figure 9.2 (continued) Movement trajectories of 15 radio-collared polar bears during the period from April 1995 to December 1997 (females: c: 11 – 6361, 12 – 6407, 13 – 7892; d: 14 – 6352, 15 – 6356).

Table 9.2 Major Periods of Sea Ice Activity Changes in the Region of Russian Arctic Seas

Period	Barents Sea	Barents–Kara Seas	Kara Sea	Laptev Sea	East-Siberian Sea
1	30.09.1995	30.09.1995	05.09.1995	15.09.1995	
	20.09.1996	22.09.1996	13.09.1996	28.08.1996	16.08.1996
	12.09.1997	29.09.1997	29.09.1997	20.09.1997	
2	12.10.1995	10.10.1995	03.10.1995	10.10.1995	
	10.10.1996	02.10.1996	30.09.1996	15.09.1996	15.10.1996
	13.10.1997	13.10.1997	06.10.1997	01.10.1997	
3	07.01.1996	31.12.1995	27.12.1995	25.11.1995	
	16.01.1997	05.01.1997	25.12.1996	05.10.1996	1.01.1997
4	17.05.1995	26.05.1995	05.06.1995	18.06.1995	
	03.06.1996	05.06.1996	20.06.1996	05.07.1996	1.05.1997
	12.05.1997	20.05.1997	25.05.1997	10.06.1997	

Table 9.3 Comparison of Mean Daily Movement Rates within Sea Ice Condition Categories for the Bears in Russia Arctic Seas, 1995–1997

Region	Population	Daily movement rate (km/24h)			
		1	2	3	4
Barents Sea	Spitsbergen	13.73 8.95–18.5[a] 10[b]	26.93 19.3–34.6 42	16.78 14.9–18.7 84	15.76 12.8–18.7 79
Barents–Kara Seas	Barents–Kara	9.06 2.5 – 15.7 4	28.59 21.5 – 35.7 60	20.73 15.1–26.3 115	14.40 12.1 – 16.7 75
Kara Sea	Franz-Josef and Novaya Zemlya	7.77 4.9 – 10.7 18	19.37 16.3 – 22.4 148	8.99 8.0 – 9.97 285	11.33 9.97–12.69 173
Laptev and Kara Seas	Laptev	17.35 0.0– 34.8 3	10.19 4.4 – 16.0 11	9.04 6.9 – 11.1 74	10.97 8.1–13.8 15
Chukchi Sea	Chukchi	15.46 6.9 – 24.0 4	20.43 5.7 – 35.202 9	8.897 6.4 – 11.4 21	11.019 9.4–12.6 20

Note: Mean daily movement rate parameters are shown with 95% confidence intervals ([a]) and sample number ([b]).

9.3.3 Comparative Analysis of Movement Rates and Patterns of Different Population Groups

Seasonal variation in observed daily movement rates was found in every group (Spitsbergen, K–W: $H = 12.71$, $df = 3$, $P = 5.2288 \times 10^{-3}$; Barents–Kara, K–W: $H = 13.115$, $df = 3$, $P = 4.015 \times 10^{-3}$; Franz-Josef/Novaya Zemlya, K–W: $H = 56.991$, $df = 3$, $P = 2.571 \times 10^{-3}$). Except Chukchi (K–W: $H = 5.194$, $df = 3$, $P = 0.1581$) and Laptev groups, when there were insufficient data for the analysis (Table 9.3).

The daily movement rates during the advancing ice period for individuals from three populations (Spitsbergen, Franz-Joseph/Novaya Zemlya and Barents–Kara region) were higher than in the other three periods (Table 9.3). For example, an analysis of daily movement rates for the Spitsbergen group shows that these parameters have a maximum in the fast sea ice edge movement period (period 2 vs. 1, 3, 4 using the Student's *t*-test: 2 vs. 1, $t = 2.5817, P = 0.015$; 2 vs. 3, $t = 2.20, P = 0.03$; 2 vs. 4, $t = 1.74, P = 0.022$). The movement rates during the maximum, receding and minimum sea ice periods were very similar (for example, comparison of mean daily rates of the movement using the *t*-test suggested this conclusion, 3 vs. 1: $t = 1.1261$, $P = 0.263$; 3 vs. 4: $t = 0.494$, $P = 0.622$).

This conclusion is not true for the daily movement rates for the groups of animals from the Laptev's and Chukchi's populations. One of the main reasons is that a small number and different functional groups of radio-collared bears were used for this analysis. Here we suggested that the classification scheme of individuals into functional groups is important for result interpretations. In the population ecology very useful classification of individuals into functional groups (transitive and resident) would incorporate both individual response to environmental changes and the way of adaptation during different (dynamic or static) habitat type selection. Particularly, the spatial and temporal individual distribution analysis and movements of the Laptev's animals allows one to find two small and different functional female groups: resident which uses mostly fast-sea ice near the coastline, and transitive which actively uses drifting pack sea ice. The first functional group (#6361, 7892) is at the fast-ice around Severnaya Zemlya islands and the continent coastline. The second one (#6407, 7622) is related to the moving pack sea ice and quickly reacts to the sea ice condition changes in the East-Siberian and Laptev Seas. It is obvious that comparative analysis must take into account an ecological specialization and the animal functional groups. For these groups the sample size was small, therefore no efforts were made to determine the statistical significance between their rates of movements.

An annual variation of seasonal activity was found in three groups (Spitsbergen, W–K: $H = 19.749$, $df = 3$, $P = 1.913 \times 10^{-4}$; Barents–Kara, $H = 11.9904$, $df = 3$, $P = 7.4160 \times 10^{-3}$; Franz-Joseph/Novaya Zemlya, $H = 10.9601$, $df = 3$, $P = 0.01174$). For the Laptev and Chuckchi groups radio-collared data were limited in order to test annual variation in seasonal activity data. Seasonal activity estimates (Table 9.4) allowed finding maximum activity periods (advancing and receding sea ice). The same maximum activity seasons for three ecogeographical populations show that polar bears choose a movement strategy that maximizes their possibility to contact and interact with the ice edge, i.e. with a buffer kind of food in extreme periods of sea ice activity changes.

The regional comparative analysis shows that seasonal activity parameters of Spitsbergen groups are at their maximum during the advancing sea ice period and they are higher than other groups values at different significance levels (paired *t*-test estimates: Franz-Josef/Novaya Zemlya–Spitsbergen: $t = 1.363$, $P = 0.19$; Barents–Kara–Spitsbergen: $t = 2.159$, $P = 0.043$). Unfortunately, for other groups of individuals (Laptev and Chukchi) radio-collared data were limited for comparative analysis. Nevertheless females of those ecogeographical groups can be said to require more space. Although many questions about seasonal activity parameters of different groups of polar bears remain, it can be said taking into account the verbal description in (Garner et al., 1995) that the Spitsbergen group's seasonal activity peaks in the advancing and receding sea ice periods. This fact is related to the strong linking of transitive females to drifting pack ice fields; to have the maximum interaction with the ice edge in the Barents Sea, they have to cover a great distance during the season.

Table 9.4 Comparison of Mean Seasonal Activity Parameters within Sea Ice Condition Categories for Polar Bears in Russia Arctic Seas, 1995–1997

Region	Population	Seasonal activity, km			
		1	2	3	4
Barents Sea	Spitsbergen	88.54 27.5–149.5a 5; 7[a]	798.92 582.2–1015.6 6; 8	350.65 222.3–479.7 7; 14	474.53 367.2 – 581.9 7; 15
Barents–Kara Seas	Barents–Kara	76.38 46.3 – 106.5 3; 4	534.72 427.2–642.3 8; 16	427.32 307.3 – 546.8 8; 20	468.1 374.0–562.2 8; 14
Kara Sea	Franz-Joseph and Novaya Zemlya	121.38 47.8 – 195.0 8; 8	581.0 362.6 – 799.5 7; 14	300.05 192.9–407.2 12; 25	401.41 307.5–495.3 11; 21
Laptev and Kara Seas	Laptev	113.63 0.00 – 294.0 3; 3	194.1 42.0 – 346.2 3; 4	213.11 47.6 – 378.5 3; 4	327.9 149.9–505.9 3; 15
Chukchi Sea	Chukchi	394.8 144.3–645.3 3; 3	586.49 160.9 – 1012.1 3; 3	175.16 25.21 – 325.1 2; 4	425.91 180.0–671.9 3; 5

Note: The seasonal activity parameters of different populations are shown with 95% confidence interval

[a] number of individuals

[b] total number of the periods for 1995–1997

Table 9.5 Comparison of Mean Annual Activity Parameters of Polar Bears in Russia Arctic Seas, 1995–1997

Region	Population	Sample number (animals /year)	Mean annual activity (km)	Total distance past for year (km)	Navigation efficiency coefficient
Barents Sea	Spitsbergen	10	975.48 773.6–1177.4	3867.04 2868.1–4866.0	0.302 0.219–0.386
Kara Sea	Franz-Joseph and Novaya Zemlya	14	772.06 544.5–999.6	3173.35 2188.2–4158.5	0.345 0.236–0.448
Barents–Kara Seas	Barents–Kara	15	974.27 796.6–1152.0	3641.29 2828.1–4454.5	0.298 0.247–0.349
Kara and Laptev Seas	Laptev	4	373.7 245.5–501.8	1232.58 2.2–2463.0	0.429 0.106–0.752
Chukchi Sea	Chukchi	3	706.15 580.1–832.2	2574.54 1010.9–4138.1	0.29 0.112–0.468

The annual activity parameter estimate results suggested that females from the Spitsbergen and Barents–Kara populations are most mobile only at the significance level P = 0.065 (Table 9.5; W–K: $H = 8.82$, $df = 3$). These annual group parameters differ significantly only from the Laptev group's parameters $P = 0.006$ ($t = 3.348$). The mean total annual distance (S) inside the individual home range and navigation coefficient (N) of these groups are not different (comparative estimates using the t-test for S and N are as follows: S, Spitsbergen and Barents–Kara groups: $t = 0.068$, $P = 0.95$; N, Spitsbergen and

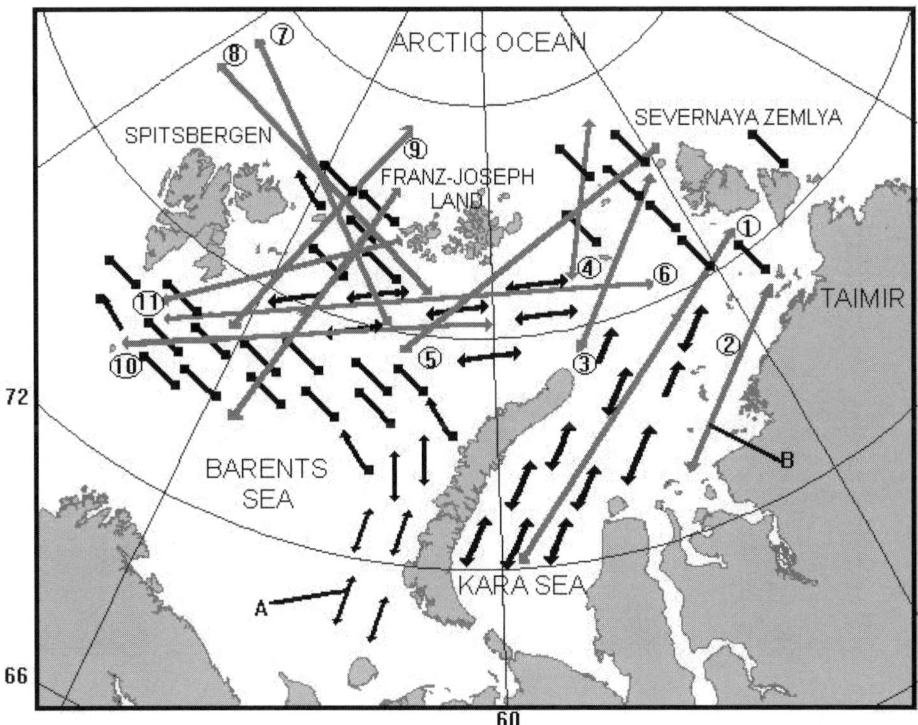

Figure 9.3 Mean annual drifting pack sea ice movement directions during the freezing/melting period (A) and polar bear females' movements (B) in the Barents and Kara Seas during the period from April 1995 to December 1997.

Barents–Kara groups: $t = 0.0019$, $P = 0.985$). It can be said that, in terms of annual activity, these groups of females are in similar ecological conditions, i.e., changes in ice in their habitats exhibit similar spatial–temporal scenarios. The mentioned groups differ from the Franz-Joseph/Novaya Zemlya groups parameters S and N, the values $P > 0.21$ (S) and $P > 0.49$ (N) at different significance levels.

Relatively high navigation coefficients for Franz-Joseph/Novaya Zemlya can be explained probably with better female orientation near islands and the continent coastline. However, the analysis of more radio-collared data is required to substantiate these preliminary subjective inferences.

9.3.4 Determination of Individual/Group Complementary Movement Patterns of Different Population Groups

Mean drifting pack ice movement directions (while freezing and melting) and radio-marked polar bear females in the Barents, Kara and Laptev Seas are shown in Figures 9.3 and 9 4. The space monitoring data analysis shows that polar bear seasonal and annual movement directions exist in the Arctic sea regions. The polar bear movement directions are defined by the specific spatial and temporal sea-ice condition changes in habitats. For example, all females in the Kara Sea region can be divided into two groups by their

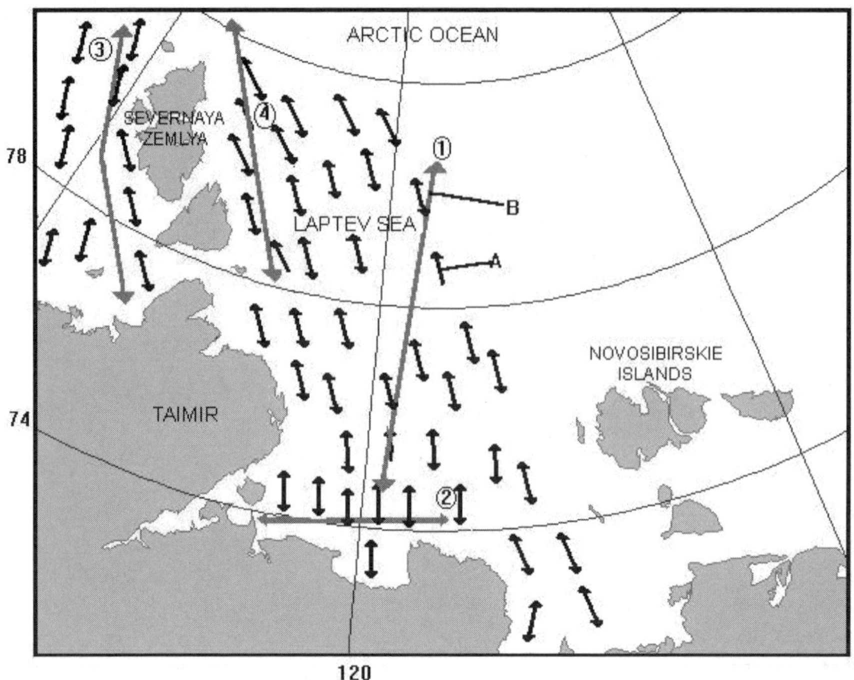

Figure 9.4 Mean annual sea ice movement directions during the freezing/melting period (A) and polar bear females' movements (B) in the Laptev Sea during the period from April 1995 to December 1997.

movement strategy. The first group was probably linked to drifting pack ice and is sensitive to the sea ice edge movement (directions 1, 2, 3, 4; Figure 9.3). The representatives of the second group move along the parallels. Detailed analysis of the second group wasn't performed because of the lack of data. Major seasonal and annual migration directions of the Spitsbergen (directions 7–12; Figure 9.3), Barents–Kara (directions 5, 6; Figure 9.3) and Laptev group's (directions 1–4, Figure 9.4) allows one to say that the complementary principle is true for polar bear females. There are definite movement directions for every region and concrete functional animal group. It is evident that transitive animals minimize significantly energy resources using this movement strategy. Although an interpretation of our figures is somewhat subjective, we believe that it is important and should be strongly evaluated in future.

9.3.5 Analysis of Polar Bear Habitat Preference

Table 9.6 presents the location distribution parameters on selected habitat types in the Barents–Kara Seas and the adjacent parts of the Arctic Ocean. The statistical χ^2-test (χ^2 = 204,73, DF = 6; P = 0,0018) indicates the existence of statistically significant difference differentiation between observed and theoretical frequencies. That is the polar bears do not use each habitat category (islands, the continent coastline, the constant multiyear sea ice and mixed ice types — open water, drifting FY ice, MY ice, etc.) in proportion to its

Table 9.6 Observed and Expected Values for Randomly Distributed Points and Polar Bear Locations in Different Habitat Types in the Barents and Kara Seas in 1995–1997

Habitat #	Habitat	Number of polar bear locations	Expected number of locations assuming uniform distribution	Proportion of locations in every type	Proportion of locations assuming uniform distribution	Conf. interval lower bound	Conf. interval upper bound
1	Spitsbergen	3	24	0.0018	0.0198	0.0083	0.0276
2	Franz-Joseph	45	19	0.0274	0.0157	–0.0266	0.0032
3	Severnaya Zemlya	15	23	0.0091	0.0189	–0.0017	0.0213
4	Novaya Zemlya	26	29	0.0158	0.0239	–0.0057	0.0219
5	Taimir and Kola Peninsula	26	34	0.0158	0.0280	–0.0022	0.0266
6	Multiyear ice	122	287	0.0742	0.2364	0.1280	0.1965
7	Open water and mixed sea ice types	1408	798	0.8559	0.6573	–0.2396	–0.1576

occurrence (Figures 9.5 and 9.6). For example, polar bear females use Spitsbergen significantly less than they do the contour area of home range. It could seem at first sight that the tolerance area by several habitat factors doesn't include Spitsbergen Island. However, additional biotelemetry data analysis (Wiig, 1995) shows that resident polar bear females from local groups use Spitsbergen Island with some buffer area. Probably, the interpopulation concurrence of females is observed in this case.

The confidence interval analysis indicates that polar bears use the lands areas along the continent coastline and Franz Joseph Islands, Novaya Zemlya and Severnaya Zemlya proportionally to their area inside the contour area (Table 9.6). This conclusion seems to be of great importance for habitat selection and radio-collaring goals, particularly with the aim of studying the keystone population ecological parameters.

The statement about the minor use of the constant MY sea ice of the Arctic Ocean by polar bear females means that mostly definite very transitive polar bear females use this type of ice.

The statistical estimates indicate that one of the preferred habitat types is a mosaic habitat (open water, drifting FY and MY sea ice, etc.). The detailed study and finding of polar bear preferred habitats in terms of sea ice types and concentrations requires additional research using simultaneously radio-tracking and remote sensing data.

9.4 CONCLUSION

These studies demonstrate how satellite microwave data and telemetry allow the estimation of polar bear habitat selection parameters and home range of relatively discrete populations. Particularly, these studies show the potential use of integrated remote sensing and telemetry databases for estimating regional variability of daily, seasonal and annual movement rates, major periods of sea ice activity change, functional structure of

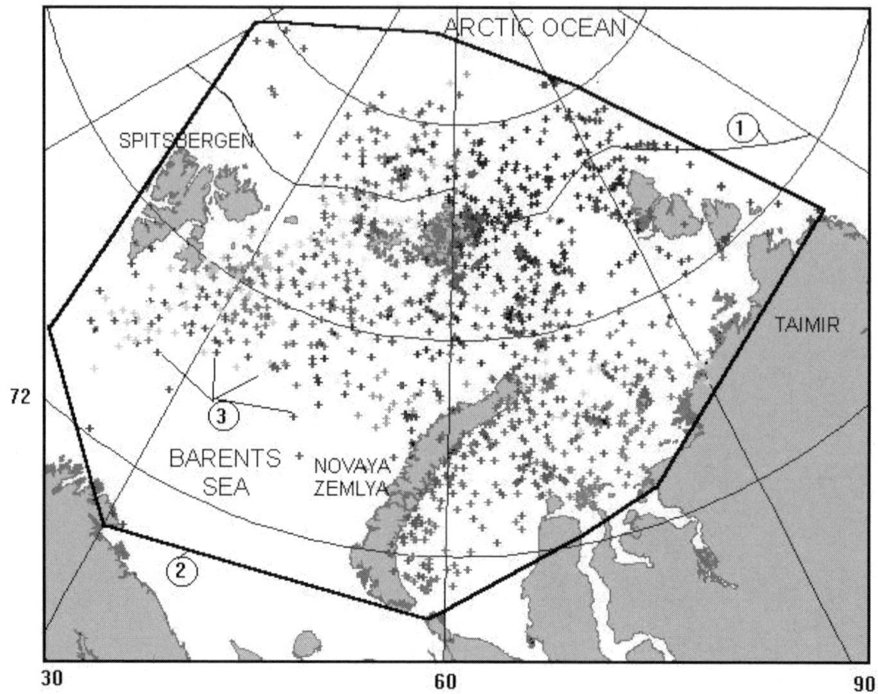

Figure 9.5 Observed location distributions of 19 polar bear females according to radio telemetry monitoring data (1 = constant multiyear sea ice edge, 2 = contour of home range, 3 = observed locations of polar bears).

ecogeographical groups of individuals; regional specific complimentary animal movement directions, and habitat preference.

The satellite monitoring has a great potential for checking the several hypotheses explaining the mechanism of polar bear adaptation to extreme Arctic conditions, for example, the mechanism of the complementary principle of the transitive animals' movements. The migration patterns presented above have not yet been sufficiently evaluated due to the restricted number of radio-collared individuals used during very short periods of time. However, it seems appropriate to assume that all resident females moved less than transitive bears, and their movements were consistent with a strategy of energy conservation.

The comparative analysis shows that daily movements during the advancing ice period for the three populations was higher than the other three periods. The calculated lower level of daily rate of movement for all populations and their large difference from the existing estimates (Uspensky, 1989) allow one to state that approaches utilized are useful for advancing Arctic ecological studies. The same maximum seasonal activity periods for three ecogeographical populations were detected. Although the differences between seasonal activity parameters for different ecogeographical groups were not fully explained, nevertheless the detection of the seasonal strategy of moving females was important.

The suggested complementary principle specifies the existing statement about the presence of "bear migration routes" (Uspensky, 1989). However, our studies show that there are complementary animal movement directions in the Arctic Seas. Unfortunately,

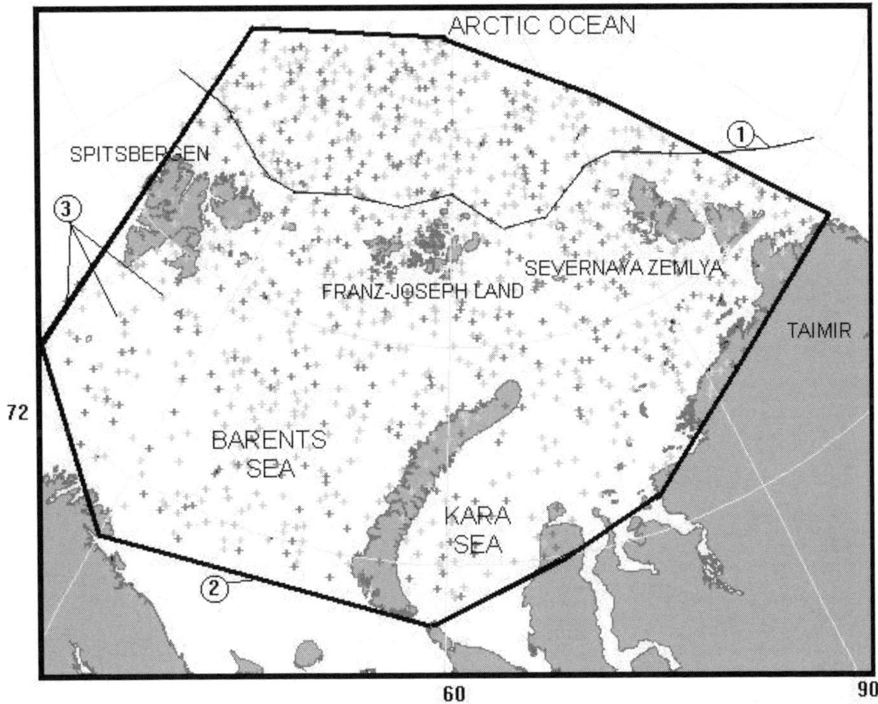

Figure 9.6 Randomly distributed points over the study area (1 = constant multiyear sea ice edge, 2 = contour of home range, 3 = expected locations of polar bears).

because of lack of data, full analysis of the navigation coefficient used to determine the female orientation character was not performed. The manner in which polar bears navigate is unknown, but obtained results allow one to state that probably the resident female orientation near islands and the continent coastline seems to be higher compared with that of transitive females, which use drifting pack ice to a greater degree.

The analysis of the habitat use parameters indicated that females do not use each habitat category in proportion to its occurrence. The preferred habitat types (mosaic habitats) do not disprove the hypothesis about the effective use of keystone ecological species parameters as a reliable direct indicator to detect regional and global climate and sea ice condition changes over time.

In these studies, the composition of all available habitat types is determined as a constant. However, mosaic habitat parameters and composition change and these changes may have essential effects on an animal's habitat selection. One way to address the problems of habitat variability is to define habitat availability separately for each observation of habitat use.

An efficient method of detailed comparison of data on habitat selection and availability that allows the availability to differ among observations is presented in Arthur et al. (1996). This method is applicable when habitats change over time and animals are unable to move throughout a predetermined study area between observations. It allows quantitative comparisons among types and it is not affected by arbitrary decisions about which habitats to include in the study. The authors of this work illustrate the efficiency of this methodology for studying polar bear sea ice habitat selection and use in the Bering and Chukchi Seas during 1990. The results indicated that the use of ice concentration categories by bears was nonrandom, and the pattern of use differed between spring and late summer seasons.

The next objectives should be to include application of this methodology (Arthur et al., 1996) for estimating polar bear habitat selection in the Barents–Kara–Laptev and East-Siberian Seas and adjacent parts of the Arctic Ocean based on multiple observations of habitat availability and use. At present, significant satellite telemetry data are accumulated to study various ecogeographical populations of polar bear. Therefore questions of satellite microwave data processing and interpretation methods should be devoted a lot of attention.

REFERENCES

Alldredge, J. R. and Ratti, J. T., 1986. Comparison of some statistical techniques for analysis of resource selection. *Journal of Wildlife Management*, Vol. 50, pp. 157–165.

Alldredge, J. R. and Ratti, J. T., 1992. Further comparison of some statistical techniques for analysis of resource selection. Journal of Wildlife management. *Journal of Wildlife Management*, Vol. 56, pp. 1–9.

Amstrup, S. C. and Durner, G., 1995. Polar bear research in the Beaufort Sea. Proc. 11th Working meeting of the IUCN/SSC Polar Bear Specialists Group. Gland Cambridge, 145–153.

Amstrup, S. C. and Gardner G. 1994. Polar bears maternity denning in the Beaufort Sea. *Journal of Wildlife Management*, Vol. 58 (1), pp. 1–10.

Amstrup, S. C., Stirling, I., and Lentfer, J. W., 1986. Past and present status of polar bears in Alaska. *Wildl. Soc. Bull.*, 14, pp. 241–254.

Argos, 1984. Location and data collection satellite system user's guide. Service Argos. Toulouse, 36 p.

Arthur, S. M., Manly, B. F. J, McDonald, L. L. and Garner, G. W., 1996. Assessing habitat selection when availability changes. *Ecology*, Vol. 77(1), pp. 215–227.

Belchansky, G. I., Ovchinnikov, G. K., Petrosyan, V. G., Pank, L. and Douglas, D. C., 1992. Processing of space monitoring data for studying large mammals in the Arctic environment. *Remote Sensing from Space. Russian Academy of Science*, Vol. 2, pp. 75–81.

Belikov, S. E., Wiig, O., Garner, G. W. and Arthur, S. M., 1995. Research on polar bears in Severnaya Zemlya (Russia) 1991–92. Proceedings of the 11th Working meeting IUCN/SSC Polar Bear Specialists Group. Gland Cambridge, pp. 167–173.

Calvert, W., Stirling, I., Taylor, M., Ramsay, M. A., Kolenosky, G. B., Crete, M., Kearney, S. and Luttich S., 1995. Research on polar bears in Canada 1988–1992. Proc. 11-th Working meeting of the IUCN/SSC Polar Bear Specialists Group. Gland. Cambridge, pp. 33–57.

Chesson, J., 1983. The estimation and analysis of preference and its relationship to foraging models. *Ecology*, 64, pp. 1297–1304.

Conover, W. J., 1980. Practical nonparametric statistics. 2nd. ed. New York: John Wiley and Sons. Inc. 493 p.

DeMaster, D. P., Kingsley, M. C. S., and Sterling, I., 1980. A multiple mark and recapture estimate applied to polar bears. *Can. J. Zool.*, 58, pp. 633–638.

DeMaster, D. P. and Sterling, I., 1984. *Ursus maritimus*. Mammal Species, 145. 7 p.

Derocher, A. E. and Sterling, I., 1990. Distribution of polar bears (*Ursus maritimus*) during the ice-free period in western Hudson Bay. *Can. Journal of Zoology*, 68, pp.1395–1403.

Dobrovolsky, A. D. and Zaligin, B. S., 1982. USSR Seas. Moscow. MSU, 230 p.

Douglas, D. C., Pank, L. F. and Greslin, J. C., 1992. Satellite telemetry and Geographic Information System: powerful tools for wildlife research and management. U.S. Fish and Wildlife Serv. Resource Publ. Washington, D.C., pp. 83–93.

Fancy, S. G., Pank, L. F. and Douglas, D. C., 1988. Satellite telemetry: a new tool for wildlife research and management. U.S. Fish and Wildlife Serv. Resource Publ. Washington, D.C. 54 p.

Garner, G. W., Belikov, S. E., Stishov, M. S., Barnes, V.G. and Arthur, S. M., 1994. Dispersal patterns of maternal polar bears from the denning concentration on Wrangel Island. International Conference on Bears Research and Management. 9, pp. 401–410.

Garner, G. W., Belikov, S. E., Stishov, M. S. and Arthur, S. M., 1995. Research on polar bears in western Alaska and Eastern Russia 1988–1992. Proc. 11-th Working meeting of the IUCN/SSC Polar Bear Specialists Group. Gland. Cambridge. pp. 155–164.

Garner, G. W., Knick, S. T. and Douglas, D. C., 1990. Seasonal movements of adult female polar bears in the Bering and Chukchi seas. International Conference on Bears Research and Management. 8, pp. 219–226.

Johnson, D. H., 1980. The comparison of usage and availability measurements for evaluating resource preference. *Ecology*, 61, pp. 65–71.

Manly, B. F. J., McDonald, L. L. and Thomas, D. L., 1993. Resource selection by animals: statistical design and analysis for field studies. Chapman & Hall. London. pp. 150.

Marcum, C. L. and Loftsgaarden, D. O., 1980. A nonmapping technique for studying habitat preferences. *Journal of Wildlife Management*, 44 (4), pp. 963–968.

Messier, F., Taylor., M. K. and Ramsay, M., 1992. Seasonal activity patterns of female polar bears (*Ursus maritimus*) In Canadian Arctic as revealed by satellite telemetry. *Zoology (London)*, 218, pp. 219–229.

Mordvintsev, I. N. and Petrosyan, V. G., 1994. Application of the satellite telemetry and geographic information system for large mammals ecology studding. *Remote Sensing from Space*, 2, pp. 119–124.

Neu, C. W., Byers, C. R. and Peek, J. M., 1974. A technique for analysis of utilization-availability data. *Journal of Wildlife Management*, 38, pp. 541–545.

Pank, L. F., Regelin, W. L., Beaty, D. and Curatolo, J. A., 1985. Performance of a prototype satellite tracking system for caribou. In: R. W. Weeks and F. M. Long (eds.), Proc. Fifth. Intern. Conf. Wildl. Biotelemetry. Chicago, Illinois, pp. 36–42.

Polar Bears., 1995. Proc. 11-th Working meeting of the IUCN/SSC Polar Bear Specialists Group. Gland. Cambridge, 192 p.

Thomas, D. L. and Taylor, E. J., 1990. Study designs and tests for comparing resource use and availability. *Journal of Wildlife Management*, 54 (2). pp. 322–330.

Uspensky, S. M., 1989. Polar Bear. Moscow. Agropromizdat: 189 pp.

Walsh, J. E., Hibler III, W. D. and Ross, B., 1985. Numerical simulation of north hemisphere sea ice variability. *Journal of Geophysical Research*, 90, pp. 4847–4865.

White, G. C. and Garrott, R. A., 1990. Analysis of wildlife radio-tracking data. Academic Press. San Diego: 123 p.

Wiig, O., 1995. Status of polar bears in Norway 1993. Proc. 11-th Working meeting of the IUCN/SSC Polar Bear Specialists Group. Gland. Cambridge. pp. 109–114.

Wright, S. P., 1992. Adjusted P-values for simultaneous inference. *Biometrics*, 48, pp. 1005–1013.

10

Detection of Sensitive Boreal Forest Types to Monitor and Assess Potential Impact of Climate Change

10.1 INTRODUCTION

A number of simulation models have been developed to assess the climatic impact of atmospheric carbon dioxide increasing (Mitchell, 1983; Harrington, 1987). These studies have indicated that disproportionate warming will occur at higher latitudes (45°N–65°N), with more pronounced effects in the continental interiors. Other studies have suggested that these changes could modify the floristic composition of forest types and, in general, the ecological function of boreal forests (Davis and Botkin, 1985; Sokolov, 1990; Solbrig, 1992). Additionally, changes in temperature and soil moisture regimes could alter the exposure and response of boreal forests to intermittent natural disturbances such as fire frequency.

Detection and investigation of sensitive forest types located near the southern ecotone of the boreal forest biome are critical in monitoring and assessing potential impacts of climate change on boreal forest structure. Sensitive forest types, which can be considered preliminary direct indicators, are very important to the development and validation of remote sensing approaches in order to transfer our understanding of the above processes from the local scales to regional and global scales. The need for better definition of directional changes of canopy structure, vegetation composition and above ground biomass under cloudy conditions for forest community types dictate multiangle (e.g., visible, near-infrared, thermal-infrared and microwave) approaches to allow effective interpretation of vegetation dynamics in time (Belchansky et al., 1992). In this chapter, we argue that, by classifying the forest community and identifying biological sensitivity of vegetation types for each local area, we achieve an intermediate level of detail that is critical for remote sensing approaches in the context of global change and in predicting the response to and effects on ecosystem processes.

Thus, the primary objectives of these studies included: (1) a syntaxonomic classification of Visimsky Nature Reserve (VNR) forest community types; (2) an evaluation of biodiversity among the forest types; and (3) a comparative micro-climatic analysis to investigate biophysical sensitivity of forest community types.

10.2 STUDY AREA

The Visimsky Nature Reserve (approximate size 13,000 ha) is located in central Russia (57°27′N, 59°35′E) in a mixed conifer/broadleaf forest ecosystem (Figure 10.1). The Reserve has a temperate continental climate with an extended cold winter and a growing season of approximately 152 days. Annual precipitation averages around 555 mm. Approximately 95.3% of the VNR land area is occupied by forest. The vegetation is characterized by *Picea obovata* Ledb. (51.1% by area), *Abies sibirica* Ledb. (0.1%), *Pinus sibirica* Du Tour (0.2%), *Pinus sylvestris* L. (5.0%), *Populus tremula* L. (0.4%), *Larix sibirica* Ledb., *Betula pendula* Roth. and *B. pubescens* Ehrh. (43.2%), *Tilia cordata* Mill., *Ulmus glabra* Huds., and *Sorbus aucuparia* L.

10.3 FIELD DATA

Data were collected at 325 intensive study areas, of which 245 were located in forest communities. They were chosen to obtain forest phytocoenoses data for all representative terrain, soil, and moisture regimes across a maximum variety of community types.

Study areas were established at sites that appeared to have homogeneous vegetation, based on archived forest taxonomic data from 1949, 1976 and 1986. Size of the study areas varied from 9 m^2 to 4900 m^2, as determined using the Dietvorst approach (Dietvorst et al., 1982). Near the center of each study area, four to 18 sample plots were established where quantitative data were collected. Percent cover of vascular plant species (total $n = 261$) was quantified at each plot using 100 pin placements on a vertical point-frame.

Descriptive data, including preliminary vegetation type, species composition of trees, percent cover of grass and moss layers, soil type, elevation, moisture regime, and terrain type were also collected during the summer periods, 1985–1993. Taxonomic, ecological, geographical and phytocoenotic characteristics were described for all vascular plants. Historical microclimate data on surface air temperature, surface air moisture and soil moisture at 30 cm depth (Kolesnikov and Kirsanov, 1979) during the growing seasons 1973–1976 were also used for this study (Figure 10.2).

10.4 FOREST COMMUNITY CLASSIFICATION

Initially, three groups of hierarchical clustering techniques were compared using the VNR field data:

1. Euclidean square-distance strategies (Duran and Odell, 1977; Enslein et al., 1986)
2. Unweighted pair-group using arithmetic averages (UPGA) based on quantitative/qualitative similarity indices such as Sorensen–Chekanovsky and Jaccard coefficients (Orloci and Wildi, 1988; Gauch and Whittaker, 1981; Dietvorst et al., 1982)
3. Minimization within-group dispersion using standardized and unstandardized Euclidean distance (Gauch and Whittaker, 1981)

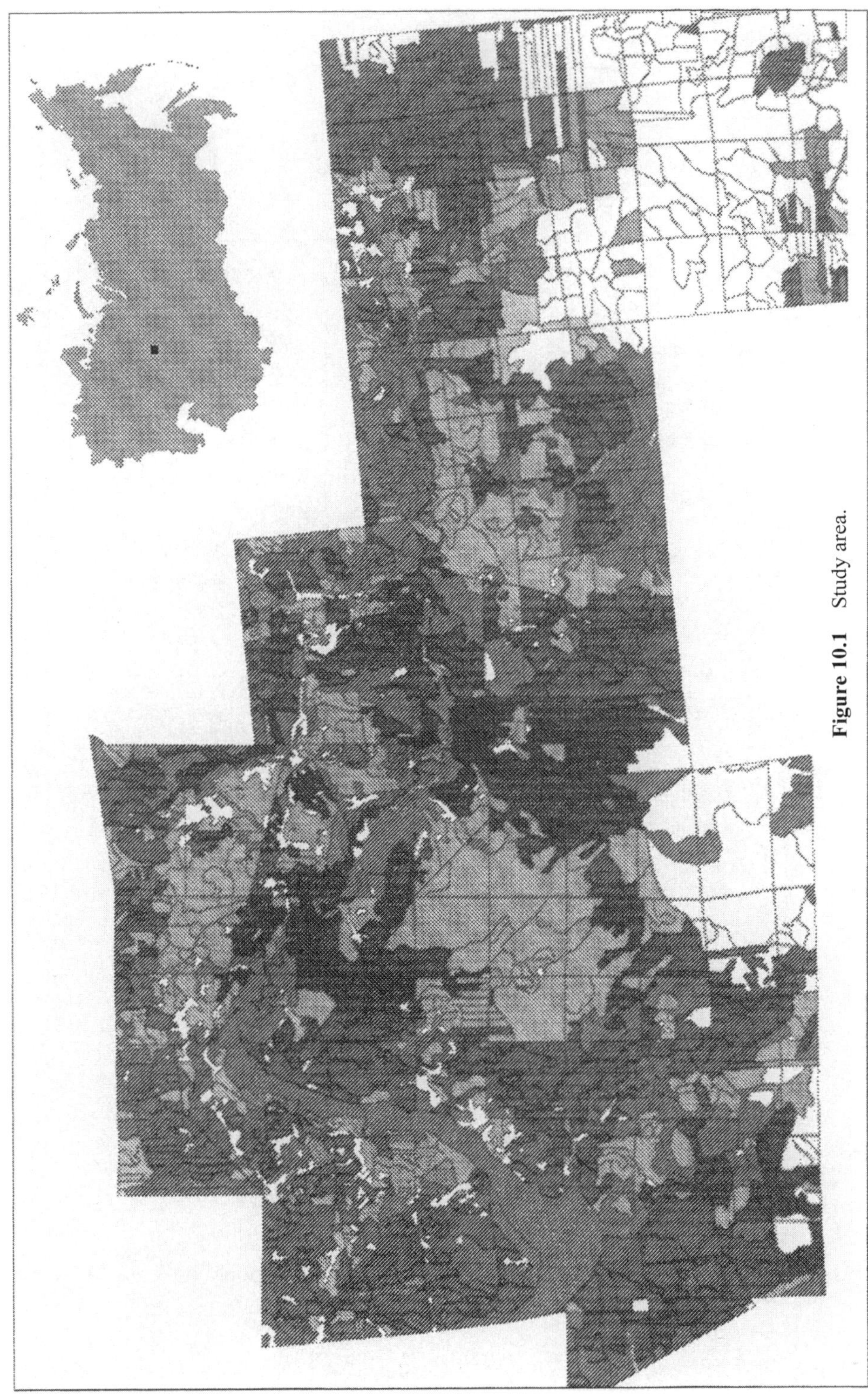

Figure 10.1 Study area.

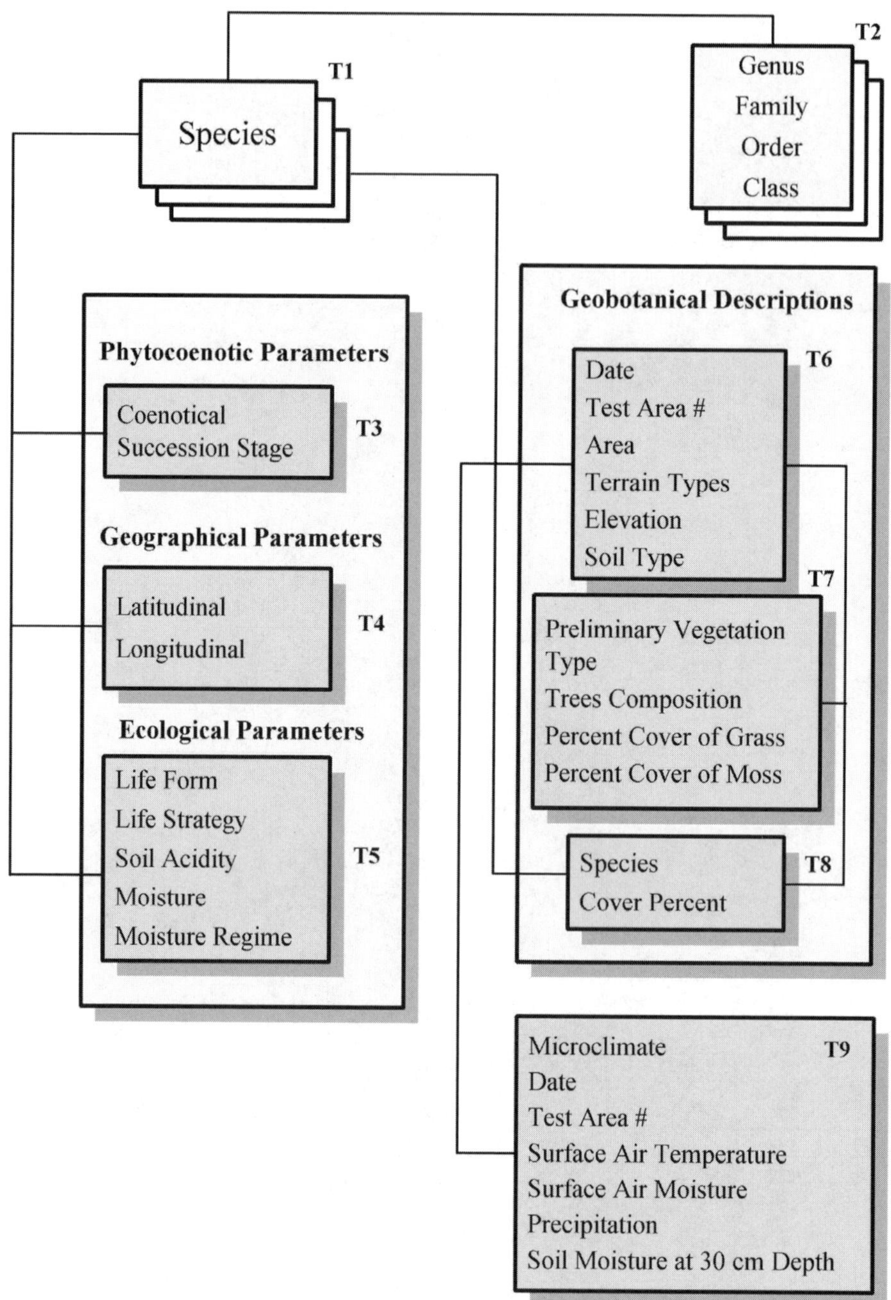

Figure 10.2 Conceptual model of database. Functional relationships (1:1 – one to one, 1:M – one to many) between tables: T1–T2 (1:1), T1–T3 (1:1), T1–T5 (1:1), T1–T4 (1:1), T1–T8 (1:1), T6–T8 (1:M), T7–T8 (1:M), T6–T9 (1:M).

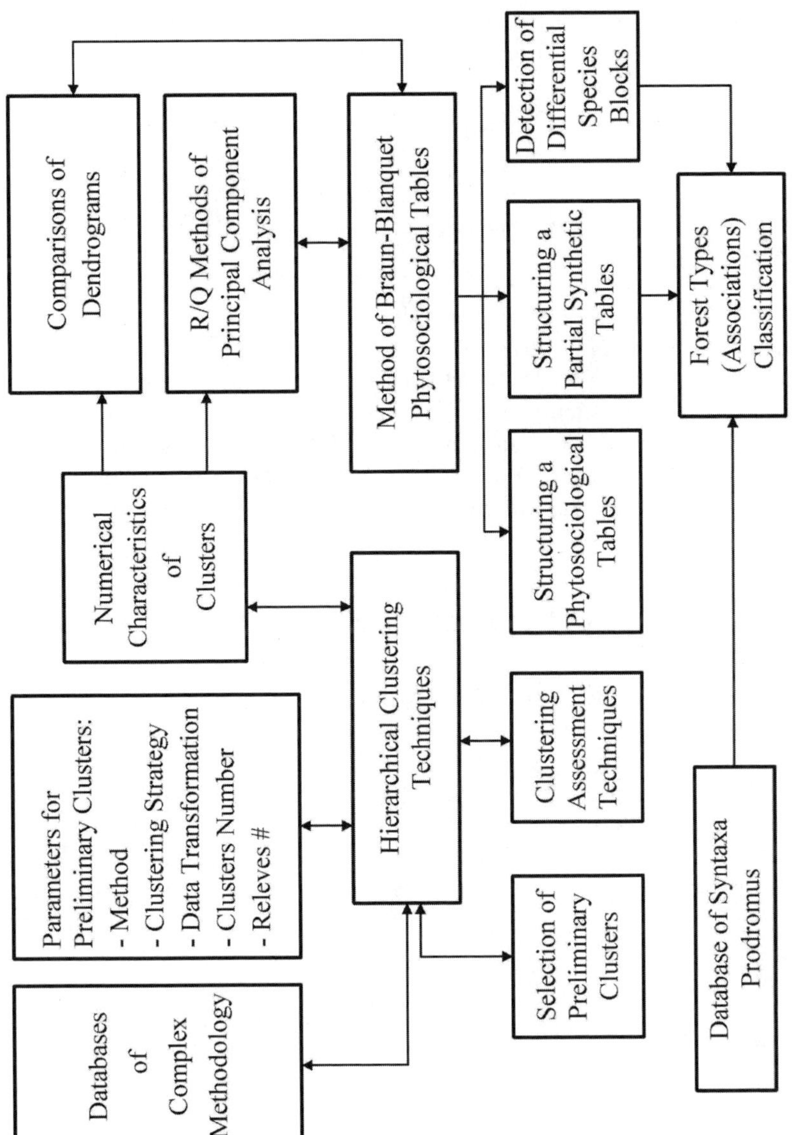

Figure 10.3 General functional scheme of forest community classification.

The comparisons indicated that the UPGA method was more effective and compatible with a Braun–Blanquet approach for identifying preliminary clusters within the forest community database. Subsequent methods were developed on the basis of combined hierarchical clustering techniques and Braun–Blanquet phytosociological tables.

Three steps were involved:

1. Detection of preliminary clusters using UPGA methods based on quantitative Sorensen–Chekanovsky similarity indices
2. Structuring a phytosociological table using results from step 1 and *R/Q* (relevés/species) methods of principal component analysis and chi-square statistics (Maarel et al., 1978)
3. Classification of final phytocoenons based on results from the first two steps: phytocoenons detected, partial synthetic tables and differential species/relevés blocks (Figure 10.3).

Steps 1 and 3 were conducted with a threshold value of 0.45, a fusion limit of 0.60, and a constancy limit of 70 (Maarel et al., 1978).

Ecological interpretation of the classification scheme was aided by evaluating different measures of biodiversity (simple species richness index and *Q*-statistics) as well as various models of species distribution such as exponential, hyperbolic, geometric-series, log-series, log-normal, MacArthur's broken-stick model and taxonomic, ecological and geographical parameters (Magurran, 1992; Pielou, 1977).

10.5 COMPUTER PROGRAMS AND DATABASES

The Interactive Information System (IIS) was programmed using PARADOX relational database software (Borland International, Inc., Scotts Valley, CA).

The IIS databases include relevés and their ecological parameters of stands, taxonomic, geographical, phytocoenotic vascular plant characteristics and microclimatic parameters (Belchansky et al., 1990).

Methods of forest community classification and assessment, based on combinations of hierarchical clustering techniques and Braun–Blanquet phytosociological tables (Gauch and Whittaker, 1981; Enslein et al., 1986), as well as methods for calculating biodiversity indices (Whittaker, 1975; Magurran, 1992; Pielou, 1977) were programmed into the IIS using the PARADOX software language.

The IIS displayed data in tabular and graphics format at the computer terminal using an interactive interface. Summarized and processed data can be output in report format, including estimates of common statistical parameters and biodiversity indices (Figure 10.4).

10.6 COMPARATIVE FLORISTIC AND MICRO-CLIMATE ANALYSIS TO IDENTIFY SENSITIVE FOREST TYPES

A dendrogram of classification results for 14 forest types (associations), based on the Sorensen–Chekanovsky dissimilarity index, is presented in Figure 10.5. The forest types

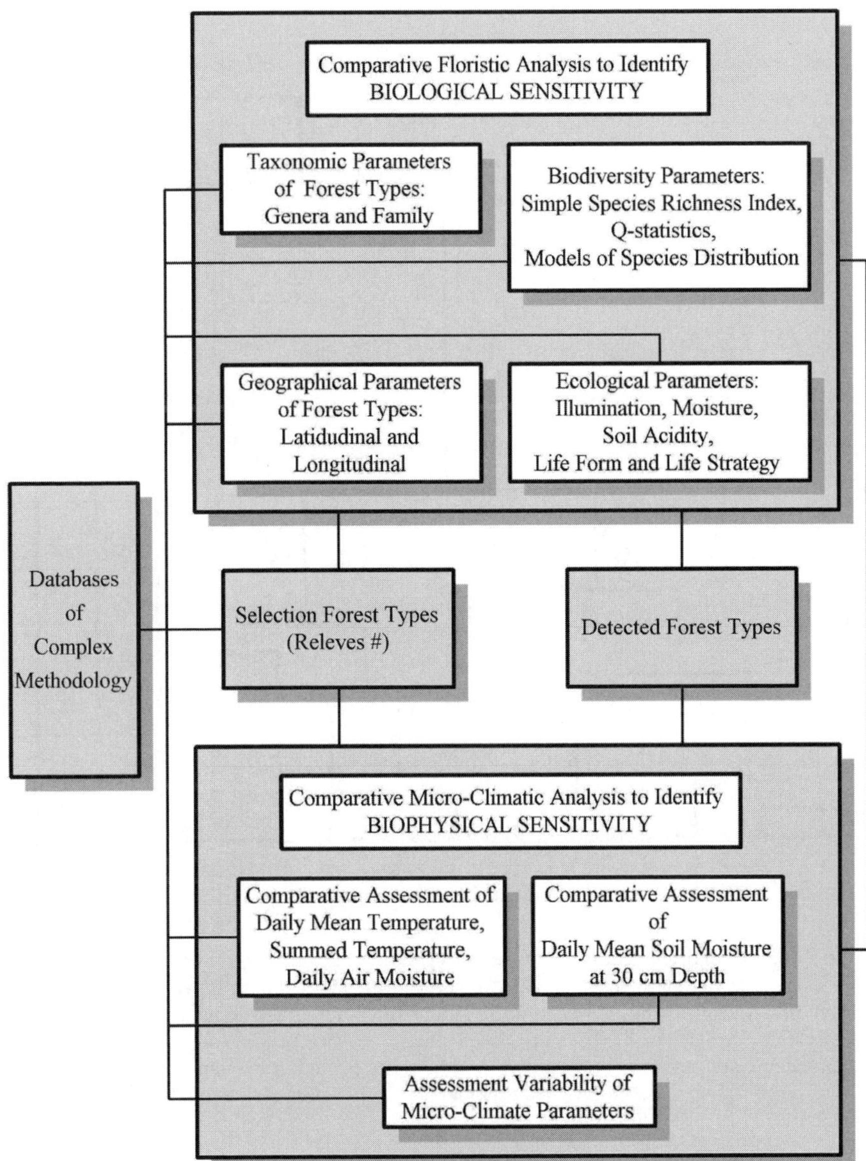

Figure 10.4 Conceptual scheme of forest types detection.

were classified to a high level of taxonomic relationship in accordance with the USSR vegetation syntaxa prodromus (Korotkov et al., 1991).

Forest types #1–#7 belong to the alliance Milio-Abietion sibiricae (mixed nemoral conifer/tall grass forest) of the Class Querco-Fagetea; associations #8–#10 belong to the suballiance Eu-Piceenion; #11–#12 belong to the suballiance Sphagno-Piceenion of Class Vaccinio-Piceetea (boreal conifer moss forest); #13–#14 belong to the Class Vaccinietea uliginosi (swamped sphagnum with sparse growth of trees). The last three associations

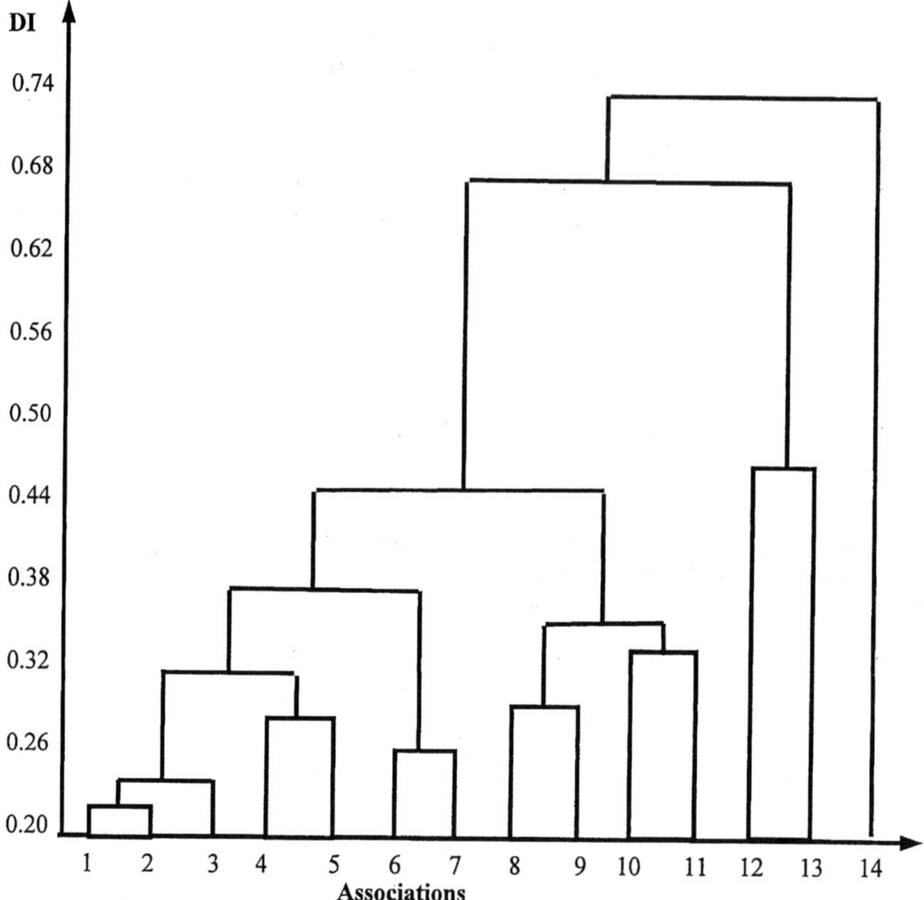

Figure 10.5 Dendrogram of classification results based on the dissimilarity index (DI). Forest types: 1: fir–spruce/tall grass, 2: fir–spruce/lime/tall grass, 3: fir–spruce/lime–large fern; 4: fir–spruce/lime/birch–sedge; 5: fir–spruce/birch–tall grass–fern; 6: birch/ tall grass; 7: birch/motley grass; 8: fir–spruce/large fern; 9: fir–spruce/birch–small fern; 10: fir/spruce–green moss–short grass; 11: spruce–birch/horse-tail–sphagnum; 12: spruce–birch/sedge; 13: spruce/sedge–sphagnum; 14: spruce/horse-tail–green moss–sphagnum.

must be classified more exactly in the future because they were represented with only three or four relevés (other types had 6 to 30 relevés).

We could not describe some of our phytocoenons within existing low-level syntaxonomic units (associations/subassociations). They could be described, probably, as new. Working names were used for these phytocoenons in this regard (Figure 10.5).

Sensitivity results of forest types to elevation gradients are presented in Figure 10.6. Results indicated that most forest types can be easily sorted along an altitudinal gradient.

There are two zones, a lower (350–450 m) boreal forest zone and a higher (450–600 m) mixed nemoral/boreal forest. However, the derivative tall grass and mixed grass birch forest types (#6 and #7) were exceptions. Decreasing elevation was associated with

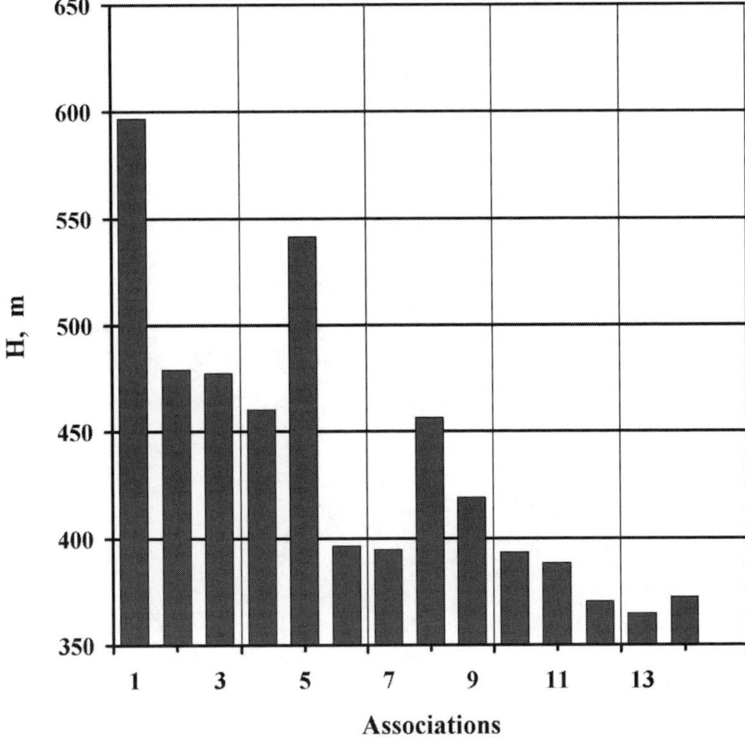

Figure 10.6 Altitude (H, m) profile of forest types.

increasing proportions of boreal and hypoarctic species, and decreasing proportions of nemoral and mixed nemoral/boreal species.

Q-statistics and species richness parameters indicated that forest community classification could be interpreted in a biodiversity context.

Phytocoenons were divided into three groups based on the *Q*-statistic parameters (Figure 10.7).

The first group had high biodiversity and included mostly native or conditional-native mixed nemoral/conifer associations (types #1–#5) as well as the derivative tall grass and motley grass birch forest associations (types #6 and #7).

The second group had intermediate biodiversity and included the native and derivative dark-conifer boreal forest associations (types #8–#12).

The third group comprised the derivative swamped sphagnum associations (types #13 and #14) which had sparse growth of trees and low biodiversity.

A similar division of phytocoenons was achieved using species richness parameters.

This division corroborates the concept that higher biodiversity occurs in derivative and mixed nemoral/conifer forests in comparison with dark-conifer boreal forests.

Results of species distribution models indicated that exponential and hyperbolic theoretical models best fit the raw (empirical) data. Correlations between the theoretical models and raw data were improved from 0.85 up to 0.97 at the 95% confidence level. Similar results were reported by Magurran (1992) for boreal forests in Ireland. Analysis of

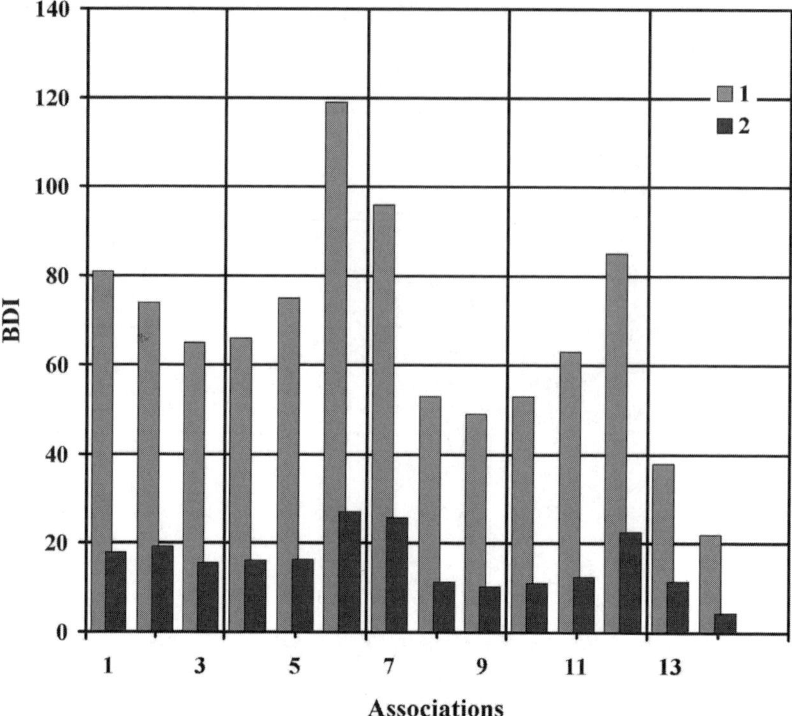

Figure 10.7 Forest type biodiversity indices (BDI): 1 – species richness, 2 – Q-statistic.

taxonomic (genera and family), geographical (latitudinal and longitudinal species distribution) and ecological parameters (illumination, moisture, soil acidity and life form) suggested that the 245 relevés could be objectively divided into the 14 forest types. For example, using fractional percentages of different latitudinal species, the 14 associations can be divided into the same three groups (#1–#7, #8–#12 and #13–#14) that were previously described (Figure 10.8).

Comparative analysis of microclimatic parameters used available archives data from five forest types (#2, #5, #8, #10 and #13). Although the microclimate database was limited in space (five forest types) and time (4 years), it reflected important aspects of the thermal and moisture regimes for three primary classes: #2 and #5 (Class Querco-Fagetea); #8 and #10 (Class Vaccinio-Piceetea); and #13 (Class Vaccinietea uliginosi).

Comparative assessments of daily mean temperature, summed temperature and daily mean soil and air moisture indicated that forest type #2 had a more moderate microclimate (Table 10.1). In addition, variations in daily mean temperature and daily mean soil moisture were more stable for forest type #2. Median air temperature in forest type #2 was significantly higher than types #5, #8, and #13 ($P < 0.001$, Wilcoxon signed-rank test); no difference was detected between it and type #10 ($P = 0.17$).

Classification results and the comparative analyses of biodiversity indices and microclimate parameters indicated that two boreal forest types: Abieto-Piceeta tilieto-altiherbosa (APTA, #2) and Abieto-Piceeta hylocomioso-nanoherbosa (APHN, #10) are more sensitive to global warming.

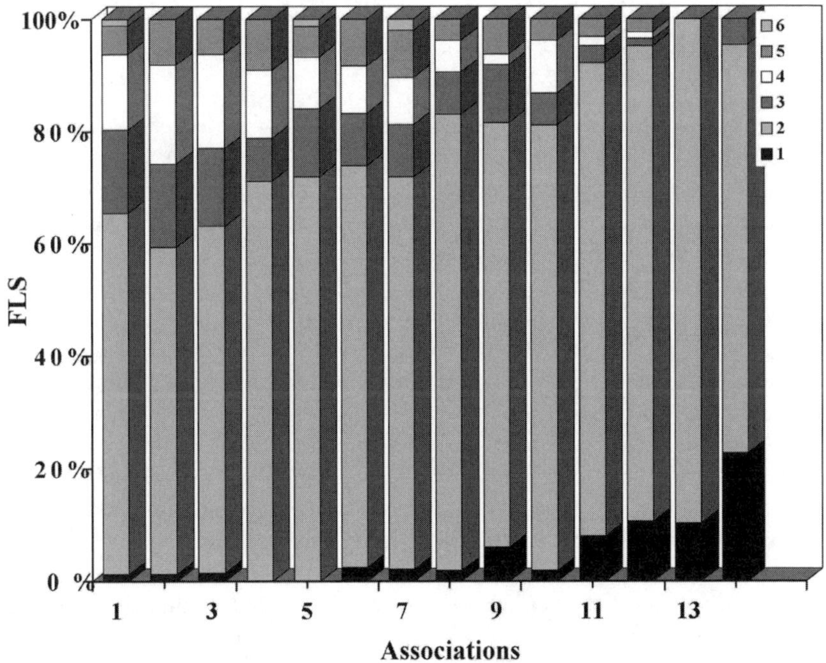

Figure 10.8 Diagram of geographical structure (fraction of different latitudinal species (FLS)) of forest types: 1: plural zonal and hypo-arctic; 2: properly boreal; 3: nemoral/boreal;. 4: properly nemoral; 5: boreal/nemoral; 6: boreal/forest–steppe and forest–steppe.

The APTA forest type has low natural variability of primary microclimatic parameters and high maximum daily mean temperature during the growing season, which help to distinguish it from other mixed nemoral/conifer forest types. The fractional percentage of latitudinal species for APTA is minimal for boreal forest associations (58%) and maximal for mixed nemoral/conifer associations (32.5%) (Figure 10.8). Furthermore, APTA is a climax type and the lime constant is present in both overstory and understory APTA communities. The APHN forest type has high mean daily temperature in the early dry summer period and simultaneously low soil moisture, thus increasing its susceptibility to fire.

Table 10.1 Mean Multiannual Daily Air Temperature (T), Soil Moisture (SM) and Their Coefficients of Variation CV-T and CV-SM, Respectively, for Different Forest Types in the Period 1973–1976

FOREST TYPE	T	CV-T	SM	CV-SM
#6	10.8	29.26	56 02	25.03
#3	11.95	26.53	37.00	19.56
#8	10.76	26.80	32.36	25.91
#10	11.42	29.26	41.04	26.39
#11	10.87	28.38	238.98	35.76

10.7 GLOBAL CHANGE AND BOREAL FOREST COMMUNITY DYNAMICS

This study used data from 245 relevés to identify and describe vegetation structure in the southern ecotone of the boreal forest in middle Ural, Russia. A principal vegetation gradient was described across an ecological transition from higher elevation nemoral communities to lower elevation boreal forest associations. Plant biodiversity generally decreased across this gradient, with notable exceptions of high diversity in forest types associated with riparian areas.

Microclimate analyses demonstrated the potential use of integrated comparisons for assessing forest sensitivity in the context of global climate change, however, analyses of more VNR climate data are required to substantiate these preliminary inferences.

One potentially sensitive forest type, "fir–spruce/tall grass/lime" (type #2), had high maximum daily mean air temperature during the growing season, together with low natural variability in both air temperature and soil moisture, which helped distinguish it from other mixed nemoral/conifer forest types. The fractional percentage of latitudinal species in type #2 was minimal for boreal forest (58%) and maximal for mixed nemoral/conifer associations (32.5%) compared with other forest types. Also, type #2 is a climax forest community. Assuming that long-term global warming could modify the composition of heat-loving nemoral grass and shrub species that have short reproductive cycles, such changes might be realized earlier in warmer, more stable ecotopes of mixed nemoral/conifer forest types, for example type #2.

A second forest type, "fir–spruce/moss–short grass" (type #10), showed higher mean daily temperatures during the early summer period and simultaneously low soil moisture, possibly increasing its susceptibility to fire if future climate conditions become warmer and drier. Although ecological indicator species of natural fire cycles occur in forest type #10 (pine and characteristic moss species), concerns regarding increased fire susceptibility are valid (Gorshkov and Gorshkova, 1992), because pine is a poorly renewing species in the previously described and closely related Abieto-Piceeta hylocomioso-nanoherbosa forest type (Kolesnikov and Kirsanov, 1979).

This study has made an initial contribution toward automating the wealth of long-term ecological information that has been collected for several decades by Russian Nature Reserves in a variety of ecosystems. Databases of long-term ecological monitoring are uncommon. The value of Russian Nature Reserve data is not only important for local resource management, but also to the global community of ecologists studying topics like biodiversity and global climate change using ground and satellite observations. The IIS, developed for the Visimsky Nature Reserve as part of this study, established a database structure that can be easily adapted for application at other reserves. For the Visimsky Reserve and boreal forest studies in general, further database development needs to include expanded information about mosses and microclimate data. Also, further research needs to identify key parameters for monitoring the phytocoenotic dynamics of boreal forest communities.

REFERENCES

Belchansky, G. I., Petrosyan, V. G. and Boukvareva, E. N., 1990. Interactive Information system: Database elaboration and Biodiversity Research. International Conference on the Role of the Polar Regions in Global Change. Fairbanks. Alaska. Vol. 1. pp. 112–118.

Belchansky, G. I., Kondratyev, K. Ya. and Petrosyan, V. G., 1992. Conceptual aspects of space biodiversity monitoring. *Proceeding of Russian Academy of Science*, V. 332, No. 1, pp. 200–203.

Belchansky, G. I. and Petrosyan, V. G., 1995. Complex methodology of detecting forest types sensitive to global climate change. *J. Progress in Current Biology*, Vol. 115. No. 2, pp. 5–12.

Davis, M. B. and Botkin, D. B., 1985. Sensitivity of cool temperature forests and their fossil pollen to rapid climatic change. *Quaternary Research*, 23, pp. 327–340.

Dietvorst, P., E. van der Maarel and H. van der Putten, 1982. A new approach to the minimal area of plant community. *Vegetatio*, 50(1), pp. 77–91.

Duran, B. and Odell, P., 1977. Clustering analysis. (In Russian), Moscow. Statistics. 128 p.

Enslein, K., Ralson, A. and Wilf, H. S., 1986. Statistical methods for digital computers. Moscow. MIR. 464 p.

Gauch, H. G. and Whittaker, R. H. 1981. Hierarchical classification of community data. *J. Ecol.*, 69, pp. 537–557.

Gorshkov, V. V. and Gorshkova, V. G. 1992. Recovery characteristics of forest ecosystems after fire. Botanical Institute of Russian Academy of Science RAS Press. 39 p.

Harrington, J. B., 1987. Climatic change: our view of causes. *Can. J. of Forest Res.*, 11, pp. 1313–1339.

Kolesnikov, B. P. and Kirsanov V. A., eds., 1979. Dark conifer forest of middle Ural. AS USSR Press. Sverdlovsk, 144.

Korotkov, K. O., Morosova, O. V. and Belonovskaja, E. A., 1991. The USSR vegetation syntaxa prodromus, Publ. by G. Vilchek. Moscow: RAS Press. 346 p.

Maarel, E. J., Jansen, G. M. and Louppen, J. M. W., 1978. TABORD, a program for structuring phytosociological tables. *Vegetatio*, 38(3). pp. 143–156.

Magurran, A. E., 1992. Ecological diversity and its measurement. Moscow. MIR. 142 p.

Mitchell, J. F., 1983. The seasonal response of a general circulation model to changes in CO_2 and sea temperature. *Quart. J. Royal. Mot. Soc.*, 109, pp. 113–152.

Orloci, L. and Wildi, O., 1988. Mulva-4 a package for multivariate analysis of vegetation data. Swiss Federal Institute of Forestry Research. 121 p.

Pielou, E. C., 1977. Mathematical Ecology. John Wiley & Sons, Inc. New York. 385 p.

Sato, N., Sellers, P. J., Randall, D. A., Schneider, E. K., Shukla, J., Kinter III, J. L., Hou, Y. T. and Ablertazzi, E., 1989. Effects of implementing the Simple Biosphere Model (SIB) in a general circulation model. *J. Atmos. Sci.*, 46, pp. 2757–2782.

Sokolov, V. E., 1990. Biodiversity and sustainable development. Conference report: Sustainable development, science and policy. Bergen. 8–12 May. pp. 217–229.

Solbrig, O. T., 1992. The IUBS–SCOPE–UNESCO program of research in biodiversity. *J. Ecological Applications*, 2(2), pp. 131–138.

Tans, P. P., Fung, I. Y. and Takalashi, T., 1990. Observational constraints on the global atmospheric CO_2 budget. *Science*, 247, pp. 1431–1438.

Whittaker, R. H., 1977. Evolution of species diversity in land communities. *Evol. Biol.*, 10, pp. 1–67.

SUBJECT INDEX